FINDING
W.H. HUDSON

December 2023

For John

[signature]

Hudson's birthplace and childhood home. It was revived by his admirers, long after it had fallen into disrepair. Image courtesy of the Friends of Hudson Ecological Park Association.

FINDING W.H. HUDSON

The Writer Who Came to Britain to Save the Birds

CONOR MARK JAMESON

PELAGIC PUBLISHING

First published in 2023 by
Pelagic Publishing
20–22 Wenlock Road
London N1 7GU, UK

www.pelagicpublishing.com

Finding W.H. Hudson: The Writer Who Came to Britain to Save the Birds

https://doi.org/10.53061/YGVN4872

British Library Cataloguing in Publication Data
A catalogue record for this book is available from the British Library

ISBN 978-1-78427-328-6 Pbk
ISBN 978-1-78427-329-3 ePub
ISBN 978-1-78427-330-9 PDF

Typeset in Goudy Oldstyle Std by S4Carlisle Publishing Services Chennai, India

To the memory of Hudson and his friends, in particular the women who founded modern conservation campaigning; who turned wildlife from something to be owned and exhibited into something to be nurtured and protected. And to underdogs and chasers of lost causes, everywhere.

Photograph by the author.

William Henry (aka Guillermo Enrique) Hudson age 26 or 27. Image courtesy of the Smithsonian Institution, Washington DC, to whom it was sent by Hudson in 1868.

'He had unfailing sympathy with young and inexperienced workers, and gave them every encouragement. When I was young he helped me over many difficult places connected with my position as Honorary Secretary of the Society for the Protection of Birds. His influence was in a curious way both potent, and permanent.' **Etta Lemon**

'There was no one I thought more highly of as a man, or respected as a genius. That he was a genius, I think all his real admirers know. Some day the world will become aware of it.' **Robert Cunninghame Graham**

'Hudson's writing is like grass that the good God made to grow, and when it is there you cannot tell how it came.' **Joseph Conrad**

'Mr Wells, Mr Bennett, and Mr Galsworthy ... while we thank them for a thousand gifts, we reserve our unconditional gratitude for Mr Hardy, for Mr Conrad, and in a much lesser degree for the Mr Hudson of *The Purple Land, Green Mansions*, and *Far Away and Long Ago*.' **Virginia Woolf**

'There was no one – no writer – who did not acknowledge without question that Hudson was the greatest living writer of English ... I have never heard a writer speak of him with anything but reverence that was given to no other human being. For as a writer he was a magician.' **Ford Madox Ford**

'I'm not one of you damned writers – I'm a naturalist from La Plata.'

'I have no style. I simply write what I think.' **W.H. Hudson**

Contents

Preface

For most of the 25 years that I worked for the Royal Society for the Protection of Birds (RSPB), at The Lodge, Sandy, Bedfordshire, W.H. Hudson was just 'the man above the fireplace', looking back at the world from a painting: pensive, mute. Then I got to know him better.

A story told by Hudson in his 1901 book *Birds and Man* captured my imagination. He describes staying in a big house one night, sitting by a fireplace, as a storm raged. He fell asleep and had a nightmare, that when he died he would be stuffed by the taxidermists of hell, like so much of the wildlife he loved, and condemned to eternity with no voice. And it struck me that here he is today, beside a fireplace, voiceless. A mission began to give him back his voice. The more I found out about Hudson, the more perplexed I became that he had faded into obscurity.

In an odd way, this has felt like tracing family, as well as the soul of the organisation for which I worked for so many years. Hudson was contemporary with my grandfathers' grandfathers. I don't know much about them, or my great-great-grandmothers, other than that they were essentially rural people, like so many others living through a time of displacement – from the land to the city, from the outdoors to the indoors, from Scotland to Ireland and Ireland to America, with all the loss and regret that accompanied those advances in civilisation. I have recognised in Hudson what feel like family traits, and a kindred spirit in his love for wild nature, and his emigration.[1]

The Pampas origins of Hudson, of course, add a whole other dimension of intrigue. His poet friend Edward Thomas captured it rather neatly: 'W.H. Hudson began by doing an eccentric thing for an English naturalist. He was born in South America.' I love that he brings emigration full circle, via Ireland and the USA. He came back – to England – with many stories to tell. He made sense of Britain, in a way only the fresh perspective of 'half-foreign' senses could. He also dared to challenge its norms and sacred cows: not having, nor knowing, his place.

This book is an attempt to give Hudson his voice back, to help it be heard; in a sense to revive him and his milieu. It seems that everyone

Detail of the Hudson portrait by Frank Brooks that hangs above the fireplace at RSPB headquarters.

wanted to keep the letters Hudson wrote to them – and he wrote *a lot* of letters. Hudson, meanwhile, burned almost all of those he received, and is said to have requested that his friends do the same. In many cases they went against his apparent wishes, enabling me to piece together the friend and colleague they so obviously loved, for all his 'native lack of delicacy',[2] and his complexity: for indeed he was complex.

Hudson lived with fragile health and an almost constant dread of death. He veered between euphoria and gloom, between love and compassion for people and fellow creatures, and bitterness and scorn. People better qualified than me might today search for labels for his mental health.

In any event, while writing this book I have not found there to be any shortage of material to work from in forming a sense of Hudson's circumstances, character and motivations. There are several published books of his letters – to Edward Garnett, Morley Roberts, George Gissing (and many others) and Robert 'Don Roberto' Cunninghame Graham – and published compilations of his correspondence to a range of friends and colleagues. The notes and references are grouped by chapter as endnotes.

Hudson said that he wished to be forgotten, and discouraged his contemporaries from picking over his life. It didn't deter Morley Roberts, one of his enduring friends, from immediately writing a 'portrait' of him, with around ten further biographies to follow. What has struck me is that no one has focused on Hudson the field naturalist and campaigner – his efforts for his beloved Bird Society and the cause of bird protection have usually been at most a subplot, and usually incidental. But Hudson was first and foremost a spirit of nature and a champion of its protection. As he himself once declared to his literary friends: 'I'm not one of you damned writers: I'm a naturalist from La Plata.'[3]

One way or another, Hudson's importance to the history of nature conservation has been overlooked and forgotten in his adopted homeland. I hope, by telling this story, to help put that right. If this book inspires others to find out more about Hudson, his life and times, it will have done its job. I will be happy for this work to be continued and augmented, and for any of my lines of enquiry to be explored further.

Conor Mark Jameson
Cambridgeshire, June 2023

The Bird Society – a note on terminology

W.H. Hudson always called it the Bird Society, the organisation we know today as the RSPB. Following his lead, and his instinct for readability, I have used this name throughout. It means the organisation that began life as the Society for the Protection of Birds (SPB, before the Royal R was added in 1904 to make it RSPB) and which merged with the Fur, Fin and Feather Folk in the early 1890s.

Acknowledgements

Special thanks to Sara Evans, Diana Donald, Jane Bransby, Tessa Boase, Peter Dance, Lachie Munro, Cecilia Hecht, Jason Wilson (author of *Living in the Sound of the Wind*) and Nigel Collar. Also sincere thanks and abrazos to Andie Filadora and Maria Laura Josens, for their kind hospitality and companionship on our tour of the Pampas. To Francisco Gonzalez Taboas, my guide in finding Hudson's childhood home, and to Atilio Martinez, Ruben Ravera and the Friends of Hudson Ecological Park Association for welcoming me there and showing me round their beautifully curated Museo. Nigel Massen, David Hawkins and all at Pelagic Publishing, as well as Jeremy Complin for his indexing expertise. Elizabeth George, Lizzie Sparrow, Tara-Lee Platt, Emma Whiteway and the RSPB library team. To the Society of Authors, as the literary representative of the Estate of W.H. Hudson. And in particular to Justine Palmer and the trustees of the Royal Literary Fund for their kind words and support at a vital time.

Special thanks for photographs to the Smithsonian Institution (cover photo), the Brooke Charity, the Estate of Helen Thomas, with special thanks to Jeremy Mitchell, the Edward Thomas Fellowship and the resources at the Edward Thomas Study Centre in Petersfield Museum and Art Gallery, the National Portrait Gallery, the Friends of Hudson Ecological Park Association, and Nick Brooks.

I would also like to acknowledge and thank the following for their interest and support over the last few years: Brigid Allen, Felipe Arocena, Melissa Ashley, Simon Aumonier, Richard Bashford, Sarah Baxter at the Society of Authors, Ewen Binns, John Birkett, Tim Birkhead, Bryan Bland, Hugo Blomfield, Libby Bodio, Stephen Bodio, Ana Ines Larre Borges, Andres Bosso, Frank Bowles, Sarah Broadhurst at the Zoological Society of London, Derek Brockway, Pat Brockway, Jason Brooke and the Brooke Trust, Nick Brooks, Ronnie Buchanan, Peter Buchanan, Leah Burke, Jimmy Burns of the British-Spanish Society, Emma Campbell at King's Lynn Library, Hernan Casanas of Aves Argentinas, Clare Chadderton, Charles Chadwyck-Healey, Ruth Chambers, Jenny Childs of Birmingham University Archive, Brian Clews, Mark Cocker, Mark Cousins, Robin

Cunninghame Graham, Bill Curtis, John Davis, Luke Davies, Geoff Dibb, Jane Dismore, Mary Done at Kew, Lisa Dowdeswell at the Society of Authors, Philip Eade, Caroline Evans, Joseph Farrell of the Cunninghame Graham Society, Helen Fisher at Birmingham University Archive, Simon Fletcher, Suzanne Foster, Roger Gaitley, Jocelyn Galsworthy, John Glaves-Smith, Roxana Glenn at Kew, Georgie Green, Kat Harrington at Kew, Melissa Knox-Raab, Blake Lee Harwood, David Harrison, Alister Hayes of The Royal Parks, Anthony Heaton-Armstrong, Gaye Henson, Richard Hines, Judith Frances Hubbard, Julian Hughes, Rob Hume, Caroline Jackson-Houlston, Kay Jameson, Kevin Jameson, Jamie Jauncey, David Lindo, Amanda Long at the RSPB, Tommy MacKay, Eric Marland, Stephen Mason, John T. Masterson, Nancy Masterson, Angela McAllister, Gerry McGarvey, Mary Mclernon, Liz Merchant at the Anglo-Argentine Society, Stephen Menzie at *British Birds* journal, Matthew Merritt at *Bird-watching* magazine, Patrick Minne, Allison Moorhead, Isabel Moorhead, Edward Morant, Franny Moyle, Margaret Murray, Lynne Myfanwy Jones, Jeremy Mynott, Fatima Nollen-Reardon of the Anglo-Argentine Society, Matthew Oates, Justine Palmer, David Payne, Malcolm Peaker, Sarah Pook, Richard Porter, Caroline Pridham, Mike Pringle, Rachel Ridealgh at Norfolk Library and Information Service, Niall Robertson, Nilda Mercedes Rodriguez, Andrew Rossabi, Jean Saunders of the Richard Jefferies Society, Helen Smith, Sue Steptoe, Jim Stevenson, Sean Street, Matthew Sturgis, Michael Tinker, Martin Todd, Javier Uriarte, Andy Varley, Elizabeth Velluet of the Richmond Local History Society, Ray 'Badger' Walker, Ian Ward, Jane Winter and Jonny Wright.

PART 1

LATE VICTORIANS

CHAPTER 1

Smelling England

'The land of my desire'[1]

William Henry Hudson first set foot in England early on a bright, chill morning in May 1874, after spending the whole of April crossing the Atlantic Ocean from Argentina. 'I'm not coming with you to London,' he announced to his fellow passengers, as they stepped gratefully and unsteadily off the ship. 'I want to go into the countryside and find English birds.'[2]

He checked into a hotel near the railway station, and explored Southampton. Hudson was immediately struck by an aroma that pervaded the streets, unlike anything he had known before. He enquired of passers-by what the smell might be. They must have given him funny looks, being unaware of anything unusual. It is our first hint of Hudson as a possessor of acute sensory abilities, but it is also a measure of the freshness of his impressions of his new home, and their impact on him. It was only later that he was able to identify the source – breweries.[3]

The following day, Hudson turned his attention to his first main ambition: meeting the local birds. So he hired a pony and trap, with a boy on the reins and an American from the ship who couldn't be shaken off, and they headed first to nearby Netley Common. The poor boy was wedged between Hudson and the American, with Hudson nudging him and pointing, wanting to know what bird that was, and what bird was singing this note, and that song. On the other side the American, a student of agriculture, was pestering the poor boy with questions about farming.

'What grass is that?'

'The grass what horses eat!' the exasperated boy yelled back, over the clatter of hooves and metal wheels. Sadly for Hudson, the boy was no bird-guide either. He could name for his client only two species: the skylark and 'dobbin-dishwater'. Hudson believed that they were in fact looking at a robin and not a grey wagtail, which derives this old name from its habit of frequenting river edges alongside washerwomen and children.

Hudson's friend and first biographer Morley Roberts described how at the end of his reminiscence Hudson issued 'a terrific cackle of laughter'. It seems Hudson had a peculiarly high-pitched laugh for a man of his imposing stature, like a green woodpecker that he no doubt heard but the identity of which he could only guess at on that introductory sortie into the English countryside. It's got so I can't hear a green woodpecker (or yaffle, as they used to be known) without thinking of him. The far-carrying sound haunts me, but in a good way. It's possible to reconstruct vignettes of Hudson's life such as this one from some of the more detailed of his anecdotes that are related in letters and articles, in which he recalls snatches of conversations and interactions with local people, usually in rural settings.

Hudson would always be vague about his early days, and years, in England, but luckily, two years before he died, Morley Roberts had the idea of interviewing him and transcribing his responses. For whatever reason, Hudson broke with habit and consented – providing, from memory, nearly 50 years after the event, the above picture of his first day in England, that precious and defining moment.

Hudson wasn't asked about – or wouldn't talk about – the month-long voyage that had preceded this escapade. I longed to know what his sea crossing had been like, but found that in all his books he makes few references to being on that ship. I did find this one, written shortly after he left behind an unhappy few days in the Peak District and returned to the pastoral south. The quote encapsulates the attachment Hudson developed to the southern counties of England: 'Never since I had known England, from that morning in early May when I saw the sun rise behind the white cliffs and green downs of Wight and the Hampshire shore, had it seemed so surpassingly lovely – so like a dream of some heavenly country.'

I later realised that there was something that would bring Hudson's voyage – and those first formative days here – vividly to life: a cache of letters that he had written while on the ship, and in his first days at Southampton. These were discovered long after Hudson died, a precious time capsule.

After those first few heady days among Hampshire's leafy lanes and wild birds, Hudson took a seat for his first ever railway journey, bound for London, 80 miles to the north-east, where fortunes might be sought.

Lurid London

I've tried to imagine the young immigrant's inner turmoil, his excitement and trepidation as the train approached the great city, as he emerged from Waterloo Station clutching a suitcase, among a throng of other passengers, his 6-foot 3-inch frame head-and-shoulders above most in this bubbling river of bonnets and hats. I wonder if he was prepared for the scale of the place, the smoky air, the grime. In the winter just passed, a cattle show in the city had been enveloped in a smog so thick and noxious that it left the livestock gasping for breath, many of them collapsing and dying.

This was all in marked contrast to the fresh, green, budding countryside he had just passed through in the train as he craned to see every bird, perhaps scribbling notes as he went, thinking about his ancestral origins. Hudson's paternal grandfather was said to be a Devon man. I can picture him turning over in his mind and sharing with fellow passengers the dream he was following, to become a professional ornithologist, if not a poet, in whatever form that might take.

He had for some years been corresponding with scientists at the Zoological Society of London, as well as the Smithsonian Institution in Washington, and had established some credentials as their bird man in Argentina. He provided specimens of birds and descriptions that were advancing the sparse knowledge of the avifauna of that region, including two species new to science, and which would in time come to bear his name.

But is there a living to be made in this? a fellow passenger might have asked him, voicing a question no doubt echoing in his own thoughts. Hudson might have smiled thinly and raised his eyebrows in a gesture of hope and effortful optimism. Perhaps he articulated the one basic reason

Hudson's canastero, one of two species new to science that Hudson discovered.
Image courtesy of Nick Brooks.

for seeking kindred minds so far from his Pampas home: 'Not in all the years of my life in the Pampas did I ever have the happiness to meet with anyone to share my interest in the wild bird life of the country I was born in.'[4]

Perhaps he was having early notions of challenging the trade in bird plumage that was beginning to decimate the birds he had known and loved. He might have described some of the appalling cruelty he had witnessed – young birds left to starve in the nest as adult birds were netted and clubbed in their breeding colonies. His beloved hummingbirds too: tiny, miraculous fliers, killed and embalmed, to be turned into adornments for dresses and hats, or jaded trophies, mounted behind glass.

Hudson later recalled his first impressions of London, the vast, mushrooming city, already the largest in the world:

> I put up at a city hotel, and on the following day went out to explore, and walked at random, never enquiring my way of any person, and not knowing whether I was going east or west. After rambling about for some three or four hours, I came to a vast wooded place where few persons were about. It was a wet, cold morning in early May, after a night of incessant rain; but when I reached this unknown place the sun came out and made the air warm and fragrant and the grass and trees sparkle with innumerable raindrops. Never grass and trees in their early spring foliage looked so vividly green, while above the sky was clear and blue as if I had left London leagues behind.
>
> As I advanced further into this wooded space the dull sounds of traffic became fainter, while ahead the continuous noise of many cawing rooks became louder and louder. I was soon under the rookery listening to and watching the birds as they wrangled with one another, and passed in and out among the trees or soared above their tops. How intensely black they looked amidst the fresh brilliant green of the sunlit foliage![5]

Besides visiting London Zoo, which had opened to the public 20 years earlier, and the Crystal Palace, setting for the Great Exhibition, showcasing technological and scientific achievements, he would have admired the scale of the recently completed Houses of Parliament and the Natural History Museum under construction.

Buenos Aires was also rapidly growing at this time, but Hudson would never have seen anything on this scale.[6] London had almost 5 million inhabitants, many of them – like him – recently arrived. The population had multiplied fivefold since 1800, and almost 2 million more people would live there by the end of the century. In weeks to come, he would peer through pea-souper fogs, hold his finely tuned nose at the stink of the filthy River Thames. He called it 'lurid London', and wrote of the 'rank steam of slums'.[7] But it was here that the country boy from a far-off land would have to stay.

For now, Hudson had a career to discuss, and an appointment with the top man at the Zoological Society of London at Hanover Square, an individual and an address that would come to figure prominently in his story, and the campaign to save the birds.

Creative tensions with Dr Sclater

One person more than any other held the key to Hudson's future in Britain: Dr Philip Lutley Sclater, one of the era's pre-eminent figures in natural history. Posterity records him as having identified the six main zoogeographic regions of the world, including the Neotropics, from which Hudson had come, and the Palearctic, in which he now found himself. The two had corresponded for several years, as Hudson supplied bird-skins and notes on bird habits and habitats to Sclater at the Zoological Society, where he had risen to occupy the top position of secretary. Hudson's long-distance correspondence had been published in the society's journal.

Hudson could now pursue his publishing project with Sclater to produce the definitive textbook on Argentine birds. He had already provided much of the new material and first-hand knowledge of these species, and descriptive powers to bring them to life on the page. Sclater was the empiricist, the scientific mind. One way or another, despite Hudson's need of favour from his main contact in scientific circles and the imbalance of power between them, they were unable to get along well: 'All the antagonism of poet to pedant, and of outdoor to indoor naturalist, pervaded Hudson's work with Sclater.'[8]

But although Hudson didn't warm to his collaborator, and it wasn't in his nature to suffer fools gladly, he had to grin and bear it. The two-volume

Argentine Ornithology would take the best part of 20 years to produce. And it would not have taken Hudson long to realise that this wasn't going to put food on his table any time soon.

Morley Roberts recalled Hudson's lukewarm view of his collaborator. 'Judging from his description this eminent professor lacked charm,[9] to say the least of it, but balanced the lack by a keen regard for royalties somewhat repugnant to a much poorer collaborator.'[10] That they didn't become best friends is perhaps less surprising than that they could collaborate at all, and over so long a period. Sclater was in no hurry – he didn't need to be. And of course he was very busy, and a man of considerable standing. His father owned the Hoddington Estate in Hampshire, and his elder brother was a Conservative Member of Parliament (MP).

It is a pity in a way that they didn't meet at Hoddington soon after Hudson stepped off the *Ebro* at Southampton, to bring into focus the contrast in backgrounds and status of the two men. In fact they met at the Zoological Society's offices in Hanover Square, where Sclater also lived, in a plush apartment that came with the job. The irony was not lost on Hudson that 'two men who have not one thought or taste in common should be associated together in writing a big book'. And as for what they *did* have in common? 'We are both big ugly men, and there all resemblance ends.'[11]

If working with Sclater was tedious, there was the consolation of free tickets to the zoological gardens. 'They look at us with a strange friendliness in them as it they knew what we, after thousands of years of thinking, have only just found out', wrote Hudson after meeting a lemur at London Zoo.[12] He remained preoccupied with evolution, and had issues with some of the assumptions in the theory. It would lead to issues with another, even more eminent, figure in natural history. In fact, it already had.

Issues with Darwin

For reasons best known to himself, Hudson liked to let his friends and colleagues believe that he had arrived in England in 1869, five years before he actually did. This error continued to be repeated for several decades after his death, which suggests that the letters he had been sending to Dr Sclater in the years leading up to his emigration from Argentina in 1874 were not generally known about. I assumed these letters must have been long lost, and not kept by the Zoological Society of London.[13] Then I discovered that someone had been well ahead of me, not here in the UK, but in America. The bundle of letters to Dr Sclater had been sought out and transcribed by a Scottish journalist called David R. Dewar, and published by Cornell University Press in 1951.

I was intrigued as to why writers on Hudson had for so many years never thought to enquire of the Society about any such correspondence, as some letters were published in their journal and sight of them would have set the record straight on his arrival date in Britain, and also opened up the question – the mystery – of why Hudson wanted people to believe he spent those five years, 1869 to 1874, in Britain rather than in Argentina. My mind raced. Did he need an alibi for some reason? Or did traumatic events happen in the period before he left that he needed to erase? What's also curious is that if Hudson liked to let people believe he was younger than he was, adding five years to his time in England hardly helped with that.

I was able to buy a copy of this small bound volume of seven hitherto unpublished Hudson letters, together with five that were printed in the society's 'Proceedings', which had been overlooked by Hudson's biographers for decades. Perhaps this speaks to the fact that these Hudson scribes were less interested in him as a naturalist than in other aspects of his life and personality. This is curious, as Hudson defined himself first and foremost as a field naturalist. Saving the birds was his mission; writing was a means to that end.

The 1951 volume is dedicated 'To Lovers of W.H. Hudson everywhere', while recognising that this constituency may already be diminished in number – citing a recent article in the New York Times that reported 'the general public has forgotten Hudson, but, with his friend Joseph Conrad, he will come back into favor'. They were right about Conrad, at least. The letters reveal that Hudson dared to find fault with Charles Darwin and – worse still – aired his criticisms in public. While broadly accepting the principle of natural selection, Hudson found errors about South American woodpeckers in Darwin's book,[14] and he wasn't shy about pointing them out.

Darwin had travelled in South America, first arriving in 1832. In On the Origin of Species by Means of Natural Selection, first published in 1859, he had described the behaviour of the campo woodpecker, a species that Hudson knew well from his days in La Plata. Darwin used this as an example of a species adapting to a new environment, one that he said was almost treeless, saying that now the species never climbed trees. Hudson dared to state in the pages of ZSL's Proceedings, in 1870:

> so great a deviation from the truth in this instance
> might give opponents of his book a reason for
> considering other statements in it erroneous or

> exaggerated … The perusal of the passage I have
> quoted from, to one acquainted with the bird referred
> to, and its habitat, might induce him to believe that
> the author purposely wrested the truths of Nature to
> prove his theory.[15]

Ouch. Not surprisingly, Hudson's accusation prompted a lengthy response from Darwin, which was also published in the *Proceedings*, in which he said: 'I should be loath to think that there are many naturalists who, without any evidence, would accuse a fellow worker of telling a deliberate falsehood to prove his theory.'[16]

In the 1872 sixth edition of *On the Origin of Species*, Darwin adjusted the wording, recognising that it was wrong to say this woodpecker species never climbed trees, although it might, in a treeless situation, be able to live without them, and nest in earth banks where suitable. Like the green woodpecker in Eurasia, it was adept at foraging on the ground. In any event, this tells us something about Hudson's lack of airs and graces, and his forthrightness of views and approach.

Hudson's public criticism of Darwin may not have helped his aspirations to move in academic and scientific circles, but it doesn't seem to have done him any harm in the eyes and esteem of Alfred Russel Wallace, who became a later ally.

Like many people with his background, Hudson was brought up reading the family Bible. This was a form of recreation, as well as no doubt being encouraged – if not required – by his parents. But they were also liberal-minded. There is no more sense that they forced him to attend church than they thought formal schooling necessary for their six children growing up on the edge of the Pampas. But one way or another Hudson was sufficiently God-fearing – or God-believing – to have his world rocked when an older brother returned from a trip to a big city brandishing a copy of Darwin's theory, first published in 1859 when Hudson was 18 years old:

> The idea of a beneficent Being who designed it all –
> did not come to me from reading, nor from teachers,
> since I had none, but was thrust upon me by nature
> itself. In spite, however, of its having come in that
> sharp way, I, like many another, succeeded in putting
> the painful question from me and keeping to the old
> traditions. The noise of the battle of Evolution, which
> had been going on for years, hardly reached me; it was

but a faintly heard murmur, as of storms immeasurably far away 'on alien shores'. This could not last.

One day an elder brother, on his return from travel in distant lands, put a copy of the famous *Origin of Species* in my hands and advised me to read it. When I had done so, he asked me what I thought of it. 'It's false!' I exclaimed in a passion, and he laughed, little knowing how important a matter this was to me, and told me I should have the book if I liked. I took it without thanks and read it again and thought a good deal about it, and was nevertheless able to resist its teachings for years, solely because I could not endure to part with a philosophy of life, if I may be so describe it, which could not logically be held, if Darwin was right, and without which life would not be worth having.[17]

So it is clear where Hudson stood. He was briefly in denial. After that, and perhaps reluctant to throw his lot in completely with the new theory, he became a Darwin challenger: 'Humanity angry at the intolerable insult implied in the Darwinian notion. But we have now so

Sparkling violetear hummingbird (photographed in Ecuador).
Image courtesy of Nick Brooks.

far outgrown that feeling that it is no longer an offence for the zoologist to tell us not only that we are related to the lemur … but even such a creature as the bat!'[18]

As far as courting favour with influential figures was concerned, Hudson had made a shaky start. He may have burned his boats with Darwin, and was walking a bit of a tightrope with Sclater, but he wasn't finished yet.

John Gould and the indoor naturalists

Perhaps it was Dr Sclater who suggested Hudson should try his luck with the businessman artist John Gould in his search for paid employment. Some people, if not Hudson, might have been a little starstruck on meeting Gould, another of the big names in ornithology. He was by common consensus a difficult man. Edward Lear called him harsh, unfeeling and violent, and it was Lear, along with Gould's wife, who did much of the painting.[19] I have prints of 'Gould' toucans hanging on my walls, but I hadn't realised that he had little time for creating the actual artwork himself. He might be better described as an art director, shrewd and hard-nosed in business, ambitious and enterprising. Hudson's visit to him was perhaps inevitably ill-starred. Morley Roberts recorded that Hudson had found his dealings with Gould so unpalatable that he didn't like to be reminded of them. But recall them he did, years later:

> I shall never forget the first sight I had of the late
> Mr Gould's collection of humming-birds, shown to
> me by the naturalist himself, who evidently took
> considerable pride in the work of his hands. I had just
> left tropical nature behind me across the Atlantic,
> and the unexpected meeting with a transcript of it
> in a dusty room in Bedford Square gave me a distinct
> shock. Those pellets of dead feathers, which had long
> ceased to sparkle and shine, stuck with wires – not
> invisible – over blossoming cloth and tinsel bushes,
> how melancholy they made me feel.[20]

I can picture Gould standing proudly beside his trophies, jaded and moth-eaten on their wire nooses, dangling pathetically in the musty room. There were thousands of them. Five thousand, in fact. Perhaps Hudson's mind flashed back to his South American childhood, and these often tiny, delicate birds glinting, iridescent, buzzing and darting, forward, back, sideways, in the warm sunlight, amid vivid blooms dripping with sweet nectar.

Gould also tried to import hummingbirds alive, believing that he could simulate, perhaps in a greenhouse, the conditions they needed to survive. They lived for a time on his ship bound for Britain, 'until they arrived within the influence of the climate of Europe. Off the western coast of Ireland symptoms of drooping unmistakeably exhibited themselves.' Unsurprisingly, 'they never fully rallied'. One of the birds made it alive to his house in London, but then died within two days.[21]

Gould was evidently suffering from some physical affliction that added to his bad humour, and perhaps amplified his moans and groans. Hudson was unsympathetic, and thought Gould had 'as little real illness as he had manners'. He thought of him as a 'necrologist' and an 'embalmer of nature',[22] what John Ruskin called 'the skin-and-bone man, the layer up of mouldering skins and empty eggshells'. Morley Roberts, meanwhile, thought Gould a 'pretentious and unscientific ornithologist'.[23]

If they discussed the global trade in dead birds and their plumage for the fashion industry, Gould might have told him that he wasn't keen on any legislation that would restrict imports; that might make it more difficult for people like him to source specimens. Maybe Gould scorned Hudson's knowledge of British birds, which in these early days was rudimentary. Roberts recalled how Hudson satirised his meeting with Gould in a published sketch, although the title of the magazine in which it was published eluded him. 'That Gould should regard Hudson as some astounding intruder who dared to believe he knew anything of birds, and should be intensely rude between groans, was a joke that did not pass unavenged', he wrote.[24]

Hudson's experience with Gould was probably a turning point for him, such was his dismay, and the creeping realisation that the world of these great scientists, artists and ornithologists might not be what he had imagined.[25] His ambition of becoming if not a professional scientist then at least an ornithologist scraping a living – 'my one dream',[26] he called it – was rapidly fading. His face, his accent, his dress, his lack of education … who was he kidding?

I can picture Hudson making his way through the streets of London, the beggars and street urchins of the East End, the crowded, filthy streets and, despite his own parlous state, being kind to children who were begging, in the guise of selling matches. Then I see him sitting on a bed in a comfortless, dreary, spartan boarding house, an alien lost in a vast, smoky, noisy city, in which he had no other reason to be living but pursuing life as a naturalist, a dream in danger of fading fast. Things were about to get even worse.

Down and out

After the whirlwind of his early explorations, meetings and occasional excursions, there is something of a blank canvas on which to paint Hudson's early years in London. Morley Roberts wrote of 'the irritation he always felt if anything recalled to him his time of bitterest poverty and stress. It is indeed due to this that I, or others, know so little.'[27]

We do know that Hudson spent a lot of time in the British Museum library, reading voraciously, especially on the natural history of his new homeland, and scouring *The Times* and other newspapers to keep abreast of current affairs. It was on his 1875 library readers ticket that he described himself as a field naturalist, the self-identification he would stick to throughout his life.[28]

But for now, he was broke, with no obvious way to monetise the knowledge and expertise he had. He would have to seek alternative employment. Somehow or other he ended up at the office door of a man called Chester Waters, who was trying to make money by offering an ancestry research service mainly to visiting Americans who were keen to trace their family histories in the British Isles. It's possible Hudson first went there to investigate some of his own heritage, to find out more about his Irish paternal grandmother and his parents' claimed link to the Pilgrim Fathers, who left Plymouth on the *Mayflower* in 1620, among the earliest European colonists of North America.

Hudson was hired, and this might have looked like a pretty good job on paper, but it was to be short-lived and ill-starred. His employer was heavily in debt, and one of Hudson's main tasks was to use his imposing physique to keep angry debt-collectors at bay. Sometimes he couldn't get out of the office – lest these bailiffs got in – or get back in himself. There seems to have been a system of fetching lunch into the building using pulleys from an upstairs window. Waters's wife and daughter had abandoned him.[29]

Imprisoned in this office, Hudson daydreamed of world travel. 'A beautiful dream all this, like that of the poor little pale-faced quill-driver at his desk, summing up columns of figures, who falls to thinking what his life would be with ten thousand a year.'[30]

He soon became one of these debt seekers himself, as he was owed his wages. 'It is no use asking for money. I haven't any,'[31] his feckless employer would plead. Following one particularly heated row with Waters, Hudson threw a batch of paperwork in his astonished face. His dalliance with office drudgery and deception had been short-lived. He retired 'penniless and defeated',[32] in Roberts' assessment.

One day, as he wandered in London, Hudson revisited the park with the hundreds of elms, and the rookery that had so impressed him on his first days there. He was shocked to find nothing but logs and sawdust, and men sawing the tree trunks that remained, creating space for new buildings and industry, and infrastructure for the growing network of overground and underground railways. It was a sight he would never forget, and which later served to drive his passion for saving nature.

So began the 'wilderness period' for Hudson, and at times things seemed desperate. He had no income, or fixed abode, and what funds he had brought with him were dwindling. To save money he took to sleeping in the shrubbery in Hyde Park – 'kipping down',[33] as he once called it. This was perhaps less of a hardship than it sounds – at least on warmer, drier nights – as he was a man well used to lying with his head on a saddle, under the Pampas sky.

He was consoled by the proximity to nature this gave him. He shared his bread rolls with the birds, near where he slept. He loved London's sparrows, a familiar companion to him in almost every place he visited, despite the soot and grime. In an urban environment still dominated by horse transport, the humble sparrow was still very much at home. He scribbled notes and poems on scraps of paper:

> I from such worlds removed to this sad world
> Of London we inhabit now together,
> O Sparrow, often in my loneliness,
> No other friend remaining, turn to thee…[34]

If the tuneless chirping of sparrows kept his spirits up through some dark times, the voice of a woman who gave him what he thought would be temporary shelter, signalled salvation.

CHAPTER 2

Salvations

Voice in the darkness – Emily Wingrave

What would have become of the by-now homesick, jobless and bewildered immigrant had he not met Emily Wingrave? No photographs of her survive, and even the date of her birth is uncertain: it isn't given on her gravestone. If descriptions of her as less than five feet tall are correct, she barely reached Hudson's elbow. She had masses of what had at one time been curly red hair, and a powerful voice. She taught music and singing, and would lead evening choruses around the piano, which Hudson loved, even if he himself was no Caruso. It was in part because of her singing voice that Hudson fell for Emily.

We know almost nothing about Hudson and Emily as a courting couple. At some stage he might have told her about his heart condition, contracted in youth through serious illness (doctors left him with a pessimistic view of his chances of living much longer) and his mishap with a faulty pistol in Patagonia, when he shot himself in the knee. If they told each other about past relationships with significant others, it is more than he ever reveals to his readers.

Emily managed a boarding house in south-west London, and Hudson had been a paying guest there. Within two years of his arrival in England, they were married. She was perhaps as much as 15 years his senior, which would make her around 50 when they wed. She had been a singer of some repute, and appeared on stage with Adelina Patti, one of the great divas who dominated opera in this period.

Although Hudson could have amassed hundreds of letters from and to his wife over the course of the years to follow, he destroyed virtually all of them. The few that do survive are therefore precious. How any survived at all is not clear, but my sense is that he allowed them to. I came across one surviving letter of particular interest. Emily and Will, as she called him, never had children. He never seems to discuss it, so the casual reference to it in a letter of September 1890 is entirely unexpected.[1] She has evidently had news of the birth of a child in her wider family, and he remarks: 'Doesn't it make YOU wish that you could have a dear little baby?' Emily was probably approaching 65 by this time, and Hudson in his fiftieth year. Perhaps significantly, the rest of the letter describes the drama he has just experienced while rescuing two young girls and a young woman from the sea.

Although Emily may literally have saved her husband through what can often seem like a marriage of convenience, and spared him from having to make the return migration to South America, their days of uncertainty – and sometimes poverty – were far from over. Her boarding house businesses were precarious, and they were forced to move several times to new addresses. At one point their financial situation was so parlous they lived for a week on the one tin of cocoa they had left in the larder.

Their poverty meant they were trapped. 'It was impossible to escape the immense unfriendly wilderness of London', Hudson later reflected, as it was 'the only place in the wide world where our poor little talents could earn us a few shillings a week to live on. Music and literature! But I fancy the nearest crossing-sweeper did better, and could afford to give himself a more generous dinner every day.'[2]

In 1884, Will and Emily Hudson moved to a small apartment that was redeemed by the fact that it overlooked a green, tree-filled place called Ravenscourt Park. 'I lived for a long time beside it in sad days,' he recalled, 'when the constant sight of such a green and shady wilderness from my window was a great consolation.'[3] It even cheered him in the winter months, when he was at his most gloomy.

The park was small, at 32 acres, but it was a lifeline. Hudson thought of its location as a stepping-stone for birds between the large inner London parks and the open countryside beyond the city to the west. He called it 'the one remnant of unadulterated nature in the metropolis'. Among the birds he could enjoy there he noted 'missel-thrush', spotted flycatcher and 'willow-wren'.[4]

The park had been private but was opened to the public in 1888, when 'a noisy torrent of humanity poured in'. Hudson noticed, he thought, a diminishing of birdlife when this happened – 'for a time it was almost birdless'[5] – but he thought the birds returned when they became accustomed to sharing the park with crowds of people. 'Now there is as much bird life as in the old days',[6] he concluded.

By this time the Hudsons had moved to what would become their permanent address, in St Luke's Road in Bayswater. Emily inherited her sister's house, in which they had lodged for a time. It was run down, 'stale and gloomy' and 'melancholy in its disrepair',[7] and there was still a mortgage to pay on it, but it must have eased their worries at least slightly. It was called Tower House, with a belfry-like room at the top, and Hudson dubbed it his Illimani, after one of the highest peaks in the Andes. He mentions in a letter that he has just 'been flitting', and is now east of his former position.[8] This suggests the move to Tower House was 1886; the letter is presumably wrongly dated as 21 October 1888.

The Hudsons were occasionally able to escape the city and visit the surrounding countryside, with walking holidays taken two or three times a year. 'In very lean years there was but one outing', Hudson recalled, describing their preference for lodging with locals:

> They were mostly poor people, cottagers in small
> remote villages; and we, too, were poor, often footsore,
> in need of their ministrations … We walked day after
> day, without map or guidebook as our custom was, not
> knowing where the evening would find us … This
> method of seeing the country made us more intimate
> with the people we met and stayed with.[9]

But of course mostly they contented themselves with regular walks in London's parks and green spaces, during which Hudson relayed to his wife colourful tales from his past, such as the one about the bookish sheep he had known. The Hudson family's small library had been all the book-learning the young boy and his five siblings had needed. But this informal education was for a time threatened by a rogue ewe that had decided it preferred the comforts of the family home to life among its fellows out on the Pampas. It developed a taste for the trappings of humanoid domestic life – specifically tobacco and books. The former could be difficult to find, 'but books were often left lying about on tables and chairs and were easily got at'.[10]

'She knew very well that it was wrong and that if detected she would have to suffer.' When caught in the act, she would race off with the book, pages flapping, evading child pursuers, to a safe distance, then place one hoof on her prey and begin to tear out and chew its innards. This couldn't go on, and eventually the sheep was banished once more to the flock. 'A sheep cannot "follow his own genius," so to speak', Hudson reflected, perhaps a little ruefully, 'without infringing the laws we have made for his kind.' He took a political message from the experience: 'His condition in this respect is similar to that of human beings under a purely socialistic form of government.'

Hudson had known an English settler in Patagonia who had similar problems with a guanaco (similar to a llama) that developed a taste for cotton and linen – as long as it was white. It stole his shirt while he was in the bath, and a five-mile pursuit on horseback ensued, involving dogs and neighbours. Guanacos, it should be noted, are fast. By the time it was caught, only the hard-starched front of the shirt had survived and could be retrieved.

Handfuls of gold

For a decade and more Hudson persevered with his efforts to make money from writing, while repeatedly receiving rejection letters from publishers. He had just one noteworthy item published in ten years – 'Wanted: A Lullaby' – in a women's magazine in 1876. Curiously, he wrote this article under the nom de plume Maud Merryweather. It was the first of his early, meagre publishing successes.

> It occasionally happened that an article sent to some magazines was not returned, and always after so many rejections to have one accepted and paid for with a cheque worth several pounds was a cause of astonishment, and was as truly a miracle as if the angel of the sun had compassionately thrown us down a handful of gold.[11]

Hudson's first foray into the writing of fiction was driven partly in reaction to the long-drawn-out frustrations of co-authoring *Argentine Ornithology*, which, ten years after his arrival in Britain, was still in progress, a labour of love but a grind, and one that he knew would never put bread on the table. His first published book finally appeared in 1885, a novel with the cumbersome title *The Purple Land that England Lost: Travels and Adventures in the Banda Oriental, South America*. It is based on his adventures in and around what we know today as Uruguay. Unsurprisingly it was not a bestseller. One critic quoted by Hudson had evidently called it 'a farrago of indecent nonsense and lies'. Meanwhile, *The Academy* reviewed it favourably – 'the first taste of praise I ever had'.[12] Notwithstanding this welcome endorsement, he would now try something completely different.

Dabbling in science fiction

Hudson's next project saw him dabbling in science fiction with *A Crystal Age*, which he chose to publish anonymously. It was accepted by Fisher Unwin, a new, avant-garde publisher that was interested in radical ideas and willing to take risks with new and experimental writers such as H.G. Wells, Joseph Conrad and Somerset Maugham. *A Crystal Age* tells the story of an explorer-naturalist who has 'gone somewhere on a botanising expedition, but whether at home or abroad I don't know'.[13] After a fall into an underworld, a blow to the head and period of unconsciousness, he finds himself in a future land.

Hudson calls on episodes from his own recent experience to describe being an alien in a foreign land, arriving like a bird over the ocean, not fitting with a language and culture, having to adapt to be accepted, unfamiliar landscapes and species. In the absence of many other clues about Hudson's emotional life as a young adult who uprooted himself to Britain from South America, *A Crystal Age* can be read as obliquely auto-biographical, offering insights into how the gauche outsider felt in his new surroundings. It also gives a picture of what he wore and carried with him, and how he felt at the end of a long ramble: the narrator and hero Smith describes how his

> tourist suit of rough Scotch homespun had not suffered
> much harm, although the cloth exuded a damp,
> mouldy smell, also that my thick-soled climbing boots
> had assumed a cracked rusty appearance as if I had
> been engaged in some brick-field operations; while
> my felt hat was in such a discoloured and battered
> condition that I felt almost ashamed to put it back on
> my head. My watch was gone; perhaps I had not been
> wearing it, but my pocket-book in which I had my
> money was safe in my breast pocket.[14]

Smith explores this evidently very scenic rural place of the future, with its eye-catching birds, and a curious absence of modernity, and it meets with his approval. 'Looking on this scene I could hardly refrain from shouting with joy … a more tranquil and soul-satisfying scene could not be imagined.'[15] Smith finds a great stone mansion, and witnesses the inhabitants engaged in a funeral procession. He watches unseen until one of the group – a young woman whose beauty has caught his eye – spots him. 'Oh father, look there! Who is that strange-looking man watching us from behind the bushes?"[16] It is tempting to imagine something similar having happened in the gardens outside Emily Wingrave's boarding house, perhaps after he first overheard her singing.

It becomes clear, as he gets to know them better, that these inhabitants aren't like other people Smith has known. They have no knowledge of cities or money, or of any of the famous names he throws at them, current or historic. They are well dressed, like Pre-Raphaelites, making him self-conscious about his own appearance. He worries that this might be hampering his chances of getting better acquainted with the beautiful young woman:

> Her glances, beginning at my face, would wander down
> to my legs, and her lips would twitch and curl a little,
> seeming to express disgust and amusement at the same
> time. I was beginning to hate my legs, or rather my
> trousers, for I considered that under them I had as good
> a pair of calves as any man in the company.[17]

There is a matter-of-factness about the description of this experience that is akin to someone recounting a mystery tour to which he has been taken blindfold. At times the novel seems tongue-in-cheek. The protagonist can be irksome, spending a lot of time somewhat pointlessly describing the plumage detail of fictitious birds. He seems to have a low opinion of the society from which he has come:

> The old life, which I had lived in cities, was less in my
> thoughts on each succeeding day; it came to me now
> like the memory of a repulsive dream, which I was only
> too glad to forget. How I had ever found that listless,
> worn-out, luxurious, do-nothing existence endurable,
> seemed a greater mystery every morning.[18]

Smith's rendition of an old song as his party piece results in a hilarious response from his usually undemonstrative and dispassionate hosts. They are clearly greatly discomfited:

> When I remembered my own brutal Bull of Bashan
> performance, my face, there in the dark, was on fire
> with shame; and I cursed the ignorant, presumptu-
> ous folly I had been guilty of in roaring out that

abominable *Vicar of Bray* ballad, which had now
become as hateful to me as my trousers or boots.[19]

Some of the themes are based on his previous life: horse riding, injury
sustained in the hills, convalescing from serious illness and the delirium
of fever, vegetarianism, and witnessing the illness of the mother of the
household and her demise: 'There came to me a remembrance of my
beloved mother, whose early death was my first great grief in boyhood. All
the songs I had ever heard her sing came back to me, ringing in my mind
with a wonderful joy, but ever ending in a strange, funereal sadness.'[20]

The society he describes has reverted to a simpler life, organised like
a beehive, with the mother figure like a queen bee and one male drone
elevated to the role of king. In a fascinating twist, however, the one
high-tech element that has been allowed to persist is the brass globe that
plays ambient music: 'Listening with closed eyes to this mysterious, soul-
stirring concert, I was affected to tears, and almost feared that I had been
snatched away into some supra-mundane region inhabited by beings of an
angelic or half-angelic order.'[21]

A *Crystal Age* may have been in part inspired by the Crystal Palace
in Hyde Park, the enormous glass edifice built for the Great Exhibition
of 1851 – to show off all that was Great about industrial Britain. John
Gould's vast display of stuffed hummingbirds was said to have been one of
the most popular exhibits, housed in a special structure in the Zoological
Gardens. Although the book is often described as a utopian vision,
and was labelled as such by its publisher, Hudson was not advocating
his crystal age as a utopian future in any uncritical or simplistic way.
Recognising the novel's flaws, he agreed with the publisher that any later
editions would need adjustment. He also hoped for a new cover, thinking
the first edition's 'somewhat satanic'.[22] In the preface to a later edition,
dated 1906, Hudson describes being motivated by his sense of dissatisfac-
tion with the existing order of things, combined with a vague faith in or
hope of a better one to come.

The themes of future societies and gender relationships are familiar
today; they were developed in H.G. Wells's celebrated short story *The
Time Machine*, published eight years later, and since then adapted for
cinema more than once, as well as Aldous Huxley's *Brave New World*, a
set text for schools since it was published in 1932, and also adapted for
cinema. While these books and authors remain household names, Hudson
is all but forgotten – this early novel certainly so, not helped by the fact
that he chose to publish the first edition anonymously.

Within months of *A Crystal Age* appearing, another famous and enduring name in literature was passing away, tragically young – younger even than Hudson, at this time. But his influence had already been felt, and he would continue to prey on Hudson's mind for some time to come. His name was Richard Jefferies.

Richard Jefferies

The legacy of English nature writer Richard Jefferies lives on in large part thanks to a learned society that bears his name. He was an influence on Hudson, and his name recurs throughout Hudson's life. Paradoxically, Jefferies belongs to an earlier literary era, despite the fact that Hudson was the older man – by seven years.

Jefferies had published more than 20 books by the time his life was tragically curtailed in 1887 at the age of 38. Hudson, meanwhile, published almost nothing of note until 1885, when he was 44. Jefferies' bleak vision of the future, *After London*, appeared in the same year. Two years later came Hudson's futuristic novel, *A Crystal Age*. It seems inevitable that Hudson was familiar with Jefferies' book, though his treatment of the future is less apocalyptic and more philosophical.

Another area of overlap seems to have been the Selborne Society, which has claimed both men as members, or patrons, around this time.[23] It has not been possible to trace any more specific evidence of this or how such a shared allegiance might have brought Hudson to Jefferies' attention. In any case, I think we can be fairly sure they never corresponded, and there are three main reasons for believing this: I can find no reference by Hudson to doing so; Jefferies was in poor health by the time he might have known of Hudson; and Hudson was never 'forward' in these things, even when he had made his name.

In the summer of 1887, Hudson received a scrapbook of cuttings from his publisher that contained around 35 reviews of *A Crystal Age*, most of them, in Hudson's view, 'more or less favourable'. The most complimentary, to his mind, was the 'extremely generous' one snipped from the *Glasgow Herald*.[24] Having established the identity of the author, the writer of that admiring review, William Canton, also contacted him directly.

'Unfortunately the book has not sold well, and is still full of imperfections and mistakes',[25] Hudson wrote to Canton. 'One thing in your notice specially pleased my wife', Hudson added. 'I mean your mention of Richard Jefferies at the end, for she has always maintained that I resembled him when writing about nature, and in Natural History articles, and your words proved to her at once that she "was right"; which was a great satisfaction, as up to then she had not found anyone to agree with her. Poor Jefferies!'[26]

By the time Hudson's grateful and typically self-deprecating reply had reached Canton, Jefferies had died at his home in Goring, Sussex, in August 1887 – 'slain before his time by hateful destiny',[27] as Hudson put it.

Their tenures on Earth overlapped briefly, and so too did their spaces. Hudson lived at various addresses in south-west London, and for a time, in the late 1870s, Jefferies was just 10 miles away, in Surbiton. Jefferies was at this time already a prolific author, while the recently arrived immigrant Hudson was reading doggedly, exploring London's parks and green fringes, and making occasional forays beyond the city when time, money and Emily's schedule would allow.

The earliest reference to Jefferies I have found in Hudson's writing comes in an 1894 letter to the prominent humanist and animal rights activist Henry Salt. It gives a hint of Hudson's distaste for and resentment of the majority of writers on natural history, as practised in that period: 'The last place one would look for the feeling for nature, which gives flavour to the writings of Thoreau, Jefferies, Melville etc, would be in the works of the naturalists – so-called. Robert Mudie, the ornithologist, is an exception.'[28]

It is tempting to conceive of the chronology between the two writers as a seamless passing of a pen-shaped baton, with the briefest of overlaps. One thing that didn't pass was the gun, and the end of Jefferies and start of Hudson perhaps marks some kind of threshold between two eras in Britain, between armed and unarmed naturalists. Hudson may have done more than anyone to challenge the cultural norm that nature should be approached with a firearm, collected and taken home.

I've seen nothing to suggest Hudson ever carried or fired a gun in the near half-century of his life in England. This is especially interesting considering that he may have been seldom without one when living in South America, where he had latterly been a collector of bird specimens, lived the life of a gaucho and served in the Argentine army, when frontiers were fiercely contested and summary justice routinely dispensed. Although he never said as much, perhaps he had seen more than enough killing by the time he emigrated and felt it a relief to walk unarmed. We didn't have names for trauma back then, but I often wonder what effect the brutalities he witnessed in early life as a conscript in the Argentine army and living among the lawless gauchos of the Pampas might have had on a mind so acutely attuned to the beauties of nature.

In 1881, Richard Jefferies had described the rapidly changing countryside around London, and the new breed of armed men taking advantage of developments in gun technology and availability. The 'idlest workman' could now afford 'a double gun which to our grandfathers would have seemed a marvellous work of art'.[29]

> As soon as the first frosts have begun to shorten the
> supply of food for wild birds, and urged them to seek
> it in more public and exposed places, you may see the
> hedgerow sportsman sallying forth for his day's sport.
> On a Sunday morning you may hear the country
> for miles round London and other populous towns
> resounding with an almost constant fusillade from the
> hedgerow brigade.[30]

Hudson summed up his own attitude to guns by the time he was established in Britain:

> I know good naturalists who have come to hate
> the very sight of a gun, simply because that useful
> instrument has become associated in their case with
> the thought and the memory of the degrading or
> disturbing effect on the mind of killing the creatures
> we love, whose secrets we wish to find out.
> Abstention from killing had made me a better
> observer and a happier being, on account of the new
> or different feeling towards animal life which it had
> engendered.[31]

The saga of Hudson's *Argentine Ornithology* project with Dr Sclater was finally over in 1888, with the book published in two volumes. For all the creative tension with Sclater and Hudson's impetuousness, he had managed not to burn his bridges with this influential figure. Had he done so, the story of the Bird Society might have been quite different. Not long after this, Hudson's life would take a dramatic turn as he became acquainted with the women who held the key to his future.

Butterflies in a church

Frustratingly, there is no specific record of how Hudson came to know the women who founded the Society for the Protection of Birds, but the following scenario – based on what Hudson wrote – is not implausible.

During the hard winter of 1887–8, gulls had sought sanctuary in the otherwise uninviting surroundings of inner London. A sun-kissed English country church stood in marked contrast, with primroses and daffodils riffling in the breeze, chaffinches in song, rooks cawing and jackdaws tumbling against the egg-blue sky and towers of cloud. It was Easter Sunday. Hudson shuffled into church among a gaggle of chattering men and women, a wool-capped head and shoulders above the smartly dressed parishioners around him. They didn't recognise the mysterious figure – a foreigner of some kind, who mumbled a polite 'good morning' with traces of an American or Hispanic accent. He removed his hat, ducked at the door and flipped a penny into the collection plate.

He took a pew, settling in an aisle near the door, and glanced around at the rafters, the stained glass, the worshippers. 'It came to pass that during the service the sun began to shine very brightly after several days of cloud and misty windy wet weather,' he wrote, 'and that brilliance and the warmth in it served to bring a butterfly out of hiding; then another; then a third; red admirals all.'[32] The sonorous drone of the sermonising rector faded in and out, drowned by the pattering of the wings of butterflies, newly emerged from winter rest, against stained glass, over the sound of his favourite hymn, *Lead Kindly Light* being sung by the congregation, now standing, and accompanied by the organ.

> and they were seen through all the prayers, and
> psalms, and hymns, and lessons, and the sermon
> preached by the white-haired Rector, fluttering
> against the translucent glass, wanting to be out in that
> splendour and renew their life after so long a period
> of suspension. But the glass was between them and
> their world of blue heavens and woods and meadow
> flowers.[33]

The glass eyes of mummified birds of paradise gazed blankly at Hudson from bonnets in the pews in front of him:

> Among the women (ladies) on either side of and before
> me there were no fewer than five wearing aigrettes

of egret and bird-of-paradise plumes in their hats or
bonnets, and these five all remained to take part in
that ceremony of eating bread and drinking wine in
remembrance of an event supposed to be of importance
to their souls, here and hereafter.[34]

Service and rituals over, congregation milling around by the doorway,
Hudson's attention returned to the butterflies, and thoughts of freeing
them, but he 'soon reflected that to release them it would be necessary to
capture them first, and that that could not be done without a ladder and
a butterfly net'.[35]

With a wistful look, Hudson put his cap back on, and left the church:

It saddened me to leave my poor red admirals in their
prison, beating their red wings against the coloured
glass – to leave them too in such company, where
the aigrette wearers were worshipping a little god of
their own imaginations, who did not create and does
not regard the swallow and dove and white egret and

Etta Lemon's first recollections of Hudson pre-date the first meeting of the Bird Society.
They would be comrades in campaigning for well over three decades.

bird-of-paradise, and who was therefore not my god
and whose will as they understood it was nothing
to me.[36]

Around the same time, a teenager called Etta Smith was sitting in
a crowded church in Blackheath, Surrey.[37] The hymn *All Things Bright
and Beautiful* was ringing out around her, lustily rendered by organist and
congregation. Etta mouthed along to the words:

He made their glowing colours,
He made their tiny wings…

As she sang, she scribbled names discreetly in a notebook: names of the
dead birds and parts of dead birds (and other fauna) dangling from the hats
of those around her; names of the wearers…

Back at home, she sat down at her writing desk and began to compose
letters: 'Dear Mrs …' She wrote carefully until she had a small pile of neatly
folded envelopes in front of her before sitting back, breathing deeply, a
determined glint in her eye. Later, she set off to post the envelopes, one
by one, through the doors of the relevant local churchgoers, along the way
passing a boutique with window dummies bedecked in millinery plumes.

Queen Victoria had been on the throne for 50 years and feathers in
costume were all the rage, not least during the celebrations to mark her
golden jubilee. Changing attitudes to plumage use was no small ambition.

It is tempting to imagine Hudson in the same pews as the teenager
Etta Smith as she jotted in her notebook the names of the ladies around
her who were wearing bird plumes in their hats and dresses, for later
admonishment. It's not so far-fetched an idea. That young woman later
married and became Etta Lemon, who would go on to lead the SPB/RSPB
for many decades. Lemon recorded in a tribute to Hudson that her first
recollections of him were from before 1889, and the first meeting of the
Bird Society.[38] One way or another, these converging lives were about to
intersect, and they would be comrades in campaigning for three decades
to come, alongside Eliza Phillips and many others.

CHAPTER 3

The Bird Society

Campaigning women of Manchester and London

History has recorded that the Royal Society for the Protection of Birds (RSPB), which today has more than a million members, and manages over 200 nature reserves, began life in 1889 in two places simultaneously – Manchester in the north of England and Croydon in the south.[1] The groundswell of activism against the wanton killing of wild birds for fashion accessories and other indulgences sprang up in more than one place. But perhaps Manchester more than anywhere else lays claim to being the birthplace of the organisation – not least because Emily Williamson, the founder, called her movement the Society for the Protection of Birds (SPB), the name that stuck.

Meanwhile, in the south, the campaign was coalescing around a drawing room table and presided over by Eliza Phillips. She initially identified her informal group as the Fur, Fin and Feather Folk, which scores high for alliteration but was never likely to be chosen as the name

Emily Williamson, founder of the Society for the Protection of Birds in Manchester, in 1889. She would later recall the wisdom, moderation, zeal and energy of those early years, and the vital support of many famous names.

when the groups later merged. It was dropped when the two groups merged in 1892.

Author Tessa Boase calls such meetings 'the political threshing ground' of the age, where civic-minded women, usually with some spare time and money, could make plans for the betterment of society, unimpeded by the often oppressive presence of men.[2] The plumage industry and the fashion for wearing bird feathers, wings, tails, crests and sometimes entire skins was, in the emphatic view of Eliza Phillips, 'without doubt a woman's question'.[3] So, if this is all about sisters doing it for themselves, the presence and involvement of Hudson seems all the more remarkable. I have long been fascinated that Hudson was present from the start – the only man in the room. So how did he come to be there? First of all he had to be invited – and tolerated – and second, he had to be comfortable with being present. All this confirms him as a very unusual, counter-cultural individual.

For a long time I could only speculate on how it came about. The 'gap in the fossil record' of Hudson's first 15 years in London means there are no specific traces: no first letters to show how he came to know Eliza Phillips, for instance, and how they established mutual trust and understanding.

I had also long believed that there was little trace of Eliza Phillips. The RSPB's 1989 centenary book has just two passing references to her (one more than Emily Williamson). So, what *do* we know? Well, she was born Eliza Barron, and was an animal lover who became very active at an early stage in the welfare movement. She first married in 1847. Her husband, Robert Montgomerie Martin, doctor, botanist, writer, editor and civil servant who travelled widely in the colonies, was, among other things, private secretary to the Duke of Wellington. Eliza and her husband were based for a time at the Hampshire estate of the dukedom and at its grand London home, Apsley House on Hyde Park Corner. In this glamorous context she became a writer of some renown and distinction and mixed in the highest echelons of society. She had been half the age of her husband when they married, and he died in Surrey in 1868.[4]

In 1874, just as Hudson was arriving in England, Eliza was getting married again, this time to the Reverend Edward Phillips. They moved from Surrey to live at Culverden Castle in Kent – another highly prestigious address – fit for a king, if not an ageing vicar.

It was Etta Lemon who, in 1941,[5] recalled Hudson regularly attending Eliza Phillips's first meetings: 'He was in the habit of coming to Mrs. Edward Phillips' Fur, Fin and Feather afternoons at 2 Morland Road, Croydon … a frequent guest at these afternoons and greatly assisted in the plans then laid and the schemes discussed.' Based on this, later biographies of Hudson have placed Eliza Phillips in Croydon at the outset of the Bird Society, requiring us to imagine Hudson being invited and travelling on foot and train to get there, all of which presumes a well-established link already in place.

Intriguingly, Etta Lemon also recalled that Hudson was attending meetings hosted by Eliza Phillips in Croydon even *before* 1889. 'Mrs Phillips gave unstintingly of her literary ability, and great experience of the world, and Miss Hall of her money and sweet patience', she wrote. As a result, Croydon has always been taken as the place of origin of the southern end of the Bird Society.

But there is a snag. Lemon seems to have misremembered (50 years had elapsed, to be fair). Eliza Phillips moved to this Croydon address in 1891, at least two years after the first meetings had taken place. The year 1891 was also when her second husband passed away, which is presumably linked to her moving there.[6]

Lemon mentions that meetings were also hosted by Miss Catherine Hall at her home in London. If my analysis is correct, it would make Lancaster Lodge in Notting Hill the birthplace of the organisation in the south, and not Croydon. This London address brings us to within a kilometre – or ten minutes' walk – of Hudson's Tower House in Westbourne Park, and I think this helps solve the mystery of how he came to know the women founders and to be involved in their first meetings, at least in practical terms.

The business of these meetings was followed by meals, which, according to Lemon, 'were daintily served and included dishes and wines which we knew were his favourites', when Hudson 'talked most freely and showed us what was in his heart and on his mind'. Lemon also gives the following insight: 'he was no public speaker, but a fascinating talker on subjects about which he cared, and yet the enthusiasm for bird protection which he inspired was not so much attributable to what he said, or even to what he wrote, as to what he *was*'.[7]

It's another testament to the peculiar charisma of Hudson, his *je ne sais quoi* (*no se que* in Spanish, more appropriately). Whatever this magnetism, Hudson's aversion to public speaking (mentioned by Lemon) in particular and the limelight in general would come back to haunt him.

I used to assume that by the time Hudson joined the nascent Bird Society he had already acquired some minor celebrity status, and that might help to explain why his presence was tolerated, if not welcomed. But this theory collapses for two reasons – the first being that he wasn't yet a name; he had only just published *Argentine Ornithology*, his collaboration with Dr Sclater, and in any case this was hardly likely to have brought him much renown. His futuristic fantasy *A Crystal Age* wouldn't have been known about unless he chose to mention it, as he'd published anonymously. And his racy gaucho adventure yarn *The Purple Land* was hardly typical material for a drawing room table. And the point is that plenty of other male figures could have been invited, or involved, but they weren't.

There was something about Hudson.

For a long time I thought that his path might have crossed with that of Eliza Phillips through their shared attendance at or involvement with another animal welfare organisation. This could, for example, have been the Selborne Society, which claims Hudson among its early patrons. I made enquiries and did some research online, as far as this was possible, but couldn't confirm that Hudson actually *was* on their books. That society seems to have had the backing of some very big names – poet laureate Lord Tennyson, artist and author William Morris and poet Robert Browning, among others. Patrons weren't expected to do much, or even attend meetings, so if Hudson had agreed to let his name be cited as a backer, there might not be much sign of him beyond that.

For all her standing in society, no photographs are known to exist of Eliza Phillips. No photograph survives of her Vaughan House in Croydon, but I found one of a boys' home that stood nearby, which gives some idea of the imposing scale of the place and its appearance.

It was Hannah Poland, a local SPB branch secretary in London, who was the catalyst in bringing Phillips and the Manchester-headquartered SPB to each other's notice. Poland saw something that Phillips had written, and wrote to tell her about the SPB, in 1891, the same year that Phillips was widowed and moved from Kent to Croydon.[8]

Using her connections, Phillips brought on board her close associate the Duchess of Wellington as a supporter and perhaps through her the Duchess of Portland as president. Interestingly, more than a decade later, in October 1901, while exploring Hampshire on his bicycle, Hudson found his way to the Duke of Wellington's estate. He wrote to Eliza Phillips about this place she had known so well in earlier life, when married to the duke's private secretary. 'The Duke's great house is empty of people and has an air of quiet and loneliness about it that is almost like desolation. I went round it and through the park and met no person. I was all the time thinking how familiar to you must be everything that I saw there.'[9]

The first of Emily Williamson's monthly gatherings in Didsbury, Manchester, was held on 17 February 1889. The June 1891 transfer of the centre of operations from Manchester to London was brokered by the Royal Society for the Prevention of Cruelty to Animals (RSPCA). Great credit is due to the RSPCA for recognising the value in creating other organisations dedicated to specific objects of campaigning rather than trying to do everything itself. Williamson was content to see her organisation's name adopted and London become the base for future progress, c/o the RSPCA in its early years, at Jermyn Street in Westminster, close to the corridors of government power.

Hannah Poland was just 15 at the time the Manchester end of the Bird Society was forming in 1889. She must have had something about her for one so young, because her peers appointed her the first honorary secretary when the London group merged with the Manchester group in 1892.

SPB founder Emily Williamson travelled to London for discussions with Eliza Phillips about merging the two organisations. For what now seem like obvious reasons, with the benefit of hindsight, the name of Williamson's group was retained for the amalgamated body and retains the name (with the Royal prefix added in 1904) to this day. She chose well. I wonder if there was much debate in that meeting room at the offices of the RSPCA. Or did Eliza Phillips recognise that Fur, Fin and Feather Folk was just a little bit twee to meet the needs of the turbulent years ahead, with the massed ranks of the plumage industry and the guns and nets of the bird killers that would be ranged against them, disparaging them for their 'sentimentalism'?

The name SPB would have its critics too, of course. Some onlookers considered the name grandiose. 'To assume such a very ambitious title as "The Society for the Protection of Birds" for a band of ladies who do nothing but abstain from personal iniquity in the matter of bonnets, may give occasion for the unrighteous to scoff.'[10] Others considered the new society 'so modest and apparently insignificant that it was greeted with smiles of amusement rather than of sympathy'.[11]

Another intriguing twist is that Hudson seems to have been acquainted with Hannah Poland since she was an infant, having lodged with the Poland family when he first arrived in Britain, 15 years earlier.[12] It seems reasonable to assume that he stayed in touch with Poland throughout that time, unless this later reunion through a shared interest in bird protection was purely coincidental. Perhaps he had helped to nurture her interest in the cause of saving birds? If so, it's frustrating to find no mention of it anywhere.

I have found a reference to Hannah Poland in one of Hudson's earliest surviving letters written in England, from October 1892.[13] As honorary secretary, Poland was in charge of producing the Bird Society's first annual report. This production seems to have gone awry; Poland having been called away unexpectedly to the south coast resort of Worthing – perhaps because she had elderly relatives there. She had, with some reluctance it seems, handed the project over to Eliza Phillips, who was serving as the first chairman for a brief period before handing over to Hudson. Phillips had explained the problem to Hudson, and he replied, satisfied that the publication was now with Phillips to complete, having had sight of what Poland had so far produced. He regretted Poland's failure to hand the work over at least two months sooner.

I think that as her limitations became clear – and this may have been down to her circumstances more than her abilities – Hannah Poland was succeeded the following year by Etta Smith (soon to become Etta Lemon) as honorary secretary. Despite the glitch with the annual report, Poland stayed on, and remained deeply engaged with the society, being elected the first honorary member in 1905.

Hannah Poland's life had its share of tragedy, and it is perhaps her story more than any other that captures the personal drama of the plumage campaign. She was born in 1874, the daughter of a fish merchant, living in Warwick Road, Paddington, close to where the Hudsons lived: their Tower House is just a mile away. After she married Isidor Lemel in 1906, Hannah lived in a street behind Kilburn station in north-west London. She gave birth to a son in 1907, but it seems that hers was a tragic household. In 1909 her husband was admitted to Bethlem Hospital (which gives us today's word bedlam).[14] It was among the most notorious institutions of its kind in the era of primitive treatment of mental illness. He was taken there after repeatedly threatening to kill his wife. Hudson later referred to her as 'that brave young girl'.[15] Well might he have done so.

Poland is perhaps the most mysterious of the founding women who held senior posts. Another mention of 'Miss Poland' in Hudson's letters comes, tantalisingly, in one of his last, just three weeks before his death – a passing reference to her mother in Worthing.[16]

We can glean a few further insights from Hudson's letters to Eliza Phillips. One incident he mentions gives a flavour of the modus operandi of this growing network of branches and secretaries and their combined efforts to spread its gospel via influential figures across the land, and beyond. He reports meeting his counterpart, Mrs Hyde-Walker, in January 1892, just returned from a mission to Devon to try to persuade the vicar's wife to give up her plumage and bird-skin-wearing habit, to no avail. She refused even to read the Bird Society pamphlets offered her, to enhance her understanding of the matter. Worst of all, the vicar was Mrs Hyde-Walker's own brother. Hudson relayed this news to Eliza Phillips, to whom he appears to have told everything.[17]

On another occasion he and Phillips conferred about a problem with Etta Lemon, in much the same way that trustees might consider how to guide their chief executive. 'I do not know that there is anything more to say to Mrs Lemon', he replied. 'You have said everything in your letter already.' The matter seems to be a communique to the network. Hudson can only urge a friendlier tone. 'If she sends out a secretarial letter, I should advise her to smooth her instructions to the B. [branch] Secretaries a little. The tone sounds sharp.'[18]

Hudson also shared with Phillips a letter from a Mrs Agnes Crane – 'one of the few women who are thoroughly scientific, but science has certainly not made her hard'.[19] His enthusiasm for spreading the word about the Bird Society can sometimes appear to border on the obsessive. In June 1892 he sent out a hundred letters and was taking another hundred with him to the country, to distribute opportunistically.[20]

He wrote to Phillips again in August 1892 after she had returned from a foreign trip with Catherine Hall. He had been on the south coast, witnessing some unseasonal storms, sympathising with his friends who were making the ferry crossing in such conditions. 'I hope that you and Miss Hall are stronger for your visit to France, but how glad you must have been to reach the grand library at Vaughn [sic] House,' he writes, adding 'What a crossing you must have had!' He signs off with 'Please give Miss Hall my best regards.'[21]

It seems that one year after her husband's death, Phillips had the company of Catherine Hall as she settled in at Croydon during her time of bereavement. Hall had been appointed the Bird Society's treasurer, as well as having apparently been its main source of funding through its founding years. Hudson mentions their large pet dog called Finn, I think one of a number of pets that they were in the habit of rescuing and adopting.[22] On another occasion he hopes that she has been able to heat their large rooms during a spell of bitter weather.

Winters could be brutal in this era. Hudson recorded an 'exceptionally long and severe frost' over the winter of 1892/3. It meant 'a memorable season in the history of the London gulls', as the birds came inland, following the Thames upriver and into the city for shelter, in desperate search for scraps of food. And the public turned out to greet them again when 'the custom of regularly feeding the gulls in London had its beginning'.[23] Each day of this prolonged freezing spell, for almost a month, hundreds of people came to the riverbanks and bridges during their lunch hours and shared some of their food with the birds. Just as the proximity and beauty of the birds had been described as a revelation to the people, so Hudson thought the welcoming and generous behaviour of the people might have been a revelation to the birds. Not only were people welcoming and nurturing the visitors, but the trigger-happy minority also stopped killing them. 'For the last time, gulls were shot on the river between the bridges, and this pastime put a stop to by the police magistrates, who fined the sportsmen for the offence of discharging firearms to the public danger.'[24]

Hudson also called it a 'late, terrible winter'. The worst of it had passed by the start of February 1893, but a letter to William Canton indicates that Hudson was still in low spirits: 'I am cursed with an eternal lassitude … I do little work – I am ashamed to say how little; and instead of seeking new friends I have been obliged to let go my hold on the very few I had.'[25] And in March the snows came back. Hudson was chairman by this time, replacing Eliza Phillips, but may have regretted agreeing to hold these reins even for a brief interim period. The author of the RSPB's 1989 centenary book *For the Love of Birds* makes brief reference to 'Hudson's ineptitude as Chairman', and regrets that Mrs Lemon 'proffers no gory details' of it. It's unclear what the evidence is that Hudson deserves to be described as inept. Whatever his shortcomings as a committee man (I suspect mainly a lack of enthusiasm for managing others, and admin), of which Hudson was all-too self-aware, it seems a remarkably uncharitable description of an important figure in a voluntary role helping the founders to establish the organisation. For the record, Phillips also wrote many of the early pamphlets as head of publications. Noticeably, the organisation's obituary for Phillips makes no reference to the Fur, Fin and Feather Folk of Croydon.

Besides the neglect of Hudson, I had often wondered at the absence from written history of the RSPB's female founders. And I have long assumed that they must have been the sorts of people who were not apt – or able – to leave sufficient traces for the piecing together of a collective biography. I have understood, from inside knowledge of the RSPB in more recent decades, that a succession of would-be authors of this story have come and gone, leaving empty-handed and frustrated at the paucity of records. Much of the record was lost in the Blitz of the Second World War. What survives is packed away in boxes, and difficult to access. Limits on time, space and resources, and higher priorities for a conservation charity can be added to the mix.

In July 2019, after the fashion of Hudson, I travelled to Croydon by train and bicycle to – unlike Hudson (who never overcame his fear of public speaking) – give two talks to the Croydon RSPB group about 'the man above the fireplace'. Even if Croydon wasn't after all the setting for the very first meetings, the fact that Eliza Phillips lived there from 1891 justifies its place in the heart and heritage of the organisation.

I checked in to a guesthouse on the other side of town, and in the living room there I met a man aged 100 – someone who had been alive when Hudson was alive. Suddenly the Hudson era didn't seem quite so long ago – the gap was bridged. The old man chewed cheerfully on Haribo gums from a bag (I'm not saying this is the secret of longevity) but was too hard of hearing to do much more than smile serenely and nod.

I found the spirit of Mrs Phillips alive and well at the Old Whitgiftian Association Clubhouse, a pavilion overlooking the cricket ground there. 'I'm glad you talked about the women founders too', one of the group said to me at the end. 'I'd have been pretty annoyed if you hadn't.' Quite right too.

Not surprisingly, this group is more invested in the origins of the society than any other I've visited, given the proud links with Vaughan House on Morland Road, although Mrs Phillips's house is no longer here; a school now occupying the space. That evening, as the parakeets screeched overheard on their way to roost, group leader John Davis took me there after the second of my talks. He also very helpfully provided me with a copy of Eliza Phillips's will, in which she left an extremely generous gift to her close and loyal friend and colleague Mr Hudson.

Not long after those first gatherings of the emerging society, Hudson met another crusading campaigner, championing an even wider range of causes, which included animal welfare.

Meeting Don Roberto Cunninghame Graham

Hudson can hardly have failed to notice the headlines about Robert Cunninghame Graham, whose nickname alone – 'the Gaucho Laird' – must have intrigued him. *The Times* called the aristocratic radical a 'cowboy-dandy'.[26] Cunninghame Graham was the first socialist to take a seat in Parliament, the odd man out among the 600 members. He was a founder of the Scottish Independent Labour Party alongside his close friend Keir Hardie. His most commonly applied nickname 'Don Roberto' reflected his mixed Scots-Spanish heritage.

Robert 'Don Roberto' and Gabriela Cunninghame Graham at their Gartmore estate, near Stirling. 'I wish to God I could go and see your capercaillies,' Hudson wrote to them, in 1896.

Don Roberto ran for Parliament in 1885 and again in 1886, the Liberal candidate (he had yet to help create the Scottish Labour Party) for North-West Lanarkshire, a constituency close to Glasgow. His platform covered a long list of progressive issues: votes for women and for men currently without a vote, home rule for Scotland as well as Ireland, the abolition of the House of Lords, free school meals, improved workers' rights and nationalisation of land and industry. He was duly elected second time around, defeating the Unionist candidate by just over 300 votes.

In his first speech in Parliament he called for an eight-hour working day. 'I came to this House straight from the prairies of America, where want is unknown, so that the sight of such misery as exists in London was brought home to my mind with exceptional force', he declared. 'As long as I have the honour of a seat in this house I shall continue to press this question.'[27]

Frustrated by tortuous procedure, Don Roberto was suspended from Parliament in 1888 for daring to accuse his opponents of a 'dishonourable trick' in procedure as they thwarted his efforts on behalf of workers on

15-hour days and pitiful wages. Such an accusation was practically unheard of. It was, the Speaker told him, 'a strictly improper and un-Parliamentary expression'. Don Roberto was twice asked to withdraw his statement, and both times he refused. 'I never withdraw,' he replied, defiantly. 'I simply said what I mean.'[28]

A year before this, he had attended a lively Trafalgar Square demonstration in support of workers' rights, freedom of assembly and Irish home rule. His wife Gabriela was watching from a hotel balcony as he was beaten to the ground by police and arrested, blood streaming from his head. 'I went simply to assert the right of free speech, and not as a Socialist or as a Radical',[29] he always maintained. He stood trial soon afterwards and was sentenced to six weeks in jail.

On his release, he declared that he would not wish jail even on his political opponents, but he seems to have been none the worse for his time behind bars, regarding it perhaps as useful research. He was soon back agitating and organising for social reform, including for prisons. Having served his suspension from the House, he arrived at Parliament once again and met Prime Minister Lord Salisbury, who asked him, with a grin, 'Well, Mr Graham, are you thinking where to put your guillotine?'[30]

Unbowed by the rough treatment and spell in jail, Don Roberto continued to make headlines, addressing Labour movement rallies from Hyde Park to Paris to Madrid, sometimes with Gabriela on the platform beside him. On one occasion he was expelled from France, as the authorities clamped down hard on demonstrators, some of whom were shot dead.[31]

I can only imagine the humble Hudson's surprise and pleasure when one morning he received a letter from Gartmore House near Stirling, Scotland. It was from Cunninghame Graham, expressing the pleasure he had derived from Hudson's writing about South America, and his own familiarity with its subject. This remarkable politician, adventurer and kindred spirit would become a great friend to Hudson, and a very useful man to know.

I became aware that correspondence between the two men had been published in America, 20 years after Hudson's death. Only 250 copies of a small volume had ever been produced. Surprisingly, I found it advertised for sale online, but less surprisingly, it would cost hundreds of pounds to buy. So, I set out to the University Library in Cambridge to have it dug out of the archives. Only Hudson's side of the correspondence survives, leaving me to guess what Don Roberto's letters had said. His first one, in March 1890, had evidently complimented Hudson on his evocation of the Pampas in an article. 'I have the feeling for it which each of us has for his own, his native land', Hudson replied, wondering what visiting gauchos would think of a London November 'pea-souper' fog.[32]

The two men enjoyed sprinkling their conversation with old Spanish and gaucho idioms, a language only they knew, among their London friends. 'The flavour of gaucho talk is lost in translating',[33] Hudson mentions in an early letter. Their shared knowledge of the Spanish language as well as the gaucho life gave a dimension to their friendship unlike any of Hudson's others. They were able to discuss subjects as diverse as the conquistador Aguirre, Basque culture and the benefits of reading *Don Quixote* in the original Spanish, wherein it retains its flavour and spirit.

It struck me, reading through Hudson's letters, how Don Roberto emerges as his alter ego. He would be the obvious partner in any gothic Western version of Hudson's life story. The correspondence provides touchstones through Hudson's life and the historic events of the era, like a cinematic voice-over, which also describes moments of reminiscence, flashbacks to the early and sometimes traumatic events of childhood and young adult life that had shaped them both, and Hudson in particular. This commentary helps at least partly to explain him and shed light on the mystery that surrounds him. Don Roberto will be a constant presence through this story.

Hudson lost no time in replying to that first letter, eager to meet this kindred spirit of the Pampas, admitting it would be 'a rare pleasure to meet with anyone at this distance with whom I can compare notes about it'.[34] He invited Don Roberto to visit Westbourne Park, 'if you can spare the time, and are not afraid to travel into such a desert as this, with macadam and mud for paja [grass] and smoke, or something dark, for cielo [sky], I should be very glad to see you any day'.[35] They met up soon afterwards, becoming firm friends immediately.

Their paths might have crossed much sooner. Don Roberto had lived on and knew the frontier that had bred Hudson, and their time there had overlapped by several years. When Hudson was exploring his South American homeland in the early 1870s, a youthful Don Roberto was in Paraguay, travelling 600 miles on horseback, following the River Parana to the Iguacu Falls, researching the life of the early Jesuit missionaries. His book about the adventure, *A Vanished Arcadia*, later influenced the 1986 film *The Mission*, with Robert de Niro and Jeremy Irons leading a star-studded cast.

With his colourful turn of phrase, flamboyant outfits and tall, imperious demeanour, in or out of Parliament Don Roberto would remain one of the most celebrated figures in late Victorian London. He rode a magnificent

Argentine stallion he called Pampa, arriving at Parliament on horseback, like an old Spanish hidalgo, and exercising it each week along with all the other well-to-do equestrians at Rotten Row, the gallops in Hyde Park. Hudson would sometimes meet him there, in part also to see the celebrity stallion.

There is a touching story behind this too. Don Roberto had rescued Pampa from a life of drudgery pulling trams in Glasgow, finding the horse miserably thin and fallen in the street, unable to rise in spite of the driver's lash. As soon as he saw the wretched animal, he resolved to buy it there and then, handing over a roll of notes, removing its harness and taking it home with him to his Gartmore estate. He restored it to its full snorting, mane-swishing, hoof-stamping, tail-switching glory. From the brand on its rump, he was able to trace its origin to the Pampas both he and Hudson had known so well.[36] While Hudson was not quite in need of such restoration, I cannot help conceiving of Don Roberto's adoption of Hudson as instrumental in putting some kind of new-found spring in his step too.

Hudson noted that the fact alone of Pampa still having a tail marked it apart from the other horses on display, with their severely docked stumps. He wept while embracing Pampa's neck, as memories of his distant youth flooded back on the warm breath of the animal.[37] Pampa reminded Hudson of 'a fiery animal I bought from an old gaucho when I was a boy – the horse I loved so much'.[38]

How Hudson must have been tempted to hoist himself on to Pampa's back, and no doubt Don Roberto encouraged him to do so as they regaled each other with their stories of the old life among the gauchos. But if Hudson ever did relive this dream, it's never mentioned by him in letter or book. There are so few mentions by him of horse-riding during his years in Britain that it might be he held a general policy against it. He mentions on at least one occasion that the cost of hiring horses was a deterrent.

The two men will soon have worked out the other parallels in their lives. They had married at around the same time; Don Roberto and his beloved wife Gabriela in a London register office just two years after the Hudsons. Like the Hudsons, and for reasons never discussed, Don Roberto and Gabriela were childless. Hudson would hear the story of how they had first met in Paris, when Don Roberto, on horseback, almost knocked over his future wife. When he dismounted to apologise, forgetting where he was he spoke to her in Spanish. To his astonishment, Gabriela responded in kind. It was love at first sight. Her full name – Gabriela de la Balmondière – reflected her exotic and prestigious background – a 'high-born' French father and Chilean mother. At least that's what she and Don Roberto liked to tell people. The truth, when it finally became public knowledge a century later, would be more prosaic. In fact, Gabriela's real name was Carrie Horsfall. She had run away from her family home in Yorkshire.[39] I can't be sure how much Don Roberto told Hudson about the family politics, but Don Roberto's mother Missy Bontine did not approve of the marriage – hence the register office wedding.

For all that they had in common, the two men were an intriguing contrast of personality types. The introverted Hudson enjoyed his friend's extroversion and adventurousness and indulged him in his whims. Their mutual friend Morley Roberts nailed it when he said Don Roberto was 'like a Velazquez come to life'.[40] Recalling the three of them together, Roberts paints a vivid picture: '[Hudson] looked down on us and told us tales and laughed at politics, and came back again to the natural life of man, of beasts and birds.'[41]

Roberts, himself a man of great artistic and scientific intellect, confessed to being a little in awe of both. He described the relationship between them: 'To Hudson, Graham was a kind of Don Quixote, a tilter at mills.'[42] For all his Scots accent Hudson thought Graham 'in essence, by some miracle, a veritable son of Spain', and 'the first and last nobleman in England'.[43] Don Roberto summed up his view of Hudson: 'there was no one I thought more highly of as a man, or respected as a genius. That he was a genius, I think all his real admirers know. Some day the world will become aware of it.'[44] He also said: 'to me his astonishing attraction is that, apart from all his other gifts, his style, quiet humour, sarcasm and pantheistic mind, he was at heart an old-time gaucho of the plains'.[45] While Don Roberto seemed so free to roam far and wide, 'a cruel world imprisoned Hudson in his London cage'.[46] Hudson would find a kind of vicarious freedom through the adventures of his new friend.

Don Roberto's world

London's Belgravia – then as now – was among the wealthiest neighbourhoods in the country, and Hudson was invited here to meet Missy Bontine Cunninghame Graham while Don Roberto was staying with his mother. Hudson's response is typical: 'I could no more dine at Chester Square with you and your friends than with Fairies and Angels. Those beings do not really dine, they sup, but let that pass. The fact is, being poor I long ago gave up going to houses dining.' To reinforce his point, he then quoted the Bible's urging against 'unequal alliances', a lesson that had obviously resonated with him since childhood.[47]

Hudson's response might have been show-stopping enough to stifle the friendship, but Don Roberto sympathised with his position, and was soothed by his suggestion that they meet instead at the Zoological Society offices in Hanover Square (into which the Bird Society would later move).

Perhaps Hudson would later accompany Missy Bontine on a visit to the zoo, because before long he had also become close friends with her, putting aside his usual reticence for the sake of his new friend, but perhaps also because he could recruit her to the cause of the Bird Society. Hudson became a regular at Bontine's apartment, where she hosted salons and helped provide her son with the setting in which to become one of the most well-connected public figures of the era.

A letter survives in which Hudson thanks her for the generous donation she had made to the Bird Society, and discusses the persecution of rooks in London, a subject that exercised them both. Hudson had become intimate with the habits of this species: 'it will dine on dung in cold weather', he once noted, 'though not cheerfully'.[48]

Inevitably, Hudson also called on her son to help the cause of the Bird Society, and no doubt Don Roberto, drawing on his parliamentary experience, advised the emerging Bird Society chairman, as he advised others, on how to structure a political campaign:

> A programme should be:
> 1. Short
> 2. Definite
> 3. It should not argue or endeavour to meet
>   objections. Nail the flag to the mast.[49]

And he would have warned Hudson that the society would have to learn patience in seeking change: 'It has been said that it takes fifty years to pass any law in England', he pronounced. 'If that be so, it may take a hundred years for any idea to make its way into an Englishman's head.'[50] He would remain a useful fount of advice to Hudson and the Bird Society on navigating parliamentary procedures and culture.

Don Roberto had inherited his mother's corvid love and had long wished to know how he might encourage these birds to establish a rookery on his own land. He had written to an eminent Scots ornithologist on the matter, who had replied at some length, saying nothing about rooks but 'pointing out the fallacies of socialism as a political creed'.[51] Hudson later recounted this story in his book *Birds in London*. Although not named in the book, the ornithologist in question recognised himself and got in touch. It was John Alexander Harvie-Brown, who luckily could see the funny side. He loved most birds (if not rooks), and had given the Bird Society a donation of £10, with which it opened its first bank account.

Hudson, more helpfully, was able to cite the example set by the parson of Morwenstow in Cornwall, who got his rooks by praying for them – every day for three years. The birds adopted the very trees where they were wanted. It is not recorded whether Don Roberto wanted rooks badly enough to emulate this approach.

Before long, Hudson encouraged Don Roberto to follow the example of his mother and actually join the Bird Society, suggesting he might even pay as much as a shilling and become an Associate. The Bird Society's first 5,000 members were almost exclusively women, but Hudson was not the only man in the room for long. 'At first the Society was composed of women only, but several gentlemen have shown their approval of the Society by joining it as Members, or by authorising the mention of their names as earnest sympathisers.'[52]

There were by now around 130 local branches, and Hudson was secretary of the North Kensington group. What would become a long-held tradition of aptly named staff (what the behavioural scientist might call nominative determinism) was initiated with the appointment of Eastbourne branch secretary Miss Lily Trotter – the humour surely not lost on Hudson at least, who will also have known the lily trotter (an aquatic South American species) by its local name, the jacana.

'One of our members has got a little bird protection act through the committee,' Hudson could soon report, to demonstrate that political inroads were being made. 'It shows that we are waking up to the fact that something must be done to preserve our bird life from destruction.' And there is an indication of his ambition and optimism for the nascent cause: 'when this Society numbers forty or fifty thousand members then I think we can safely go to Parliament and ask for a better Bill'.[53]

Don Roberto left the House of Commons, having been unsuccessful in the 1892 general election, just as Hudson was about to take on public office of a less high-profile kind with the chairmanship of the Bird Society. 'I have endured the concentrated idiocy of the Asylum for six years', Don Roberto reflected ruefully, 'now I think I may do my fooling alone.' He had left the Liberals to stand in a Glasgow constituency as a Scottish Labour Party candidate. In his last speech in the Commons he spoke of Ireland and the efforts of his friend Charles Parnell to establish home rule.[54]

Hudson regretted his friend's absence from politics when he no longer occupied a seat on the benches. He had evidently taken a seat there himself once or twice, if only in the visitors' gallery, on reconnaissance. 'One rather misses a man of independent mind', he told his friend. 'The "six-hundred gentlemen" look rather monotonous, like a vast flock of sheep, with two rams facing each other in the centre, each with his dozen or so capones grouped around him.'[55]

Don Roberto claimed to trace his ancestry to King Robert the Bruce in fourteenth-century Scotland. There were (and possibly still are) those who considered the Scottish throne to be rightfully his. And although his own politics were radical and he was no royalist, his younger brother Charles rose to command Queen Victoria's royal yacht *Osborne*. In the absence of any Don Roberto offspring, Charles's son (Don Roberto's nephew) was heir to the family estate. However, for all its illustrious past, its estates and wealth, the Cunninghame Graham family was starting to know hard times. Besides the vagaries of global capital, Don Roberto's father had squandered a fortune through gambling and other vices, his judgement perhaps impaired by a head injury. He became a danger to his family and lived for some time under supervision, at a safe distance. Missy Bontine was established in London while Don Roberto grappled with taking over and trying to save the family's principal home, Gartmore, a grand house in a spectacular setting overlooking the Lake of Menteith in central Scotland. Revenues may never have been huge on these 10,000 damp acres, and costs for the great estates were rising, hit like everyone else in Britain by years of agricultural recession in the face of cheaper imports.

Don Roberto had taken a more hands-on approach to running his estate, conscious that if he had developed it for sporting interests such as grouse shooting and otter hunting it might be generating more revenue. But these things weren't for him.[56] Nor would he countenance creating a pheasant shoot on his land, being against animal cruelty. He would challenge the moral outrage he encountered in those who thought the Spanish cruel and point out their double standards. 'There is the national crime of the bull-fight,' he would concede, 'but what of pheasant-shooting, tame deer-hunting, etc.? did you ever see a big pheasant shoot, with everyone dabbed in blood, and three thousand birds shot?' Whether the proceeds from shooting would have helped much or not, Gartmore faced ruin. The strains of agricultural recession and falling rents were starting to show.

Despite their difficulties in Scotland, having won his trust and accustomed him to Belgravia, Don Roberto and Gabriela tried to tempt Hudson to visit them at Gartmore. 'If you can stand a Spanish fonda, you can stand Gartmore for a few days,' Gabriela wrote to artist William

Rothenstein, later a close friend of Hudson. 'Don't expect anything and you won't be disappointed.'[57] Don Roberto and Gabriela were using just a few rooms of the great house. Had he visited, Hudson would have seen the strange portraits of ancestors on the walls, with the originals painted over with new, mocking faces.

Hudson declined, despite imagining the place 'a terrestrial paradise where the capercaillie exists and a guest is not expected to go to church on Sunday'. The capercaillie was one of the rare species he had recently written about for the Bird Society in his *Lost British Birds* pamphlet, as it had been hunted to extinction and reintroduced in Scotland so that it might be hunted again. Hudson knew he would never see one for himself unless he ventured to the pinewoods of the north.

When Hudson mentioned to his friends that he had visited Scotland, but just once, soon after he arrived in Britain, it can only have redoubled the Grahams' insistence that he acquaint himself properly with their homeland. They obviously persisted with the offers. In 1896 Hudson wrote again: 'I wish to God I could go and see your capercaillies.'[58]

Freed from active politics, Don Roberto prepared to embark on his latest adventure, to go in search of a long-lost goldmine in Galicia, northern Spain. Gabriela, researching her own book about sixteenth-century nun Santa Teresa de Avila, had worked out the location of the Roman mine. It may sound like a bit of an escapade but this long shot, if successful, could have been the answer to the family's financial problems.[59] Alas, the adventure provided better stories than it did hard cash. Returning to Britain, Don Roberto, with encouragement from Hudson and Gabriela, began to devote more time to writing books of his own. Not that this would be likely to featherbed a failing estate, as Hudson's experience would amply demonstrate. 'Better, I say, to live as I do, on rather less than £100 a year,' he advised his friend, while encouraging him to write, 'and be free – yes, free even from life's "pleasures".'[60] Don Roberto later commented that he had never known a man less interested than Hudson in common comforts. 'Few men in their right senses set out deliberately to live by literature',[61] Don Roberto would eventually conclude. It didn't stop him emulating his friend and going on to become a prolific and powerful writer, championing the poor and oppressed at every turn.

As well as his voracious research on British natural history and his explorations of the sprawling city, Hudson was gathering material for a novel about poverty in London. For some reason, when he finally published *Fan - The Story of a Young Girl's Life* in 1892, it was under an assumed name – Henry Harford. He was determined to remain anonymous, and achieved this for many years, perhaps because he was trying to make money rather than a reputation. He even asked the publisher never to reprint the book, even though its initial print run was just a few hundred. Hudson knew slums and deprivation from the streets close to his home. The impoverished young girl Fan, like him, seeks sanctuary among the nature of Kensington Gardens. She is taken under the wing of a piano teacher, a character inspired by Emily, his wife. Unsurprisingly, Hudson aka Harford didn't receive much mail in response to this book via the publisher, but a letter did arrive from his brother in Argentina. Edwin candidly advised him – as only brothers can – that fiction wasn't his forte. He also implored him to come home.[62]

Hudson's pen was by now in full flow. *The Naturalist in la Plata* was also published, in 1892, and sold well enough that a second edition was soon issued, then a third. It drew a glowing review from none other than Alfred Russel Wallace, discoverer of evolution by natural selection at the same time as Darwin, but with far less lasting recognition. *Idle Days in Patagonia* was published in 1893, with a similarly warm reception.[63]

The received wisdom is that Hudson was politically conservative,[64] and compared to Don Roberto he was, but such a simplistic conclusion sits uncomfortably with the facts of his close friendship with this, arguably the most radical high-profile political figure in the country. He clearly shared his friend's distaste for Britain's military campaign in South Africa, for example. On matters political in general, Hudson knew his own limits: 'If others thought as I do our order would be quickly reduced to chaos', Hudson admitted to his friend in one letter.[65]

Don Roberto wrote in sombre mood at Hogmanay 1891, words that give some sense of his mercurial character: 'The brutal lord at the Pelican club; the brutal lady at the pigeon match or pheasant battue. The brutal man of business ... the brutal treatment of wives, horses, children, dogs and all the weaker animals that God has sent ... the thought that it is all to go on unchanged till a thousand or more years are out, robs me of my gladness in the contemplation of the Glad New Year.'[66]

Learning to fly

Dr Sclater became chairman of the British Ornithologists' Club in 1892, the same year that the Bird Society chose the Society for the Protection of Birds or SPB as its official name. He is listed among the early members and supporters in the 1892 annual report. The scale of the global plumage industry was eye-watering by this time: a single 1892 order of feathers by a London dealer (either a plumassier or a milliner) included 6,000 bird of paradise, 40,000 hummingbird and 360,000 East Indian bird feathers.

The Bird Society, now fully merged and with roles assigned, and besides going to war with the plumage industry, campaigned throughout the years 1891 and 1892 for a bill in Parliament to address flaws in the Wild Birds Protection Act 1880. The early campaigners were eyed with caution by more than just the massed vested interests of the plume and fashion industries. There was a lot of money in the trade in birds' eggs and skins, much prized by collectors.

Writing years later about these early days, the society's leader Etta Lemon reflected on how 'scientific ornithologists looked askance at its aims and objects on account of its determined efforts to curb the rapacity of unscrupulous egg collectors, and the lucrative traffic by commercial dealers in rare specimens'.[67]

In late May 1892 Hudson saw in a newspaper notice of the marriage of Miss Margaretta Louisa Smith, daughter of Captain Smith of Blackheath. 'Is this your Bird Society's Margaretta?' he asked Eliza Phillips (interesting to note his use of 'your').[68] She would have confirmed that it was indeed the rising young star of the society. 'She will, I am quite sure, prove quite a pearl to the man who has been so fortunate as to win her,' Hudson declared. The lucky man was Frank Lemon, a solicitor who would in time also become a key figure in the history of the Bird Society.

The society gave Hudson a new lease of life, and another purpose for his writing, providing an outlet for his talents and passions. Alongside Phillips he wrote much of the content for the pamphlets and reports, no doubt until late into the night, by lamplight in his Tower House belfry. In autumn 1893 he dropped a bomb that was game-changing for the plumage cause, when, in a fit of anger, he dashed off a letter to *The Times* lambasting the abusers of wild birds. It caused a storm.

In the cold light of day, passions subsided. Feeling suddenly in the public spotlight, Hudson experienced a tinge of regret at what he had done. 'I sent that letter scarcely hoping to see it in print,' he explained to Phillips, who was perhaps also caught between the excitement of realising the impact they could make and wondering if he had gone too far. 'But merely as a relief to my feelings. I was almost sorry I had sent it. It was in some respects rather brutal. Only yesterday I heard it was in, from the Secretary of the British Ornithologists' Club. He offered to bet me a hundred to one that it would have no effect!'[69]

This sort of reaction no doubt reassured Hudson that he had done the right thing. 'These clever, cynical persons give us poor help,' he added.

'I was glad indeed when I found that the Thunderer itself had backed it up with such a powerful article.'[70]

The ripples spread through other media. It 'made a little stir,' Hudson noted. 'So many of the evening papers spoke about it, and favourably.' Of course, not all the papers were on side, including 'the *St James's Gazette* which abused me and the *Times* leader-writer.' Hudson collected the press cuttings to share with Phillips.[71]

It all left him feeling a little exposed. When he later ventured out, he could sense the eyes of the world upon him:

> I had to go to an afternoon meeting, and did not
> feel too comfortable about the letter, as there was a
> great crowd and a great many ladies. I retreated to an
> obscure corner, where only Mrs Little Lord Fauntleroy
> [author Frances Hodgson Barnett] and Sir Bruce
> someone … he came to me with a cup of tea and he
> said he had read the letter with great delight … Several
> other gentlemen all seemed very heartily to approve.[72]

Then there was the mailbag that resulted, and the opportunities to recruit new supporters. '*The Times* Editor sent me a letter from Lady Hooker. I have replied at some length telling her the history of the B.P. Society.' He asked Mrs Phillips to send her some of the campaign pamphlets. 'I am sorry to put this task on you,' he added, 'but I fear that Mrs Lemon may not be in working form yet.'[73]

Many more people were now taking an active interest in the war of words around plumage, even beyond London. On a rare visit to Oxford, Hudson found his hosts 'following the controversy with keen interest'.[74] Eliza Phillips, keen to seize the moment, suggested they make a pamphlet out of the *Times* letter, including their supportive leader column. Hudson once again hesitated:

> I scarcely think it would be a wise thing to reprint the
> Times letter and leader … I do not think the Society could
> well set its approval on all the wild and whirling words of
> the letter… the things I said because I know they would
> affect certain people as a red rag does to a bull.

Despite his misgivings he told Phillips she must do as she liked about it. She had told him she wanted to 'strike while the weapon is red hot'.[75]

Phillips had her way, and the letter and the editorial were duly produced as a pamphlet. The episode had a lasting impact on Hudson, having propelled him and the Bird Society into the public spotlight. With success in publishing at the same time, he became aware of a potential conflict of interest, or at least that their opponents might try to make capital out of this.[76] When Phillips proposed to include his name in a notice advertising the work of the organisation, he was quick and unequivocal in objecting. 'No one must be allowed to say that I am making use of the Society to advertise myself', he insisted.[77] He was by this time – October 1895 – enduring sleepless nights. He resigned the chairmanship soon afterwards, through a combination of this potential conflict of interest and the realisation that the role needed someone more suited to administration and oratory. Etta Lemon later called his passing of the baton 'one of the best things Mr Hudson ever did for the Society',[78] which I believe is a reflection on the suitability of his successor rather than a comment on any harm done under Hudson's chairmanship. It would free him up to continue to write the books he wanted to write, in the way he wanted to write them.

Hudson also urged a more moderate tone in some of the society communications that followed soon after his *Times* controversy, and some reining in of policy demands. He urged Etta Lemon not to issue one paper that contained 'so extravagant a proposal as that a law should be passed totally prohibiting bird nesting and bird shooting one year in every three. There is good stuff in the paper but it is all spoilt and made ridiculous by the writer suggesting such impossibilities.'[79]

They also debated whether to include pictures of birds' nests and eggs in their lecture slides. Hudson and Lemon were evidently relaxed about it, Hudson believing that the accompanying narrative could be worded in a way that dissuaded boys from robbing nests. But Phillips had misgivings, so the images were omitted.[80] It is a policy more or less followed by the society to the present day. The development of these new illustrated lectures he thought would be 'a great help' to the cause, while regretting that he could not present them himself. 'Long, long ago I said to you, and I have since said the same thing to Sir Edward Grey ... that if only I had it in my power to get up and speak, I should never write a word ... I would rather chatter like a poll parrot than write like an angel.'[81]

It is a matter of some regret that so little remains known of the RSPB's founding women as a whole, aside from what might be gleaned about their professional activities from the society's annual reports, and the surviving letters from W.H. Hudson. Considering her status, it is also curious how little has been written by or about the society's president, the Duchess of Portland Winifred Cavendish-Bentinck, who would hold this position for 63 years, from 1891 until her death in 1954.

Winifred Cavendish-Bentinck, Duchess of Portland

A hint of the duchess's presence is detectable in one Hudson letter. He and Eliza Phillips had been in touch with Richard Bowdler Sharpe of the British Museum in respect of an early policy matter, Phillips having recently attended a lecture there. Hudson got the reply, and as always was quick to share it with his colleague, as it was bird business. 'He is a careless man', Hudson concluded, both for not replying to *her* letter, and for not enclosing something he meant to send.[82]

Bowdler Sharpe asked about the Duchess of Portland, keen to understand the extent of her powers and influence and how the society intended to deploy her. 'That is going in to sublime matters about which I have no knoledge [*sic*],' Hudson remarks to Phillips. His spelling error, while not unheard of, is undoubtedly deliberate here, an ironic reinforcement of his point.[83]

Whatever was the case then, not many people have knowledge of the duchess today. Surprisingly little trace of her survives, but I did manage to gather some material to form an impression of her life, and her importance as the Bird Society's upper-class figurehead. She was born Winifred Dallas-Yorke in 1863, daughter of a Lincolnshire squire. She spent many happy childhood days at Murthly Castle in Perthshire, where her grandmother Mrs Graham lived. Her earliest memories were of the castle, but she seems to have spent most of her early years near Louth in the Lincolnshire Wolds, riding horses along the lanes. She was thought something of a tomboy. 'I had the happiest childhood anyone could have,' she later reflected.[84]

At the age of 16 she was taken to Rome for two years, to be tutored. Between there, Paris and Lausanne she became acquainted with many illustrious figures in cultural life, including the King of Italy. When he later visited London he remarked to the duke, 'Look at that lovely woman!' and was reminded that they had met before.[85] She had instinctive empathy with animals and, even when travelling abroad, she could never ignore a creature in distress. A friend remembered 'Winnie' rescuing overheating goldfish from a shop window in Switzerland, and having a wretched mule humanely euthanised in southern France after five hours spent sourcing a replacement for its owner. Her friend spoke of 'the indescribable amount of good that Winnie does'.[86]

The Duchess of Portland
as Duchess of Savoy.

Winifred Cavendish-Bentinck, Duchess of Portland, dressed as the Duchess of Savoy
for the queen's jubilee in 1897. Friends spoke of 'the indescribable amount of good that
Winnie does'. Image courtesy of the National Portrait Gallery.

Her tour over, Winnie returned to Britain, and surprised everyone by having developed into 'the most fashionable beauty in the Row' when lined up in London with her fellow debutantes. She was said to be 6 feet tall, and 'willowy'.[87] When visiting the Sussex Downs near Brighton one day the young Winifred is said to have consulted with a gypsy soothsayer, who foretold that she would win the greatest matrimonial prize in all England. Sure enough, after she had been 'out for several seasons, in the Society sense of the term',[88] she was spotted by the Duke of Portland at Carlisle railway station, as he travelled to Scotland for the shooting season and she was on her way south. He orchestrated an introduction soon after, promptly popped the question – and bought his fiancée a sable coat at great expense as an engagement present.

While the first meetings of the Bird Society were taking place around the drawing room tables of Notting Hill in London and Didsbury, Manchester, the 25-year-old Winifred was making plans for her wedding, which took place in June 1889. The future king was among the guests. The now duke and duchess would live at the Welbeck estate in Nottingham-shire, adjoining Sherwood Forest of Robin Hood legend. The duchess's gypsy palm-reader had also warned that illness would befall her, but that all would be well. It may have been no surprise therefore when, soon after she began life at Welbeck, Winifred was struck down with typhoid fever. As predicted, she pulled back from death's door, and was soon back to her old self.

I don't think we know exactly how the contact was first established – perhaps through the Duchess of Wellington – but in 1891 Winifred became president of the Bird Society. The duchess is referenced briefly and surpris-ingly seldom in Hudson's letters. There is a hint of his – shall we say – uncertainty about the commitment of their illustrious figurehead, although the longevity of her service would surely quash any suggestion that she was not wholehearted in her devotion to the cause. His first reference comes in a January 1892 letter to Eliza Phillips, on the subject of egrets. They are unpaintable, he says, like hummingbirds, especially in their exquisite breeding finery, when they sport nuptial plumes of unsurpassable delicacy. He states again what you suppose they must have agreed many times before in the three-year existence of the Bird Society – 'how dreadful to think that creatures so lovely are exterminated' (for these plumes). And then comes the aside about Winnie – 'which the Duchess of Portland says she has never worn *because she does not think them pretty!*' (his emphasis). 'I suppose if someone shot or murdered Iris she would not refuse to wear a few of his rainbow-tinted wing-feathers in her hat,' he adds. Someone must have owned a parrot (or similar exotic bird) called Iris; perhaps it was the duchess herself, if not Eliza Phillips. To make matters worse, Hudson had just read about the marriage of the Duke of Portland's brother to Lady Olivia Taylor, at which the bridesmaids had worn hats decorated with '*white birds*' (his emphasis, again). Egrets – even little ones – being too large for a bonnet, he thought these unspecified birds might be sea mews, kittiwakes today.[89]

Not all aristocrats were so slow to fall in line with the plumage campaign. In one letter, Hudson drew Phillips's attention to Lady Florence Dixie's *The Horrors of Sport*, her recently published book of this title. As she had also travelled in Patagonia, it is fair to assume Lady Dixie may have made contact with Hudson as his account of *Idle Days* there rolled off the presses in 1893. She had once been an active participant in blood sports but had seen enough.

The Duchess of Portland opened doors for the Bird Society in the highest echelons and was increasingly well connected in court circles. On two occasions the Portlands were commanded to attend Windsor Castle to dine with the queen, and in winter 1897 they hosted a shooting party for Prince Edward, although Winnie played bridge with Princess Alexandra while the men shot pheasants and deer.

Besides her love of animals and birds, the duchess is remembered as showing great compassion for the local mining community in Nottinghamshire, helping to establish new premises for disabled war veterans and injured miners, and providing cookery and sewing classes for miners' daughters. Shortly after her marriage she persuaded her horse-racing enthusiast husband to donate a portion of his substantial earnings from winning the Derby (not once but twice) to help build almshouses for the poor at Welbeck.

Winifred's life seems to have been a busy schedule of civic and charitable duties, locally and beyond. The Portland dukedom's property portfolio included a house in London's Grosvenor Square, where, in the 1890s, the duke and duchess hosted glamorous balls and danced till daylight with friends to the music of Gottlieb's or Drescher's Band. However, according to the duke in his memoir, they tired of this as it was impossible to restrict the guest list to people they actually liked, and they couldn't accommodate everyone who expected to be invited, 'however large the house'.[90] Nevertheless, they retained their love of entertaining, having 'kind hearts and plenty of leisure'.[91] Sometime after 1898, when the soon-to-be King Edward began his affair with Alice Keppel, Winnie committed a faux pas when, out of loyalty to her friend Princess Alexandra, she didn't invite Keppel to a party at Welbeck. 'The Portlands are in very bad odour',[92] the prince's friend Charles Wynn-Carrington reported. The prince had taken it as a slight to himself. Even his mother knew to invite Keppel when he visited Sandringham. While Winnie's stock may have fallen with the future king, her bond with the future queen was no doubt cemented and can have done the cause of birds no harm when her royal support was needed and sought not long afterwards.

Inevitably, some voices within the growing Bird Society were raised against the recreational shooting of birds, including at the 1896 annual meeting. As a result, the constitution was adjusted and a declaration of neutrality on the ethics of shooting birds categorised as 'game' was put in place. If sport shooting enthusiasts had been nervous about the wider agenda of this rising bird protection movement, they could now relax a little.[93]

From the late 1880s, besides forming alliances in the animal welfare movement that would coalesce as the Bird Society, Hudson was part of a small group of friends that one of their number – the author George Gissing – had dubbed 'The Quadrilateral' after an Italian fortress. The others were Morley Roberts and the painter Alfred Hartley. They would sometimes visit the Sussex coast for short breaks. On one of these occasions, in 1890, it was just Hudson and Roberts who made the trip south from London, Gissing's turbulent life having kept him away that time. This particular peaceful and sunny September morning at Shoreham-on-Sea would lurch suddenly into a life-or-death drama. 'Hudson not only saved my life, but the life of someone else,'[94] Roberts would recall.

CHAPTER 4

Branching Out

Saving lives in Sussex

Hudson and Morley Roberts were loafing on the beach. Three young girls and an elderly lady passed them, and polite greetings were exchanged. Roberts may not have known, till that day, that Hudson couldn't swim, but he was about to find out when, a short while later, as they sat smoking and sketching, their peace was harshly broken by the ear-splitting screams of the elderly woman, and the cries of the girls. 'Looking up, we saw them far beyond their depth and obviously drowning,'[1] Roberts wrote.

The pair rushed immediately to the aid of the girls. Roberts, by his own estimation a strong swimmer, ploughed straight into the water, not pausing to remove even his hat or to drop his pipe. Hudson waded straight in after him, up to his neck, using his long arms to maintain his balance in the fast-moving tide and on the loose pebbles beneath his feet. Roberts reached the first girl and was able to bring her back for Hudson to grab and relay to the shore. A young clergyman also came to their aid.

Roberts now dived to rescue the second girl, who had gone under. Whether in shock or unconscious, rigid with cold and terror, she took some prising from Roberts by Hudson as she was brought to him. The third girl was also by now near the seabed, and Roberts managed to dive once more, reach her and bring her to Hudson.

With the rescue drama over, Roberts passed out on the shore. When revived, he had to go back to the boarding house to recuperate. Hudson remained, sopping wet on the shingle, while the three traumatised girls were kept warm and consoled. A small crowd had gathered, many dry clothes or blankets fetched and offered, and wet garments peeled off, accompanied by the sounds of soft sobbing and the gentle murmur of gulls.

Afterwards, Roberts was modest about his role in the rescue mission, and generous about Hudson's, and his friend's bravery, considering his inability to swim, as well as his greater age and delicate constitution. He felt he owed his life to Hudson, and so too did the girls. Roberts also recalled that Hudson made efforts to stay in touch with the 'salvaged damsels' for years afterwards. For some reason – perhaps modesty – Hudson doesn't seem to have written about the incident in any of his books, so I was delighted when I found that letters have survived that describe the incident and its aftermath in vivid detail.[2] In one rare letter to his wife Emily, Hudson's version of events accords with Roberts's. He adds amusing detail too, about how his wool suit had afterwards shrunk while drying, and how she would have to send him some other clothes to wear before he could bear to be seen in public. And then it was all over the papers – Hudson slightly aggrieved to see it reported that he couldn't swim. Touchingly, he asked Emily if she could spare a pound to send him, so he might extend his stay. There is also a touching footnote in the next Hudson letter. A few days later the two lifesavers were invited to the home of the old lady and were there reacquainted with the two younger girls they had saved.

I began to wonder if any descendants of the rescued girls and young woman might be traceable, any record of them retained somewhere, if only in family folklore. Roberts's account offers one clue to the identity of one of the girls. He says she was the daughter of an artist he calls Aumonier. I worked out from online searching that in fact two of the girls were Aumonier sisters and they were more likely the artist's granddaughters. I found a James Aumonier, and his evocative landscape paintings called *Sheep-washing in Sussex* and *When the Tide is Out*.

I wondered if there was any other inherited knowledge of this dramatic incident, someone today who would recall their great-grandmother speaking of or passing on details of such a childhood drama, such a life-changing moment. Because Aumonier – being French – is quite an unusual name in Britain, I thought it was worth making a few speculative enquiries. I soon struck lucky. My old friend David Payne, who lives in Brighton, reminded me that not only does he know an Aumonier – Simon – but that I know him too. Simon works for a major environmental consultancy with which I collaborated on an international partnership project for migrant birds, while I was with the RSPB.

I contacted Simon and he replied promptly. Sure enough, he is related to the artist James and therefore two of the rescued girls. Simon even has two of James's paintings, and books and sculptures by others in the Aumonier dynasty. According to 1891 census records, the Aumonier family lived at Oxford Villas in New Shoreham. The daughters were called Louisa and Nancy. They went on to become a musician and a maternity nurse, and neither married.

On a breezy, sunny day in July 2021 I visited the setting for this drama, after cycling along the seafront from Brighton to Shoreham, wondering on which part of this long stretch of shingle, with its groynes extending into the waves, Hudson and Roberts had been sitting on that fateful morning in autumn 1890. The beach was empty at one place where I paused. As chance would have it an elderly lady passed, with three young children. I helped her lift a buggy over the railings onto the shingle. One girl had a large inflatable rubber ring. 'Be careful with that,' I couldn't help saying to her, while realising it might not have been my place. They were the only people on the beach. I watched for a brief while and then left, confident that no harm would befall them.

At Worthing, the next town along, I explored the pier, opened in April 1862, and found a display of Victorian photographs, including one that shows three young girls posing for the camera, and one of an elderly lady at the water's edge with a child – another eerie echo of the incident on that September day.

Among the mementoes of Hudson that lurk in files in the RSPB archive at The Lodge, there is a certificate he received in recognition of

An evocative image that helps conjure the time, place and 'salvaged damsels', when Hudson and Morley Roberts rushed to the rescue of three young swimmers, September 1890.

his bravery in helping to save three lives from the sea, although I didn't have time to search for it before lockdown. It is in a sense thanks to his part in this life-saving mission that there are people alive today who might owe their existence directly to W.H. Hudson.

This Shoreham rescue brought to my mind a well-rehearsed story passed down in my family and told to me from a very young age, of how my mother's father, as a young boy, had a near-death-by-drowning experience in a frozen lake in County Down when he dared himself to cycle on its frozen surface, with disastrous consequences. The ice broke and bike and boy went under. That he was able to drag himself out of the freezing water and back to shore is testified by my mother's existence, and the rest.

Professor Alfred Newton – godfather of ornithology

At the time the Bird Society was getting mobilised, Professor Alfred Newton had for more than two decades been Britain's biggest name in ornithology. His legacy lives on in Cambridge: the library in the Zoology Department is named after him, and I have spent many lunch-hours flicking through useful texts there.

I found at least one remaining enthusiast, my then BirdLife colleague Professor Nigel Collar, himself something of a latter-day Newton. Nigel even has a copy of – and was kind enough to loan me – Newton's biography. It was written by a former Newton pupil, A.F.R. Wollaston, and published in 1921.

As Hudson's steamship with sails, *The Ebro*, approached southern England's white cliffs back in 1874, he may have seen some of the seabirds for which the British Isles are internationally important – gannets, kittiwakes, gulls trailing in the wake of the boat, occasional shearwaters and petrels skimming the crests of waves. A crew member might also have made the point to Hudson that the birds alert them to the presence of fish, and in thick fog or darkness give useful and often life-saving warnings that land is nearby.

I don't know if Hudson was aware at this point that a few years earlier Professor Newton had cited the Isle of Wight as one of the places where seabirds were being so badly persecuted that stronger laws were needed to protect them. Newton had been the architect of the 1869 Sea Birds Protection Act, aimed at putting a stop to the unrestrained slaughter of breeding birds such as kittiwakes and gannets on our sea-cliffs. Sadly, despite the seabird law, and to underline its ineffectiveness, 130,000 guillemot eggs were recorded as taken from the cliffs at Bempton in Yorkshire in 1884.[3]

Newton was Professor of Comparative Anatomy at Cambridge. His university rooms in Magdalene College were a pivot around which the rapidly developing knowledge of birds was turning. The first meeting of the British Ornithologists' Union (BOU) was held in his rooms there, and he hosted regular gatherings to discuss all things birds with his students, and other visitors. He was also editor of the BOU journal *Ibis*.

Newton cut an imposing figure, and had the appearance and demeanour of an old country squire, old-fashioned even for his day. 'Newton of Elveden, and afterwards Prof. of Zoology, combined in himself the two souls of the conservative old-fashioned Squire and the academical professor in the University. They combined very well,' Hudson later remarked.[4]

While of the establishment, Newton was also open-minded, and an early adopter of Darwin and Wallace's theory of evolution, by no means typical of scientists at the time. He was tall, and made taller by his stove-pipe hat. He smoked a long-stemmed pipe and had an easy, leisurely manner. He moved in all-male circles, which seems to have been very much how he liked it.[5] Newton also donated funds to help the early Bird Society in its plumage campaign, but stopped short of becoming a member, preferring to maintain some distance while the organisation worked out its policies on thornier subjects on which his position might differ, such as hunting and collecting. Newton was also instrumental in shaping the Wild Birds Protection Act 1880. When an amendment to these laws was sought by some to make egg-collecting illegal, this was eventually dropped. Newton was not in favour. He wrote to Lord Walsingham: 'I do hope you will resist any attempt made by sentimental people to make egg-collecting an offence. If it were so there would be endless trouble – parents wouldn't pay the fines for their children, and the gaols would be full of boys.'[6] It may be significant that Newton's beloved elder brother was a dedicated collector of eggs, to name just one of the influential figures who weren't quite ready for any such law changes.

There were other attempts by the Bird Society to create a more effective legal framework for bird protection. In 1893 Sir Herbert Maxwell introduced a Bill aiming to enable county councils to protect the eggs of birds they deemed at risk. It passed in the Commons but was amended in the Lords, under advice from Newton, then revised in committee. Maxwell wouldn't accept the alterations, and it was dropped.

It seems unthinkable that Hudson never visited Newton in Cambridge, but for a long time I could find no trace of him ever doing so. He went once to Oxford, that I can ascertain, and not for academic reasons. It might be that he set his face against both places, as too high-brow or intimidating, perhaps. Then at last I found a reference to Cambridge in a letter of 1894,[7] in which Hudson tells Bird Society founder Eliza Phillips that he is due to visit Professor Newton in Cambridge to get his final advice on *Lost British Birds*, the society pamphlet that Hudson is compiling.

But did Hudson actually go? I have my doubts. Had he done so, he is unlikely to have written in the pamphlet that the St Kilda wren had already been wiped out by the collectors: Newton would have corrected him in advance, instead of later. The St Kilda wren is an example of evolution in action, closer to home than the Galapagos and Darwin's famous finches. Isolated from its source wren ancestors, on a remote Hebridean island, the St Kilda wren was beginning to exhibit some different characteristics. This could have spelled its doom. No sooner had it been declared to be a new, distinct type of wren than the collectors set their sights on it. That distant island was promptly 'invaded by the noble army of collectors' as Hudson put it, to add specimens and eggs to their display cabinets. Newton later advised Hudson that he had been premature in announcing the demise of this Hebridean wren. Premature or not, Hudson's description of its plight helped to save it, with the prompt action in Parliament that ensued to protect it.

For as long as he lived, Newton remained in Hudson's thoughts. He recalled him as 'the old conservative academic ... the doyen of the Ornithological world, who glared at me, an Argentine, who dared to come to England and write about English birds'. He wrote this as late as November 1921 as he was compiling his book on animal senses, and exchanging speculations with Morley Roberts on subjects such as the mysterious ability of birds to migrate.

There are also some traces of bitterness as he relates a memory:

> a long and very funny story about the great Birdist
> and poor little me. The funniest part of it was its
> culmination when he was head of a Section of the Brit.
> Ass. and in his address made a vindictive attack on my
> pamphlet *Lost British Birds*, then, to show there was no
> ill-feeling, he wrote a letter to *The Times* on some bird
> question in which he spoke very nicely of 'my friend,
> Mr. Hudson'.[8]

Although he thought that Newton disdained him – thinking him an uneducated man from the Pampas – Hudson recognised Newton's overall contribution to the cause of wild birds, and was generous about it:

> On going into the literature of the subject extending
> thirty years back or more I am amazed to find that from
> first to last Professor Newton has been the chief helper
> of anyone who has done anything in the matter ...
> That volume after volume has been gone thru in
> manuscript by him, and that half the literature on the
> subject would never have been written without his
> inspiring aid. The amount of work he has done outside
> of his own professional labour is really something to
> wonder at.[9]

In private, Hudson could also be critical of the great man at the time the Bird Society was finding its feet. I found this cryptic message he wrote to Eliza Phillips, who had just shared with Hudson a letter from the professor: 'Here he is one day preaching to us never to expect too much, to take what we can get and be thankful, and the next day showing how dissatisfied he can be with what you do!! I dare say that if you wrote a circular letter entirely satisfactory to him it would please nobody else.'[10] This must, I think, have been Newton's reaction to Hudson's provocative and outspoken letter to *The Times*, fulminating against the plumage industry. Another Newton letter followed a day later. He was evidently much exercised by the sudden escalation of the debate being conducted through the pages of this influential organ. He seems to have challenged what Hudson said about skylarks – perhaps that the species might be endangered by the trappers.

'Whether he is right or not about the larks I should rejoice to have the world regard them as I do,' Hudson railed, '– not as a bird to be eaten or put in a cage. But then I am only a sentimentalist, or, as the *St James's* [*Gazette*] calls me A Man of the New Sentiment.'[11] The *Gazette* had identified a dangerous shift in society towards greater compassion towards animals. It had also detected changes in what society considered funny, Hudson adding: 'the "new sentiment" it describes as a thousand times worse than the "New Humour".'[12]

Prompted by the professor's intervention, Hudson also mentions having just read Newton's earlier address to the British Association of 1868 on 'the Zoological Aspects of Game Laws', praising it for its brevity (not realising it was just an abstract), and asking the Zoological Society librarian to type a copy for him.[13] This would be reproduced by the Bird Society in its 1893 Annual Report.

Newton also advocated the protection of specific places and not of species – though later in his life he was magnanimous enough to admit that he might have judged this wrongly. It was the kind of difference of emphasis that meant he and the Bird Society were not always aligned. Overall, he took what he thought was a pragmatist's view – laws needed to be workable, and if your demands were too radical, or laws too difficult to enforce, you would alienate the parliamentarians and get no help from them on anything, and waste time in the courts.

Newton continued to help and advise the Bird Society when he could – maintaining the arm's length distance – until the end. He died in 1907 before things began to escalate on the plumage campaign front – there had been no plumage bills by then, but a run of them followed soon after. He was buried in All Souls Ascension Cemetery in Cambridge.

I had in mind for a long time to visit Newton's grave, and after being thwarted in earlier attempts by lockdowns and other circumstances I finally made the pilgrimage to find it. Nigel Collar kindly agreed to meet me there, to help in the search, as he had been before. I arrived early, on a murky, mid-November day, robins lilting softly among looming, dripping yew trees. Before Nigel arrived I had time to make a circuit of the small cemetery, and arrived back at the former chapel of rest, where I found a man, a stonemason/letter carver who introduced himself as Eric Marland, the chapel now serving as his workshop. 'Are you looking for anyone?' Eric asked me. 'I'm waiting for someone,' I explained. 'We're going to look at Professor Newton's grave.'

Eric asked me what my interest was in Newton, and I explained about his association with Hudson. His eyes lit up. Not only did he know about Hudson, but to my astonishment he explained that he had engraved the plaque that hangs on the wall of Hudson's London home, which was installed 30 years ago for the ceremony attended by the Argentine ambassador, to mark 150 years since Hudson's birth.

Travels in Berkshire

Hudson had witnessed the rapid growth of popularity in bicycle riding since his arrival in Britain, when it first took off. The early bikes had been hazardous, the rider being at some height off the ground. They were also difficult to mount. Villagers used to gather to watch participants in this new fad as they teetered past, often at speeds bordering on the foolhardy. Accidents were commonplace, and added greatly to the spectacle for the onlookers.[14]

The 1890s heralded a new generation of cycles, with the chain attached to the rear wheel, a lower seat and a smaller front wheel. The old-style penny-farthing cycles with the huge front wheel were going rapidly out of fashion. Hudson, with some income under his belt, was now ready to get involved. With some of his new earnings from writing he bought a state-of-the-art chain-driven Rover safety bicycle for £10.[15] He was now equipped to broaden his range, and horizons. In so doing he might try to write a book about the birds of his adopted home, and apart from anything maybe establish some credibility with the likes of Professor Newton.

Besides his 'wheel', Hudson had now also acquired a rudimentary 'binocular',[16] as he sometimes called it. He confessed to feeling a pang of guilt at the deception of watching wildlife through it, 'a beautiful hypocrisy and delightful power',[17] he called it. Five years later he would be able to afford something he called his 'glorious Freidn "binocle"',[18] with three times the magnification. He then regretted what he had been missing. Hudson also spells it Fredin in the same letter. I have been unable to find anyone who has heard of this manufacturer, presumably German, as the word *frieden* means 'peace' in that language.

The county of Berkshire retains a special place in the Hudson story as it became the focal point for his first book about the nature of his adopted homeland. He could get there by train, and bring his new cycle. There's a letter that survives from early June 1892, written to Eliza Phillips, which encapsulates the mood and moment: 'I hope to get down to the country tomorrow, and if things go well, and I find myself able to write, I hope to remain some weeks.' He gave his address as Midway Cottage in the village of Cookham Dean.[19]

He was also spreading the word about the Bird Society, reporting to Phillips that he had sent out a hundred copies of a letter introducing the cause, and taking another hundred with him to the country. His campaigning zeal is clear, if not obsessive. 'If it should result in winning one helper – someone with a name, and able to use his pen, or better still his tongue, it would be a great thing', he declares.[20]

Hudson was initially unsure of the precise nature of his mission in Berkshire, but happy to be away from the city. 'After wandering somewhat aimlessly about the country for a couple of days I stumbled by chance on just such a spot as I had been wishing to find – a rustic village not too far away; it was not more than twenty-five minutes' walk from a small station, less than one hour by rail from London.'[21]

Emily Hudson was able to join him at Cookham for a time, during the Easter public holiday. Their base was close to Winter Hill, which had the finest views of the River Thames, and 3 miles of woodland open to the public, although, as far as he could judge, nobody went there. 'The woodland birds have it pretty well to themselves',[22] he was not sorry to report. He began to compile notes with a view to writing the book, which would be called *Birds in a Village*. Hudson didn't often take friends with him on his rambles, but Morley Roberts was with him on one occasion as he began to get acquainted with the place. 'The people with whom we lodged at Cookham Dean were the simple and kindly folk whom Hudson so much preferred to inn-keepers,' Roberts later wrote.[23] Roberts recalled Hudson's delight when they were able to track down a grasshopper warbler after finally working out the source of its curiously ventriloqual 'stridulating' call, more like that of an insect than a bird. 'We heard them often but it was only after a long search that I found them in the middle of a thorny thicket. I well remember Hudson crawling into the bushes on his hands and knees.'[24]

Birds in a Village was published in 1893. Hudson called it 'my first book about bird life, with some impressions of rural scenes, in England'.[25] It helped to establish him as a writer on nature in his new homeland. Half the print run of 1,500 copies had been sold by October.

Hudson retained an affection for Cookham, and witnessed the inevitable changes in the village as he returned, first in spring 1897 with Emily. 'It was dreadful weather – wet and cold,' he reported to Roberts, 'but nightingales were singing and things looked pretty much the same down there.'[26] He was able to reacquaint himself with his hosts of a few years earlier, of whom he spoke fondly: 'Mrs. Garrett is getting fat, and Mr G. makes bricks and looks well and happy. Would I had been brought up to make bricks!' Roberts recalled Hudson's delight, 'for although he was strangely lacking in every kind of manual dexterity he loved all the primitive arts and handicrafts'.[27]

Another seven years later, in summer 1904, the Hudsons lodged with the same family, now expanded in number, and found many new houses built. But there were still magical moments with nature. It was cherry-picking season, and the villagers were busy trying to collect the harvest from their vast plantation before the birds took too many. Sheltering from a rainstorm under an oak on the common the Hudsons were joined for a time by a red-backed shrike that sat near them. They heard turtle doves purring in the woods, and found some boys there playing with a stag beetle that they called Harry Horner.[28]

Hudson was aware of his own incongruity in the context of the dwellings he often visited and stayed at. After visiting two elderly ladies in the small city of Bath, he said: 'They were, I fancy, somewhat startled at the apparition of so big a man in their small interior – one whose head came within an inch of two of the low ceiling: they seemed timid and troubled and anxious in their minds when I gave them my scrawl to decipher and copy.' They were typists, and soon warmed to him, if not his handwriting, 'and they gratified my De Quincey-like craving to know everything about the life of every person I meet'.[29]

Hudson had no doubt had a similar craving to know everything about the life of all of the wild birds he encountered, and following the modest success of *Birds in a Village* he was soon engaged in producing a 'proper' bird book: a textbook on all British species.

British Birds – textbook written to order

While he probably couldn't resist the challenge, and thought it might finally establish him as a credible British ornithologist in the eyes of his peers, Hudson found the authoring of his *British Birds* something of a chore as it was 'written to order'.[30] He was also vexed that the publisher insisted on including an introductory chapter on the anatomy of birds written by another author.[31] It taught him not to sell himself in haste to a publisher and instead to go with those who would let him 'write what I like'.[32]

Hudson's initial lack of basic knowledge of the birdlife of the British Isles makes his subsequent mastery of the subject and later achievements as a published authority on its natural history all the more remarkable. From this standing start, and within the limits of his poverty and meagre income, and for a long time without the aid of many books of his own, or optical equipment or transport, he amassed a phenomenal body of knowledge and ability to identify and describe what was around him.

'Very many men take up science as a living and, failing to get the rewards they aim at, professorships, etc., they take up the trade of writing books for educational purposes or for the man in the street. I know some of these poor fellows who get a pale wan face in the British Museum, and the libraries.'[33] When Hudson wrote this he might have had himself in mind, and his one experience of writing a textbook. He wasn't alone in his disdaining of textbooks above all else. Even Thomas Henry Huxley, as Hudson pointed out, 'protested against the multiplication of such works, and even feared that we should all be buried alive under them – the ponderous tomes which nobody reads, elephantine bodies without souls'.[34]

Hudson omitted some details of rarer species from his book, not wishing to risk furnishing the collectors with information they might use to further deplete bird populations. With this chore complete, his attention could now turn to researching a book about – or rather for – the birds of London. This would require more legwork than deskwork. In the meantime, he had important people to see. Luckily for him, and for them, these were people with whom he hit it off immediately. He could relax his rule about never mixing in high circles, again.

Sir Edward and Lady Dorothy Grey

Of all the friends that Hudson would make during his near half-century in London, Sir Edward Grey was probably the most important, for his status and influence, and for the help he – and his wife Lady Dorothy – gave Hudson at important moments. Having broken his rule for Don Roberto of not dining with the upper classes, Hudson may have found it easier to see past the titles and the status of this younger couple (Grey was 20 years his junior). He soon became comfortable in their company and actively sought them out. The Greys were fans of Hudson's first books even before they got to know him, or know him well, through the rapidly expanding Bird Society, of which they were early supporters. Both were devotees of nature, and in Hudson's writing Sir Edward took solace, as his busy political life kept him from enjoying the outdoors as much as he longed to.

Hudson later described the impression Lady Dorothy made on him when they first met:

> It was a rare pleasure, a surprise, to find one in her
> world who did not use customary phrases, who was of
> so original a mind, so transparently honest, as to make
> it a mental refreshment to converse with her. But my
> chief pleasure was in the discovery that she herself was
> a native, so to speak, of my world … where I am at
> home with my non-human fellow-creatures.[35]

Sir Edward Grey was a reluctant statesman. He was first elected Liberal MP for Berwickshire in 1887, aged just 23 (the youngest member in the House), at the same time as Hudson was finally making his breakthroughs

Sir Edward and Lady Dorothy Grey were early devotees of Hudson's writing, and later the
man himself. 'I want to write to you: first about the Hudson book,' Lady Dorothy wrote
to her husband. 'I have read a good deal; it touches very fine notes of feeling for nature.'
Images courtesy of Adrian Graves.

as an author, and Don Roberto was also taking his first seat in the House as a fellow Liberal, on paper at least.

An 1893 letter written by Lady Dorothy from her London home to her husband in the House of Commons demonstrates the importance of wild nature in their lives, and how Hudson's writing spoke to both of them.

> I want to write to you: first about the Hudson book.
> I have read a good deal; it touches very fine notes of
> feeling for nature. I felt first sad because it was such a
> long way off from what we are doing; then the feeling
> stole over me of being very faithful to the holy things
> and very firmly separated from towns. I read on and on,
> and old Haldane came in.[36]

Richard Haldane was a politician who would later achieve wider fame. He had a role alongside Grey in the Liberal government that was in power for over a decade from 1905. He is also remembered for his friendship with Oscar Wilde, whom he visited in prison and wrote about later. On this earlier occasion, Dorothy, with her head full of Hudson's evocations of nature, took the opportunity to unload all of her frustration with London life and the sacrifices she and Edward were having to make for his political career, and his sense of civic duty: 'I sort of let out, and we talked from 5 to 8 and the result is that he has gone away saying "I understand at last. You must not stay in politics. It is hurting your lives. It is bad."'[37]

Sir Edward and Lady Dorothy Grey on the carriage, with prominent fellow Liberal statesmen Winston Churchill and Lord Richard Haldane (in front).

Sir Edward and Lady Dorothy Grey, on bicycle and trailer at their Fallodon estate, to which Hudson was invited. 'There will be a horse to ride about the moors, he says, and that is a great temptation.' Image courtesy of Adrian Graves.

Dorothy feared what Edward's enforced separation from his beloved country life might do to their relationship. She told Haldane 'that if we went on crushing our natural sympathies we should probably end by destroying our married life'. She ended by promising her husband: 'I shall read more Hudson tonight and store it up for you like honey.'[38]

Sir Edward Grey joined the Bird Society that same year and, no doubt with encouragement from chairman Hudson, soon agreed to become a vice-president. Although he had always been a great outdoors man, Grey was a relative latecomer to the cause of protecting wild birds. In his wife Dorothy, and Hudson, he had very able two mentors. 'I arrived at the age of manhood knowing only the songs of two individual birds', Sir Edward wrote in *The Charm of Birds*, and no doubt told fellow members at the Bird Society's gatherings. 'One was a robin, whose tameness and persistence in singing where there is hardly further song to be heard forces anyone to know its voice.' He had more recently been learning to distinguish between the voices of blackbird and song thrush.[39]

The affinity between Hudson and the Greys is especially interesting since it seems it was only the love of nature that they had in common. Hudson never showed any interest in Grey's other great passions, fly fishing and sport. Grey was real tennis champion at the time, but sport is among the things that Hudson rarely if ever comments on. A rare exception occurs when he writes in 1902: 'I love both cricket and football as a looker-on.'[40] He only mentions sport on this occasion in response to the Rudyard Kipling's controversial poem *The Islanders*, a scathing indictment of how the British public was allowing itself to be lulled by distractions while troops were slaughtered in combat, in this case fighting the second South African (later more commonly known as the Boer) War, which ended in 1902.

In September 1894 Hudson mentions to George Gissing that he has been invited to visit Lord Grey's Fallodon estate in Northumberland, about as far north as one can travel in England. In a rare hint of an enduring love of equestrianism, he adds that 'there will be a horse to ride about the moors, he says, and that is a great temptation'.[41] Having become involved with the Bird Society the previous year, Grey was keen to get better acquainted with his new friend and colleague. The mystery is that Hudson never would never take Grey up on his offer, and never saw these other places, or the Cunninghame Grahams' arcadian home nestled in the hills of central Scotland. I found confirmation of this in the words of Etta Lemon, who later said in a brief reminiscence that 'as far as I can remember he never went to Fallodon'.[42]

Hudson will have explained that he had been in that part of the world once before, though too briefly to visit all the places he might have liked, near the border with Scotland. 'I could not spare time for the Cheviot mountains when I was there,' he later wrote to George Gissing in 1894, 'and Holy Island and the river Coquet and other enchanting scenes.'[43] And while he never took up the invitation, the desire evidently remained, as he wrote to Gissing's younger brother Algernon in July 1897: 'Every summer I dream of Northumberland, but fear now there is no prospect of such a pleasure this year.'[44]

Lady Dorothy was soon using her influence to line up a publishing deal for Hudson – a book about birds for schoolchildren, although by this time Hudson was in a position to decide his own projects, and would do so, having not enjoyed the one book he had written to order. The Bird Society's latest pamphlet, *Feathered Women*, had also just been issued. Lady Dorothy had taken the vow to shun plumage in her outfits. She wrote to Hudson: 'I have no feathers except ostrich feathers for the last ten years and have induced several people to give up aigrettes. I will take more trouble now, and get them to join the Society.'[45]

The Greys made little secret of the fact that defeat for Edward in any general election would not have disappointed them, as they would then be free once more to indulge their love of the countryside. But for all his apparent unhappiness in office, the 'liberal imperialist' Grey was enduringly popular, even among traditional conservative voters, and he kept winning elections. And while the election of July 1895 was taken by the Conservatives, under Lord Salisbury, Gladstone's Liberal party gained ground, taking 80 more seats than six years earlier. Hudson told Eliza Phillips he had been 'deafened and half-stunned with the thunder and roar of the elections' as the results started to come through.[46] He noted that Sydney Buxton of the Liberal Party had kept his seat, but that 'the other militants, radicals and extremists appear to have been swept clean away'.[47] His letter gives a strong hint that his problem with radicalism – aiming too high – was that you would be left with nothing. It was a lesson for the Bird Society to note. 'They went too far – tried to do too much, and their fate is that they will not be allowed to do anything.'[48]

His friend's result was not yet in, but Hudson predicted that 'Sir Edward Grey will almost certainly lose his seat'.[49] On this point at least he was wrong. Grey won, and was further surprised when offered the post of Under-Secretary of State for Foreign Affairs, alongside Lord Rosebery. Grey had no particular qualification for this role, nor had he up to then paid especial attention to overseas issues. Nonetheless he accepted willingly, and among his first tasks were to create a policy for adding Uganda as a colony, and examine proposals for an East African railway.

With increasing pressure and responsibility in Parliament, Grey's country cottage by the River Itchen in Hampshire, and the nature to which he could escape at weekends, became a lifeline for him and Dorothy. They had begun to keep a diary of their observations of wildlife. The following entry from 2 July gives a sense of their lives there:

> The wren sang nearly all morning. We talked about
> it while we were at breakfast the first morning, and
> thought how nice it was that we knew enough to be able
> to love it so much, and how many people there were
> who would not be aware of it, and E. said, 'Fancy if God
> came in and said, "did you notice my wren," and they
> were obliged to say they did not know it was there'.[50]

It was a rural idyll they would later share with 'Huddy', and that he would come to know intimately. Before then, there was a brief period of wider travel on Bird Society business, and a rare opportunity to make another sea crossing.

A voyage to Ireland

The Bird Society's network of local groups soon extended beyond Britain's shores, including across the Irish Sea. 'I have *not* sent the *letter* to Mrs Shane, Ballymacreese, Limerick', Hudson wrote to Eliza Phillips in June 1892, 'thinking that you would be so kind as to send it with just a line from yourself.'[51] Phillips later sent him another letter that she had received from Ireland in October 1893 that he thought 'most interesting'.[52] I wonder if it was linked to the creation of a Dublin branch of the society, or even about a legacy being left to the cause. The first such gift would indeed later come from there.

One way or another, bird and family business would take Hudson to Ireland, even if he doesn't seem to have relished getting on a boat again. Writing in July 1894 he said: 'I fear very much that I shall have to go to Ireland early in August on a visit (long promised) to some relatives over there.'[53] What his fears were, we can only guess. As for the relatives, I think it unlikely they were his, and there is some trace of a link to Emily's in-laws in a later mention.[54] Curiously, and frustratingly, Hudson left no trace of the route he took to get there – was it by train to Liverpool, and a boat from there? His experience of crossing the Irish Sea, and the seabirds they must have communed with on the way, were never described in print. If he looked up any relations of his grandmother, he never discusses it in detail. Hudson's grandmother – surname Malony – is said by earlier writers on Hudson to have been an Irish immigrant to the USA.[55] 'My mother's forbears were furiously anti-English from the very beginning of the discontent that ended in the [American] Revolution',[56] he once wrote. Yet he seems not to have explored this heritage in any depth. 'I have not penetrated far into the interior of that distant country', he wrote later.[57]

For a long time the only trace I could find of this visit to Ireland is something published 20 years later in *Birds and Man*. Hudson described his base as 'close to the Wicklow range of mountains, and Dublin and the sea in sight'.[58] He had a lot to say about the swallows there, which were nesting in the stable, late broods then joined by other birds gathering for southward migration. He was intrigued by how they mobbed him, but only if he wore a fur hat.

Then I came across a letter to George Gissing. 'We were over in Ireland in August but didn't enjoy it very much.'[59] He explains that his sister-in-law, a Mrs Brooks, was over there, and would now be coming to live in the flat downstairs at St Luke's Road. Hudson wrote from the farmhouse, called Springmount, at Rathfarnham, near Dublin. His host was a Colonel Glass, in a property large enough to have a kitchen garden, which, Hudson was pleased to discover, was partially overgrown and full of birds – a 'perfect wilderness'.[60]

He evidently made his customary rounds of the locals, in order to better acquaint himself with his neighbours and surroundings. 'The Irish peasant is a pleasant being to talk to in spite of his or her rags',[61] he observed. He reiterated this impression in a later letter when he said after reading *Irish Idylls*, by Jane Barlow: 'the Irish peasant being certainly a more humorous animal than the somewhat grim new Englander in her small agricultural village'.[62]

He was struck by the relative poverty he found in Ireland. 'What a contrast between the cottages and the bare-legged ragged people in them here', he wrote to Emma Hubbard, 'and those of Northumberland and Yorkshire',[63] where he had recently been. He later hinted at some knowledge of the Irish west coast when, for example, he likened Cornwall to Connemara, and the Cornish character to that of the Irish, but he must have based his generalised view of the Irish from his encounters in London with the diaspora there.

Hudson seemed to be resolutely opposed to Ireland ever having home rule, despite this being part of the political raison d'être of many of his associates, notably Don Roberto, and their mutual friend Wilfrid Blunt. The Irish home rule question dominated British politics for years. Don Roberto must have talked about it a lot. His friend, the Protestant landowner and reformer Charles Parnell, 'the uncrowned king of Ireland',[64] who had come so close to achieving it, and to whom Don Roberto paid tribute in his parting speech in the Commons, had died in Brighton just three years earlier. Prime Minister Gladstone resigned in the spring of 1894, to be remembered in part for introducing land reform in Ireland and attempting to introduce Home Rule, and for being undone by his failure.

Why Hudson chose to be so partisan on this matter, given his mixed family heritage, is hard to work out. He seemed generally to incline against what he may have thought was radical social change, and his political outlook was broadly that of the self-reliant frontiersman, generally suspicious of the powers-that-be, and sympathetic to the settler, all of course grounded in his experiences on the Pampas frontier and time spent as a conscript in the Argentine military.

In any case, Hudson's mind was now on other things: there were more books to write, and more wild birds to save.

CHAPTER 5

Saving London's Birds

A book with a purpose

Hudson's experience of compiling his *British Birds* textbook left him eager to write a book of his own devising, and based much more closely on his own observations: a book, as he put it, 'with a purpose'.[1] The purpose of *Birds in London* would not be to attempt to systematically document and define all of the English capital's avifauna, it was to issue a call for the better treatment of the birds – residents, visitors and others that might no longer be either, but could return.

Unlike most of the resident human population, with eyes and ears jaded by familiarity and soot, to every bird he met with in London on his travels and explorations Hudson's eye was fresh. He was hyper-alert to changes occurring around him, in the birdlife of the parks and streets. He recorded what he heard and saw as he went. There wasn't much in the way of literature to assist him. Nothing had been expressly written about the subject until James Jennings's *Ornithologia* at the turn of the nineteenth century. But Jennings's work was a poem, which Hudson did not rate. In fact he considered it 'probably the worst ever written in the English language'.[2]

Hudson would dare to be critical of the royal and other parks for the losses they had sustained in birdlife. He thought these places 'sadly mismanaged' where wild birdlife was concerned.[3] He would offer encouragement to the public authorities, and Mr Sexby, the head of parks, to do better than to think of birds and plants merely as ornaments; to provide for the 'confined Londoner' the 'glad freedom and wildness which is our best medicine'.[4] Of Hyde Park he wrote: 'is it not extraordinary that so noble a possession, the largest and most beautiful open space in the capital of the British Empire, this chief city of the world, should be degraded to something like a poultry farm, or at all events a duck-breeding establishment?'[5] 'Let us not have ducks only', he cried, 'to the exclusion of other wilder and nobler birds.'[6] He called on the park authorities to recognise that birds meant more to people than decorative borders. Providing flowerbeds for nature-starved

Londoners he likened to putting Turkish delight before a hungry man. How he hated these 'prettified and vulgarised gardened parks'.[7]

He wanted also to capture the changes that he had witnessed or could guess at, from even his own two decades or so in the city, but he soon dismissed any notion of producing what we would today call a guidebook. 'What is London?' he asks in his opening pages,[8] while explaining the pointlessness of attempting to pin down where the city's birds ended and those of neighbouring counties began. The limits of the city were unclear and fluid, and most birds are highly mobile. The city had by this time grown to 6 million inhabitants, an increase by 50 per cent even in the 20 years he had known it. It was expanding so rapidly on all sides – he likened the green spaces at the city's fringes to the edge of a volcano, 'always in a state of eruption'[9] – that any such book would immediately be out of date.

Hudson recalled a triggering moment and turning point in his life as being that warm, sunny morning in October when he set out to walk from his Tower House the mile and a half to Kensington Gardens, 'thinking how pleasant my favourite green and wooded haunt would look in the sunshine'.[10] This was the place that had so inspired him when he first arrived in and explored his new London home two decades earlier. But a terrible shock awaited him. 'Then I first saw the great destruction that had been wrought; where the grove had stood there was now a vast vacant space, many scores of felled trees lying about, and all the ground trodden and black, and variegated with innumerable yellow chips.'[11]

His mind drifted back to the early days he spent first exploring London: 'Recalling the sensations of delight I experienced then, I can now feel nothing but horror at the thought of the unspeakable barbarity the park authorities were guilty of in destroying this noble grove. Why was it destroyed?'[12] He estimated there had been 700 trees, their removal a small echo of the centuries-long process of deforestation on these islands. He met two young children, who told him that the workmen would only let them take away pieces of timber for fuel upon payment of a penny. He gave them the penny they needed, and told himself that perhaps he shouldn't 'grieve overmuch at these hackings and mutilations of the sweet places of the earth'.[13] He thought of a future time, when the dust of our 'bones had been blown about by the winds',[14] and trees would grow again on this spot, home again to squirrel, bird and bee.

Incidents like these gave him the impetus to undertake his book on London's birds, which had for some time been among what he poetically called 'the many little schemes and more or less good intentions that have flitted about my brain like summer flies in a room'.[15] The trauma of the tree devastation turned it into a mission to save the birds and the few remaining wild and semi-wild places in the sprawling, clamorous and grimy city.

Hudson's London was a very different place for birds than the one we know today. The end of summer was a melancholy time for him, but he could take consolation from the flocks of migrating swallows and martins that still gathered in the city's parks. He witnessed hundreds that remained until mid-October, before they moved south, to cross the Channel to the continent, and Africa beyond.

There were so few woodpigeons then that he counted each one – 13 of them in Lincoln's Inn Fields at the end of September, before all but an injured one disappeared. 'One spring morning, the head gardener at Buckingham Palace, full of excitement, made a hurried visit to a friend to tell him that a pair of these birds had actually built a nest on a tree in the Palace grounds.' How different from today. From where I'm sitting, writing, I can see in a north Cambridgeshire garden almost this many woodpigeons grazing on an elder tree, nibbling its early buds. In the morning there are often more woodpigeon droppings on my car than Hudson saw woodpigeons in the entire city.

Hundreds of gulls settled in the autumn in St James's Park. Understanding that a diet of bread alone would not be healthy for the birds, Hudson took to buying sprats for them, at a farthing a pound. 'No sooner would they see the little gleam of a silvery tossed-up sprat than there would be a universal scream of excitement … brushing my face with their wings'.[16] They learned to trust him and would feed from his hands. *Birds in London* would include a drawing of Hudson in St James's Park among his gulls.

Hudson was used to seeing sad, imprisoned birds for sale in London's vast, frenetic markets, but never inured to it. 'Monster London throws its nets over an exceedingly wide area', he wrote, 'capturing all rare and quaint and beautiful things for its own delight.'[17] Although preoccupied with his London book, he also made time to write the pamphlet *The Trade in Bird Feathers*. He was busy on this project throughout 1897, Diamond Jubilee year, with Queen Victoria having reached the unprecedented milestone of 60 years on the throne. The plumage campaign was gaining political traction. Questions were asked in the House of Commons on the subject of plumage use in the military. Was the government aware that these plumes were often brutally obtained from rare birds, in the breeding season? The reply came that yes, the government was aware, and that orders had already been given that alternatives should be found. This, however, was proving to be a challenge.[18] Meantime, the jubilee military procession through London would involve no feathers. Besides the plumage question Hudson had little interest in this occasion but noted that just one pair of starlings nested on the Prince Albert Memorial in Kensington Gardens, thinking it ought to have room for more.

Later in the year he was putting the final touches to his manuscript, typed up from his scrawls by Miss Beckles, 'the young lady who does typewriting for me … My Miss Beckles is not perfect', he told Algernon Gissing, 'owing I think to bad eyesight.'[19] The book was finally published in 1898.

By this time the Bird Society had amassed 20,000 members across 152 branches, with outposts including those in Washington DC and in Germany. Etta Lemon was carrying the fight overseas, speaking in French at an international ornithological conference in Provence. The society had begun to hire staff to help run the burgeoning business of saving wild birds. *The Times* remained a useful channel through which to keep the public up to date with the plumage campaign, and it covered the issue in its Christmas day leader column. The Bird Society produced its first Christmas card, although it would take Hudson some time to get used to this new fad. He also disliked postcards for their brevity, preferring longer narratives.

Meanwhile, Don Roberto was in Paris, where he witnessed US citizens celebrating the end of the Spanish–American war. Inevitably, there had been pressure on him to return to politics.

The further adventures of Don Roberto

Prominent socialist James Connolly had asked Don Roberto to stand in the 1895 general election in central Edinburgh, but he declined – citing lack of funds and not wishing to be a burden on the party. Instead, he wrote a satirical guidebook *Notes on the District of Menteith, for Tourists and Others*. He dedicated it to Trootie, a local poacher on the Gartmore estate whom Don Roberto clearly didn't begrudge a share of the lake's fish. On the title pages he remarked that all rights were reserved, except in the Republic of Paraguay.[20]

Hudson encouraged Don Roberto to write more of the edgy pieces he had started to send to the *Saturday Review*. 'I occasionally ventilate my private views and curse my enemies in that print',[21] he told him. While Hudson was lobbying beleaguered councillors in provincial towns, Don Roberto was writing impassioned letters to the American press to call out atrocities committed against indigenous people.

It was these letters that would catch the eye of the influential publisher's reader Edward Garnett, who brought Don Roberto into his writers' fold. Never shy of forming alliances with those he admired, Don Roberto introduced himself to and befriended Joseph Conrad, who would in turn become acquainted with Hudson, along with another radical and controversial landowner, Wilfrid Blunt. Irish playwright George Bernard Shaw, meanwhile, another of Don Roberto's famous friends, would remain somehow beyond Hudson's ambit.[22]

Early in 1897 Hudson walked to Rotten Row in Hyde Park to find Don Roberto exercising Pampa. Even from afar, with his keen eyesight he was able to pick out Don Roberto – tall and straight-backed in the saddle, imperious on his splendid Argentine horse – among the hundreds of horse riders exercising their mounts on the gallops. The distinctiveness of Pampa also helped, with his white facial markings; a noble, high-spirited animal with shining coat, long mane and swishing tail.

Don Roberto had been preparing for some months for his next adventure, having set his heart on reaching the mythical Moroccan city of Tarudant, which few outsiders had ever reached, and where Christians were forbidden to tread. To get there he would have to dress up in full Arab costume and hope that he didn't have to do too much talking. His Arabic was far from fluent. 'I am a heap of camels, saddles, pistols, bibles, pills, etc, etc, all suitable for the explorer', as he wrote to George Bernard Shaw.[23]

After his friend's departure for north Africa, Hudson had to rely for news of Don Roberto's adventures in the Atlas Mountains when they were published in the *Daily Chronicle*. The gallus gaucho had somehow been able to smuggle dispatches out. 'I spare you any remarks on the flora and fauna of the district, for Inshallah, I propose to inflict them on a harmless and much book-ridden public',[24] he had written to the editor. It then went quiet for some time, much to the consternation of his friends, until news came at last from Morocco in late 1897: Don Roberto had been busted – and taken into custody. Hudson was therefore mightily relieved to hear from his friend, when Don Roberto finally returned to London.

> Only yesterday I was thinking about the trouble that
> had overtaken you among the Mountains of the Moon,
> or the Atlas Mountains, I forget which, when lo!
> Your letter came to say that you were no further than
> Chester Square, S.W. If it is possible to get about the
> world at that rate of speed, why do men trouble about

flying machines and balloons, unless they want to get
to the North Pole?[25]

Hudson was not without his own problems with authority: his *Birds in London* had landed him in a spot of bother, and he wrote to tell his friend about

> another amusing item: the Barnes Conservators,
> or their Chairman, after sending me some abusive
> letters, to which I did not reply, now announces that
> he is bringing action against me for libel – for what
> I say about Barnes Common in my book. I only wish
> that the Hyde Park Ranger and all the other park
> authorities would join in putting me into gaol.[26]

Sometimes, if Hudson was proving difficult to prise from his belfry, Don Roberto would simply doorstep him. 'My impulsive friend burst in on me here because I declined to lunch with him today', Hudson reported on one occasion.[27] But his visitor lacked his usual effervescence in this instance: it was the end of days for the Gartmore estate. 'Not only is there no chance of things getting better, but, not only here but on every estate in Scotland they must get worse', Don Roberto lamented.[28] Don Roberto and Gabriela had to leave Gartmore, which was sold as the century ended. 'I feel the pain of your defeat,' Joseph Conrad consoled him. 'The fight was a good fight.'[29]

Perhaps partly inspired by his friends' publishing success, and clear of the distractions of trying in vain to save Gartmore, Don Roberto was now writing prolifically. He was completing the book about his escapade in Morocco just as Hudson was putting the final touches to his *Nature in Downland*, and naturally they were eager to exchange books. The contrast in their ranges was not lost on Hudson. 'What a poor return that will be!' he conceded. 'Travels in the Sussex Downs for travels in the Atlas.'[30]

Joseph Conrad called the Morocco book a 'glorious performance' and 'the book of travel of the century'.[31] The intrepid Don Roberto frequently inspired such awe in his friends. 'When I think of Cunninghame Graham, I feel as though I have lived all my life in a dark hole without seeing or knowing anything', Conrad remarked, no stranger to overseas adventure himself, having travelled the world as a seafaring man in earlier life.[32] He also described Hudson as 'a son of nature, and almost primitive man who was born too late.'[33]

'The book is one that one reads and does not forget', Hudson later purred. 'To my sick soul your life seems almost too full, your activities too many and great, your range on this planet too wide.'[34] A note of Hudson self-consciousness about the relative narrowness of his horizons and a hint of envy at the range and intrepidity of his friend might be detected in what he says, but my strong sense is that Hudson no longer aspired to a life of travel, or at least was entirely reconciled to the situation in which he found himself. He wished only for ten years in which to explore every English county.

Hudson was also very pleased to be able to tell his friend that he had an admirer in Lady Dorothy Grey. 'I had a delightful two hours' talk with Lady Grey about everything (politics excepted),' he wrote. 'She had a "great deal" to say about the Morocco book, and *Ipane*. She is an enthusiastic admirer of your work, and takes a particular malicious delight in the way you go for everybody and everything – especially 'sacred' things.'[35]

Don Roberto's handwriting was even worse than Hudson's, who had reassured him he could usually read him like a printed page. The speed of Don Roberto's scribbling mirrored the frenetic pace of his life and thinking, his handwriting deteriorating as his excitement mounted. 'Qué diablo!' Don Roberto on one occasion said of his own hand. 'A man who writes like this deserves a good beating.'[36] On another occasion Hudson had to tell him 'this note has been rather a floorer'. He had been able to decode part of it: 'I am invited to lunch with you and two others – distinguished men.' But in any case he declined, politely, being 'at present in bad health – and trying to medicine my ills with nature's greenness'.[37] He quoted two lines of Andrew Marvell (1621–1678) poetry:

> Annihilating all that's made
> To a green thought in a green shade.

Perhaps Don Roberto wanted to introduce him to some more of his literary friends. Morley Roberts would sometimes join the two men for lunch, including at the Café Royal, where Don Roberto had rubbed shoulders with Shaw, W.B. Yeats, H.G. Wells and Oscar Wilde. It took time, but Don Roberto would be an important catalyst in Hudson emerging from his writer's garret to embrace the new world and possibilities of Edwardian literary society. Meanwhile, his female comrade-in-arms for a decade or so from 1894 was Emma Hubbard, who also became his soulmate in nature at this time.

It's long forgotten, but Kew Gardens as it is known and loved today was under threat at the end of the nineteenth century, and Hudson and Hubbard were in the thick of the fight to save it.

Emma Hubbard – soulmate and mentor

Hubbard was 66 when they first met, 13 years Hudson's senior. She was a gifted artist, as well as poet and editor. She was widowed, and had lost two of her five children, including a daughter just three years earlier. Hubbard lived close to Kew, at Bradenham Lodge on Holmesdale Road, and was from a well-known family. Her brother was renowned archaeologist Sir

Hudson had a great meeting of minds with Emma Hubbard. 'The week seems incomplete without our talk about nature', he told her. Image courtesy of Judith Frances Hubbard.

John Evans, her daughter Frances was married to the city editor of *The Times*, Wynnard Hooper, and her son was a doctor who would minister to the sickly Hudson on occasion and counsel him on diet. It was on his advice that Hudson dined 'on bread and cheese and a glass of beer', and had 'eggs with my tea, and nothing after'. He found then that he was 'able to walk all day without getting tired'.[38]

Hubbard reminds me of a lesser-known Edith Holden, whose *Country Diary of an Edwardian Lady*, first written in 1906, would become one of the best-selling nature books of all time when it was later rediscovered and published in 1977.[39] Holden died in tragic circumstances at Kew: while collecting flowers on the Thames riverbank she fell in and was swept away by the river. She had not yet reached her 50th birthday. I wonder if this Kew Gardens connection might mean that she knew – or at least met – Hubbard, a long-forgotten nature-loving and soon to be Edwardian lady.

Although Hudson was such a faithful correspondent of Eliza Phillips, his closest ally in the Bird Society, Hubbard would become his best friend for the next decade. She was certainly his closest companion in the appreciation of nature through the 1890s. Although his circle of male literary friends was growing, it was in the main women with whom Hudson corresponded about his primary passion, and in Hubbard he found a particular kindred spirit, and to some extent a mentor. He loved their chats, and enjoyed sending her detailed notes on his explorations, hearing of her excursions by return. 'The week seems incomplete without our talk about nature, with a little, not too much, of "men, women and books" thrown in', he told her.[40]

Hubbard carefully filed all of his letters, and happily they survive. The first dates from February 1894 and is priceless – snapshotting a pivotal moment in the journey of Hudson and the Bird Society. Hubbard had evidently written to him introducing herself and with a flattering response to one of his early books, I would guess *Birds in a Village*, published the previous year. It will have resonated with her as she had connections with Berkshire and had spent many hours painting landscapes not far from there. Her enthusiasm for his writing, he said, was 'more to me than any printed review of my books'.[41] He might have had Don Roberto in mind when he mentioned that others who had written to him in this way soon became dear friends. He eagerly accepted her invitation to meet at Kew, suggesting a spring evening walk to hear the returning warblers, and listen in particular for the wood wren[42] – a species we know today as the wood warbler, greatly diminished in number and range in southern England since then.

True to form, Hudson couldn't wait until they met before asking if she knew about the Bird Society. 'I am always trying', he wrote apologetically, '– at the risk of being set down as a bore – to interest every person I come in contact with in its work and progress.'[43] He didn't mention that he was chairman, but he did make a good fist of summarising the mission of the society: 'I am convinced that if all who are bird lovers throughout the country would associate themselves together a great change could be brought about in the public mind, and that birds would be protected for their beauty and melody instead of being incessantly persecuted as is now the case.'[44]

Another two months passed before they were finally able to meet for their springtime walk at Kew to listen to returning migrant birds, particularly the warblers. Hudson wrote again in late April after his return from rambles in Kent, Surrey and Berkshire. He had heard the willow-wren everywhere he went since the start of that month: 'He is very early this year, and should be at Kew.'[45] He also reported hearing nightingales and one garden warbler, but not yet a blackcap or a whitethroat. The cuckoo and wryneck, meanwhile, had arrived a month earlier than normal. He mentioned being pleased to see that the new Wild Birds Protection Bill, to protect birds' eggs, had had its second reading in the House of Commons the day before.

They met soon after, and hit it off immediately. But the migrant-friendly weather of early spring soon deteriorated, and another month passed before they could resume their acquaintance – by which time Hubbard had duly signed up for the Bird Society. Hudson had arranged for her to be sent a copy of his pamphlet *Lost British Birds*, although he thought there could be a delay, as the organisation's twin spearhead – Etta Lemon and Eliza Phillips – had both been absent through illness.

Hubbard in turn shared with him a manuscript of a bird book she was editing, in which the author observed that the 'greatest romances' of birds took place on Sundays.[46] This was a test of Hudson's open-mindedness: there was just a tiny hint of a raised eyebrow in his reaction. No doubt Hubbard was sceptical too, for they had a meeting of minds on the matter of religion. 'Her brilliant mind could not bow to the narrow and restricted Church doctrines in which her generation had been brought up', granddaughter Frances Roper recalled. 'I was always rather in awe of the beautiful old lady. I fell completely under the spell of the family charm of personality with which she was so richly endowed. She was a wonderful artist.'[47]

Hubbard had the same effect on Hudson. They spent many hours exploring Kew and Richmond Park, as well as further afield on occasion, enjoying and discussing nature. 'I wish that you could have been there to watch the crossbills with me,'[48] he wrote at the end of 1898. And a year later: 'I am always thinking of things it would be interesting to discuss with you … in some spot so wild and desolate that it will please us both.'[49]

I can picture the pair at Kew from little written snapshots of their visits there, such as when Hudson described the two of them watching a butterfly: 'It was a hot summer's day, when the early freshness and bloom is over and the foliage takes on a deeper, almost sombre green; and it brought back to us the vivid spring feeling, the delight we had so often experienced on seeing again the orange-tip, that frail delicate flutterer, the loveliest, the most spiritual, of our butterflies.'[50]

Hubbard would set up an easel and paint her watercolours, including a series of Kew scenes. Her rural landscapes, painted mainly in the 1870s in the North Downs of Surrey, on the doorstep of Eliza Phillips's later home in Croydon, are beautifully crafted. She could also whittle, and left an extraordinary collection of carved chestnuts collected at Kew, which survive to this day. They can be seen online,[51] exquisitely detailed faces – 18 in all – peering out of the opened casings of the nuts, and framed in a box. The work is understandably treasured by the family. Hudson later commissioned his friend to provide the artwork for his *Hampshire Days* (1903). He would take or purchase photographs of relevant subjects for her on which she could base her sketches.

On another occasion in Kew Gardens Hubbard read her poetry to him, open to criticism, stopping in places to make adjustments to the wording. 'We read it aloud two or three times over to each other',[52] Hudson reported of her verse *My Moor*. He published this poem in tribute to her near the end of his life, in his 1921 collection *A Traveller in Little Things*.

It all sounds quite idyllic, but the bucolic bliss of Kew Gardens was about to be threatened by the axes, saws and shovels, and the unlikely pair would be at the forefront of the fight to repel them.

Saving Kew Gardens

Reading the latest list of Bird Society supporters, Hudson noticed that an old soldier he knew and chatted to, who stood guard at the Queen's Cottage grounds, had just signed up with the Bird Society local branch. Hudson knew he was a lover of birds – 'especially of robins'.[53] When these Cottage Grounds at Kew were given by the queen to the people to mark her 1897 Diamond Jubilee, and opened to the public the following year, Hudson wrote to *The Times* in April 1898 to urge that access be channelled in a way that maintained the value of the site for wildlife. He also underlined Kew's importance as a sanctuary for the people, calling it 'a boon to London visitors, the tired workers with hands or brain in search of refreshment for body or mind'.[54]

His suggestion was duly adopted. Commissioner of Works Mr Aretas Akers-Douglas attended Parliament to assure the House and the nation that Kew would remain 'a great sanctuary of all the wild bird life in the district'.[55] You can sense Hudson's satisfaction when he subsequently wrote, after a visit with Hubbard in spring 1899, that it now 'seems a wider, more refreshing place than ever, with crested grebes and coots at home, the shade of oaks from the sun, and the greenness of leaf and grass and fern'.[56]

Kew was certainly a safe haven for nature and nature lovers in the context of the landscape beyond, in which birds faced such relentless and untrammelled persecution. At the latest meeting of the Bird Society, they had resolved to write to the Home Secretary requesting that flagrant crimes against rare birds be properly prosecuted, in light of the shooting of three or four ospreys just arrived from Africa at their breeding grounds in Scotland. 'This last scoundrely act will probably see out this species as a member of the British avi-fauna', Hudson seethed.[57] Kew, by contrast, seemed secure. Passing migratory ospreys would even sometimes be seen fishing at nearby Pen Ponds, out of range of the rural guns.

But danger loomed. The first trace of Hudson's concern about threats to Kew becomes evident in November 1897 when Hubbard relays the news to him that trees have been felled in the Gardens. Few things triggered Hudson like the removal of what trees survived in the rapidly expanding and increasingly congested metropolis.[58] Then, early in 1900, the unthinkable happened: the government made its first announcement of an intention to build the National Physical Laboratory here. An agitated Hudson was soon writing again to *The Times*, on 14 April, about this 'great and disagreeable surprise to the inhabitants of London'. He was quick to remind readers of Her Majesty's expressed wish when she gifted the land to her people: 'The Queen earnestly trusts that this unique spot may be preserved in its present beautiful and natural conditions' had been the pledge.[59]

Sir Edward Grey got involved, perhaps at Hudson's bidding, writing to Hubbard that same month. Hubbard shared Grey's letter, and Hudson thought this encouraging. A suggestion was then made, possibly by Grey, that a copy of Hudson's letter to *The Times* might be sent to every Member of Parliament. Hudson thought the cost would be prohibitive, running to £5 for postage alone. He proposed instead to print a limited number targeted at members 'who may be supposed to take an interest in such things'. And he urged that it be reproduced verbatim. He had been advised (perhaps by Grey) to remove a passage containing what Hudson called a 'damning fact' that showed what government was capable of if the public didn't protest. Hudson saw no reason for such caution. 'I know my facts are right, and the "Works" know that too', he vowed.[60]

It seems likely that the damning fact was Hudson's reference to the cutting down of 700 elm trees in Kensington Gardens in 1880, which, when he witnessed it, had caused him so much trauma. These same trees I believe were the place he found by chance on his first rovings in London in the days after his arrival in Britain, and which I think so impressed his fresh visitor's senses, with their huge rookery in full cry. It had thrown a kind of comfort blanket over him in this strange new place. He never got over returning there for solace, and finding it torn down.[61]

At the end of April Hudson reported to Eliza Phillips that the residents of Kew, with Hubbard in the thick of it, were now agitating against the development, which would affect 4 acres. They were now taking the initiative and carrying the cost of reprinting his *Times* letter and sending it to all MPs.[62] She asked him about the feasibility of compiling the MPs' home addresses, and he pointed out that this would be tricky as many of them used temporary accommodation – hotels and hired houses – while attending Parliament. On May Day 1900 she retyped Hudson's *Times* letter while he collected signatures for the petition.[63]

While the local upwelling of resistance was welcome, with its input of resources, Hudson recognised that this alone would not suffice and that the protest must come from a wider constituency. The matter concerned all Londoners, after all, and more of them must therefore raise it with their MPs. On 10 May Etta Lemon offered to help send out more copies of the letter, so Hudson dropped 18 of them at the Bird Society offices for her to distribute to 'persons of importance'.[64]

Hudson had received a letter on 8 May from Lord Crawford, who was to ask the question in Parliament. Hudson seems to have been in the public gallery to witness the matter being raised. He was able to report that, when questioned on the proposals, Mr Akers-Douglas (who would become Home Secretary the following year) on behalf of the Conservative government had 'assumed an aggressive air', and then claimed that 'exaggerated statements had been made in the papers'.[65] Hudson knew that this can only have been in reference to his letter, as no other newspapers had covered the issue. 'If I had not written the letter the original plan would have been carried out',[66] he assured Lemon, since Kew Director Thiselton-Dyer's protest had been ignored.

Backers of the development, recognising the mounting strength of feeling against it, now revealed their alternative plan, perhaps hoping that it would look benign by comparison, and wrongfoot the protestors. They were now proposing to build on the Old Deer Park nearby instead of the site alongside the Queen's Cottage Grounds. While this might be good news for the Grounds, it had simply shifted the problem and threat to the Old Deer Park. This prompted an intervention from the famous novelist Ouida, who wrote in protest to the *Westminster Gazette*, with other letter-writers responding in turn.

Hudson wrote to Emma Hubbard immediately after a committee meeting of the Bird Society in June, at which the Deer Park threat had been discussed. A coalition of bodies including the Commons Preservation Society and the Selborne Society had now resolved to take the matter to the House.[67] Hudson escaped from London to the New Forest, in precarious health, but he remained preoccupied with the Kew campaign, even while out of town. Etta Lemon wrote to him there, about a proposed deputation to Parliament. He shared the news with Emma Hubbard. 'Mrs Lemon asks me to represent the Society – a joke on her part, I suppose.'[68] Few things could have terrified him more.

Hubbard tried to cajole him into finding the nerve to overcome his fear of public speaking. A rare surviving letter to his wife Emily is revealing, as Hudson now turned to her for moral support, she perhaps understanding the seriousness of his phobia better than anyone. He was under pressure from all sides from those who wanted him to lead the charge. He must have felt cornered, and in a quiet room somewhere in the forest he wielded his pen in self-defence. 'I have a bundle of letters saying no to write – Mrs Hubbard, Lemon, Chubb, Lord Balcarres!' He was almost in fear of his life. 'I suppose this everlasting worry about my "saying a few words" in public will only end with the silence that comes on a man once and for all', he groaned.[69]

For all his passion for this cause, Hudson was now in the grip of dread that only sufferers of social anxiety might fully understand. He made no attempt to disguise it. He wrote again from his bolt-hole in the New

Forest, this time to Eliza Phillips on 19 June, about his 'hopeless weakness' and 'helplessness',[70] pleading that they instead ask Sir Edward Grey to represent the Bird Society, in his role as a vice-president.

Of course, in the end Hudson didn't have to lead or be spokesperson for the delegation – and it is curious that, knowing him well by this stage, all of his peers would still try to push him forward.

He continued to 'strategise' while lying low, and wrote to Hubbard from Hampshire at the end of June, suggesting how they might take their protest to the Treasury, arguing that he didn't know anyone with sufficient clout to influence that key part of government, and reminding Hubbard that she was *much* better connected. Ironically, he wrote this while sitting in Highwood House, the family home of Bird Society vice-presidents Captain and Mrs Suckling, which he describes as 'magnificent', adorned as it was with portraits of generations of Sucklings dating back to Elizabethan times.

In early July *The Times* reported that a delegation of scientists had visited Mr Hanbury (I assume Robert William Hanbury, then Financial Secretary to the Treasury), with luminaries Lord Lester and Sir Michael Foster bolstering the ranks of the physicists. Hudson's scorn is searing:

> I should have thought that Lord Rayleigh and
> the other physicists could have beat their own big
> tom-toms and blown their brass rams' horns without
> assistance from others, but it's a case of let's all go
> together and help each other – and let the millions of
> Londoners go and suffocate in their slums, we don't
> care! It always seems to me that the greatest cant of the
> present time is about the scientist's love of his kind.
> But I fear I am boring you about all this.[71]

During the debate in Parliament, Major Glazebrook (Conservative, Cheshire) queried that if it has been permissible to put a golf course on part of the landscape here, why not a physical laboratory? In response to this, Hudson reasoned: 'the golfers tho' they may be fanatics are very harmless when compared with physicists who build laboratories, and their scarlet coats did not create alarm'.[72]

The kerfuffle died down for a while, in Hudson's surviving letters at least, before suddenly the conclusion of the campaign sparks vividly to life again. A scenario is described in Hudson's letter to Eliza Phillips on

25 November 1900, written after he had been to Kew once again to meet Emma Hubbard:

> When I arrived at Kew Gardens station there was
> Sir William Thiselton-Dyer [Kew Director] on the
> platform, and seeing me alight he cried out 'You've
> done the trick!' When I asked what he meant by those
> strange slangy words (so unbecoming in the mouth of
> a person in his position) he said that the Government
> had finally abandoned the project of building at Kew,
> so that danger to the Gardens and Old Deer Park was
> over.[73]

This story within the story of Hudson the campaigner not only helps bring into focus the modus operandi of the early Bird Society and his role within it, but I think also the nub of the force that drove him. Proposals to put any large building on his beloved Kew would have met with his ire and opposition, but there is something in particular about physicists being behind it, men of science who ought to know better but who, in his view, might know the price of everything and the value of nothing, that made this campaign one especially close to his heart, and soul.[74]

Much as Hudson loved and fought for the green spaces of London, and drew on them for sustenance amid the harsh urban conditions of the era, it is clear that he was never happier than when able to escape the confines of the city.

CHAPTER 6

Further Afoot

Homage to Gilbert White

Gilbert White's *Natural History of Selborne* whetted Hudson's appetite for the South Downs of Sussex. 'I still investigate that chain of majestic mountains with fresh admiration year by year,' White wrote.[1] This was about as far afield as the village curate ever ventured. As with Hudson, the Alps or even the craggy peaks of Scotland were not for him. 'There is something peculiarly sweet and amusing in the shapely-figured aspect of chalk hills,' White wrote, 'in preference to those of stone, which are rugged, broken, abrupt, and shapeless.'[2]

The Natural History and Antiquities of Selborne is widely recognised as one of the most influential early books on natural history observation

The Greys' cottage by the River Itchen, Hampshire, where the Hudsons stayed for much of the summer of 1900 and where he worked on *Hampshire Days*. 'We are quite alone,' Hudson wrote, 'and when we go lock up the cottage and leave it to the wild creatures.'

written in English, a collection of letters written by White to Thomas Pennant and Daines Barrington. It was a formative influence on the young Hudson, and he made several visits to The Wakes, White's house, locating White's nearby grave with a little difficulty on the first occasion.

The tiny Hampshire village of Selborne is just 30 miles by road from Southampton, in the direction of London. I found White's grave a little more easily than Hudson had when I made my own pilgrimage to Selborne, the site being neatly tended today. If I felt White's presence in the house there, it's because there is a life-size dummy of the man, a centrepiece of the exhibition celebrating White's life, which I bumped into when giving a talk for the Selborne Natural History Society.

Hudson made the first of two horseback rides that I can confirm in his near 50 years in Britain when he went to Delves House at Ringmer, where White's aunt Rebecca Snooke had lived.[3] He also called at a big house at Chilgrove and chatted to an 85-year-old lady whose father-in-law had been a friend of White.[4] On one occasion Hudson imagined a conversation with White beside his modest gravestone, with its inscription of his dates, 1720–1793.[5]

'I began to speculate as to the subjects about to be discussed by us,'[6] Hudson wrote. 'The chief one would doubtless relate to the birdlife of the district.'[7] Hudson contemplated how the enduring popularity of White's book had baffled some critics. 'They are astonished at Gilbert White's vitality, and cannot find a reason for it. Why does this "little cockle-shell of a book," as one of them has lately called it, come gaily down to us over a sea full of waves, where so many brave barks have foundered?'[8] Hudson answered his own question: 'I would humbly suggest that there is no mystery at all about it; that the personality of the author is the principal charm of the Letters, for in spite of his modest and extreme reticence his spirit shines in every page.'[9]

I think key to the enduring appeal of the book is the fact that these are letters, not intended for publication. Hudson conveys much of the candour of his own letters in the books he wrote. As the nineteenth century neared its end, his reputation as an author was about to reach new heights. He was soon back on the south coast, writing *Nature in Downland*, a decade after his first publishing breakthroughs and the death of Richard Jefferies.

The ghost of Richard Jefferies

'Jefferies was much in my mind just now', Hudson writes at the start of the book, 'because by chance I happen to be writing this introductory chapter in the last house he inhabited, and where he died.'[10]

While the two authors never met, a very curious thing happened to Hudson in that house at Goring, called Sea View. Hudson wrote from there to Emma Hubbard, describing his lodging as:

> A large, well-built comfortable cottage in its own grounds, ivied and pleasant to look at, with fig and apple trees in the grounds; the sea close by … My bedroom is the one Jefferies occupied, but he died downstairs in one of the sitting-rooms. In the garden there is a now ruinous summer house where he often sat to do his writing.[11]

Hudson wrote later in *Downland* of a walk he had taken near the church, approaching the house: 'A mysterious adventure befell me … My mind was full of sadness, when, hearing the crunch of gravel beneath other feet than my own, I suddenly looked up, and behold, there before me stood the man himself, back on earth in the guise of a tramp.'

'Can you spare a penny?' the stranger asked him.

Hudson obliged, and not surprisingly was troubled by this apparent visitation by Jefferies. 'I knew', he reflected, 'that those miserable eyes would continue to haunt me.'[12] But it seems the spectre didn't follow him back to the house. The following morning Hudson resumed his letter writing: 'A rainy morning! But I slept soundly, and poor R.J.'s ghost did not disturb me – they seldom come back. I wish they did.'[13]

Nature in Downland

> If the power to attain all that De Quincey craved, or pretended that he craved for, were mine, I should give it all to be able to transform myself for the space of a summer's day into one of these little creatures on the South Downs; then to return to my own form and place in nature with a clear recollection of the wonderland in which I had been.
>
> And if, in the first place, I were permitted to select my own insect, I should carefully consider them all … And after all I should make choice of the little blue butterfly, despite its smallness and frivolity, to house myself in. The knowledge of that strange fairy world it inhabits would be incommunicable, like the vision

vouchsafed to some religionist of which he has been
forbidden to speak; but the memory of it would be a
secret perennial joy.[14]

Hudson had begun to spend more time exploring Sussex, sending
reports back to Emma Hubbard, among others. Writing in June 1899,
after communing with a herring gull colony on the towering chalk cliffs
at Beachy Head, he described his dawn dip in the sea and an epic hike
across the Seven Sisters range, 'those seven gigantic downs … anxious
to get back before dark… my sixteen mile walk was as good as twenty
… I saw wheatears, males in their beautiful summer dress … about their
music there is a question, as it seems that no living ornithologist knows
the song. I should be delighted to have a letter from you.'[15]

In another letter he describes more impressive feats of rambling,
leaving a dining appointment on a gloomy evening with 5 miles of
downland to cross, the fading light and weather against him. There is a
boyish enthusiasm in his despatches, for example on an occasion when he
cycled between Kew and his home in 40 minutes, 'and would have done it
in less time if I had cared to'.[16]

Hudson had seen a recently published guidebook to Sussex, and
scoffed at the birds it still claimed for the coastal area there, including
bittern, reed-pheasant (bearded tit), bustard, stone-curlew, black grouse,
chough, guillemot, razorbill, kittiwake and shag.[17] Of the bustard, he had

The Sussex Downs, where Hudson found inspiration. 'Once we have got above
the world … then we at once experience all that sense of freedom, triumph and
elation'. CMJ.

read a description from a century before, when an old shepherd said 'there haven't been any wild turkeys either for many a year. I have heard my father say he killed two or three no great while before I was born; they used to call them bustards … Since the hills have been more broken by the plough such birds are seldom seen.'[18]

He was all-too conscious of the many other birds already missing: raven, red kite, common buzzard, honey buzzard, hen harrier, Montagu's harrier. 'It is not possible', a local told him, explaining why he had shot a buzzard, 'for any man to see a large rare bird and not "go for it". If he is not himself a collector he will be sure to have a friend or neighbour who is, and who will be delighted to have a Sussex-killed raven, spoonbill, honey buzzard, or stone curlew sent him as a present.'[19]

Hudson also rued the loss of native mammals and feared for those that remained, thinking that the fox might 50 years hence be extinct, like the wildcat, marten and wolf before it. He found badger setts and hoped that public sympathy for this species might yet save it.

Exasperated by the catalogue of loss, he was also on the tail of the local authorities while there. Wheatear trappers were still operating, and when challenged would usually offer the defence that they were after starlings, a species it was legal to trap with a landowner's permission. Revisions of the Wild Birds Protection Act in 1894 and 1896 had given county councils the ability to apply for orders to protect places and species at risk. Some 130 orders (many of them obtained at the bidding of the Bird Society) were duly in force. A new East Sussex bird protection order meant that trappers were no longer permitted to catch wheatears before 1 September. The practice of catching wheatears on their southward migration in late summer had once been a thriving industry, as the birds funnelled down to the south coast from their nesting places in the hills of the British Isles. With dwindling numbers of birds, the practice had become less lucrative.

'It is not fair that it should be killed merely to enable London stock-brokers, sporting men and other gorgeous persons who visit the coast', Hudson seethed, 'accompanied by ladies with yellow hair, to feed every day on "ortolans" at the big Brighton hotels.'[20] 'Efforts will, I trust, be made by residents on the south coast, who are anxious to preserve our wild bird life, to enforce the law,' he added. 'And I hope to be there to help them.'[21]

When news reached him of the molestation of nesting peregrines on the coast, in September 1896, Hudson advised Eliza Phillips that they should write to Sussex County Council 'to take steps to preserve their wild birds'.[22] They would run a draft past Montagu Sharpe, the Bird Society's legal brain and the person best qualified in navigating local governments. Hudson also wanted to check his facts in the Zoological Library at Hanover Square when it reopened.

Hudson's growing renown was opening doors and boosting the Bird Society's cause locally. 'Mr Robinson, editor of the Sussex paper, has just written to invite me to go and see him ... perhaps he is desirous of bringing some pressure on the West Sussex County Council.'[23] As it turned out, the editor of the *Sussex* newspaper told him that his readers weren't interested in bird protection, and it was of no consequence to him. 'What they care for is to me a weariness,' Hudson afterwards lamented to Hubbard. 'Bands and concerts principally.'[24] Despite the lukewarm reception, Hudson still managed to get two letters published in the Sussex press.

In Lewes he met the clerk of the Sussex county councils, to have a long talk about bird protection issues in the county. The clerk, evidently a fan of Hudson's books, was surprised to hear that things were so bad for the birds on the sea cliffs locally. Hudson was asked to put his thoughts in a long letter, which he wrote in duplicate in order to send a copy to Phillips and the Bird Society committee. Hudson even invited delegations from the council to accompany him to the woods and cliffs to see for themselves, and he visited at least one clerk at his home in Brighton.

He spoke to a taxidermist in Chichester, who was 'deeply hurt' at the new West Sussex order when this was brought in to protect rare birds. 'Poor fellow – how I sympathise with him!' Hudson crowed.[25] When his energies were directed at writing his Sussex book, Hudson was able to work his charm in the library at Chichester, where he was allowed to borrow books when others were not.

Even the unusual heat of that summer of 1899 couldn't lessen the pace. He wrote from Marlborough Place in Brighton, describing the pattern of his day. He had ended up in the police station. 'I have been interviewing the police here about bird-protection, and other matters,' he wrote. 'I do some writing during the morning before and after breakfast, and make visits in the afternoon; but the people visited are police officers, farmers, poulterers, bird-stuffers and such like.'[26]

He had witnessed plenty of law-breaking and told the police that they could fill the courts if they wanted, which must have surprised them as they had solemnly assured him that the law was being faithfully followed locally. 'This, I fear, is a weary subject, and is making a bore of me,'[27] Hudson again acknowledges in one letter.

Eventually, the excessive heat took its toll, and he could barely walk with illness and fatigue, but his mission was not complete. He attended Seaford Church for a Sunday service. 'A very dressy lady with a big white aigrette sat in the seat behind me, and just after the sermon she fainted and was taken out,' he reported. 'But whether it was the effect of the vicar's grating voice or of my presence so close to her, I do not know.'[28]

When he first walked in the hills of Sussex Hudson had not thought of writing a book about them, assuming such work was already well covered. But he kept his field naturalist's journal until one day the notion of compiling it all struck his mind. 'It will be, I imagine, a small unimportant book,' he wrote, 'not entertaining enough for those who read for pleasure only, nor sufficiently scientific and crammed with facts for readers who thirst after knowledge.'[29]

He was first drawn to the Downs in part to trace a poet whose verse he had known in his youth, when the hungry young reader and nature enthusiast had found fragments of nature poetry on the bookshelves at home, and in second hand stores on his occasional forays into the bustling city Buenos Aires. The poet was James Hurdis, by then fallen out of fashion and favour, but precious to Hudson, nonetheless.

He explored the Downs for three summers, and in a sense his descriptions of the heatwave of 1899 are three years rolled into one. Hudson described it as 'exceptionally hot', as he recalled the many hours he had spent on the wide spaces of these hills between the murky streets of inner London and the increasingly populous seaside resorts of the south coast. He recalled his first impressions 'from a distance of these round treeless hills that were strange to me … uninviting in their naked barren aspect'.[30]

Luckily, he had some novel methods of keeping cool. 'At this season my custom on going out on the hills is to carry a wetted pocket-handkerchief or piece of sponge in my hat: by renewing the moisture three or four times, or as often as water is found, I am able to keep my head perfectly cool during a ramble of ten or twelve hours on a cloudless day in July and August.'[31] He had learned from hard experience on the Pampas. 'Long ago, in South America, I discovered that the wet cloth was a great improvement on the cabbage-leaf, or thick fleshy leaf of some kind, which is universally used as a brain-protector.'[32]

As for clothing, 'there is nothing like tweeds of a greyish-brown indeterminate colour, with a tweed hat to match'. Hudson was against wearing straw hats, thinking them off-putting to birds. Equally, 'white or light-coloured flannels' are no good 'for those who go a-birding'. When on his wheel, he found birds 'strangely indifferent to the bicycle'.[33]

Some might have thought tweed far from the ideal fabric for heatwave rovings on the South Downs. Hudson acknowledged the limitations of his outfit of choice, and of 'my hat of an unsuitable material',[34] but it would prove an unlikely lifesaver when, one day, the unthinkable happened: he forgot his other head protection. After half a day of hiking he began to struggle.

> I experienced that most miserable feeling of a boiling brain … [like] a pot boiling on the fire, bubbling and puffing out jets of steam.[35]
>
> The day was frightfully hot. About eleven o'clock I was so consumed with thirst that I went nearly half a mile out of my way – a cart track over the downs – to interview a shepherd standing watching his flock near an old half-ruined stone building. He showed me an old well near the spot, but said that unless I had a bottle and a long string I could not get any water. However, my tweed hat fastened to the crook of my stick served as a bucket, and after drinking a hatful of cold water I felt refreshed.[36]

Hudson noted that it 'had proved directly useful in that case'.[37] He also explained why he didn't use a water bottle.

> To carry water is a precaution which I never take, because for one reason, I love not to be encumbered with anything except my clothes. Even my [field] glasses, which cannot be dispensed with, are a felt burden. Then, too, I always expect to find a cottage or farm somewhere; and the water when obtained is all the more refreshing when really wanted; and finally the people I meet are interesting, and but for thirst I should never know them.[38]

One wonders if these local providers of water were as interesting to him as he – this tall stranger with the traces of a Latin-American accent, wearing a tweed hat (and matching suit) with wet handkerchief on his head – was to them. At least he wasn't, by the time he had acclimatised to English weather and customs, wearing a cabbage leaf.

Hudson's descriptions of downland nature are often vivid:

> The air, especially in the evening of a hot spring day,
> is full of a fresh herby smell, to which many minute
> aromatic plants contribute, reminding one a little of
> the smell of bruised ground ivy ... The turf is composed
> of small grasses and clovers mixed with a great
> variety of creeping herbs, some exceedingly small. In a
> space of one square foot of ground, a dozen or twenty
> or more species may be counted, and on turning up a
> piece of turf the innumerable fibrous interwoven roots
> have the appearance of cocoa-nut matting. It is indeed
> this thick layer of interlaced fibres that gives the turf
> its springiness, and makes it so delightful to walk upon.
> It is fragrant, too.[39]

He had his own term for the very small animal life of this carpet ecosystem, woven over centuries of organic pastoralism like a Persian rug. He called it 'the fairy fauna'. He would lie among clouds of blue butterflies and watch to see where they went in the evening, finding them roosting on the grasses. Hudson evocatively described 'the sound of innumerable bees and honey-eating flies in the flowering heather',[40] rising and falling with the wind, and found five children lying on the turf, basking like butterflies:

> I experienced the blissful sensation and feeling in its
> fullness. Then a day came that was a revelation. I all at
> once had a deeper sense and more intimate knowledge
> of what summer really is to all the children of life; for
> it chanced that on that effulgent day even the human
> animal, usually regarded as outside of nature, was there
> to participate in the heavenly bounty. High up the
> larks were raining down their brightest, finest music.[41]

Hudson knew that the ploughing of the grassland often meant irreparable damage to the delicate flora. 'I seldom care to loiter long in their cultivated parts ... It has been said that if the turf is once destroyed by ploughing on the downs it never grows again.'[42] From the highest peak in the Downs, at 800 feet, on a clear day Hudson could see as far as the Isle of Wight, where Queen Victoria was staying at Osborne House, the coast where he had first arrived in England, and Hampshire, which would be the focus of his next explorations and the subject of his next county book.

In January 1900 Hudson brought his artist friend Arthur MacCormick to Sussex to make the final sketches needed to complete *Nature in Downland*. Hudson's fastidiousness is once again clear. 'I have pretty well wasted a fortnight dragging an artist about the country to get sketches, but weather was against us, and we must go back.'[43] Hudson also took him to the Natural History Museum to sketch birds for the book. Stuffed birds occasionally had their uses.

Despite wintry weather, Hudson stayed on to finish the book, and took a room at the White Horse Inn at Chichester.[44] There he found four enormous cats, a fox terrier, a jackdaw, a blackbird and a 'white owl'. He struck up a rapport with the owl. 'He is pleased at being caressed, although he always makes a great show of seizing my fingers with his beak and claws. Poor bird!'[45]

Hudson determined that the owl should be freed from its 'prison'. He was soon using his charms and powers of persuasion on the landlady. He had come prepared, having about his person a copy of the Bird Society's recently published pamphlet on barn owls, which he gave her. Happily, she consented to let him take the bird and find a way of rehabilitating it for a new life in the wild.

This new project meant that putting the finishing touches to the book would have to wait. Hudson now had a mission to find 'a farm with a very big barn and a kind family to look after the bird for a few days in the barn'.[46] He found an isolated and ancient, ivy-covered manor house, complete with moat and 17 windows on the front, and rang the bell, explaining his mission to the young, tall, handsome and gracious lady who answered, Mrs Couzins. She was enthusiastic about Hudson's plan, as both she and her husband loved barn owls above all birds, and her father had also been a bird-lover, who always preached to her and her sisters against wearing plumage. She would feed the bird herself while it learned independence.[47]

Quizzed further, she knew about the Bird Society but had not seen its pamphlets, so Hudson arranged for Eliza Phillips to send him some more, with a note from her (Hudson insisted on this) to pass on. Mrs Couzins even gave the unexpected visitor a tour of her 'strange old house',[48] and introduced him to her three young children in the nursery. Hudson made arrangements to return shortly with the owl and bade her farewell.

He returned to the White Horse with the good news that he had found the ideal new home for the owl's rehabilitation and freedom. Alas, in the meantime the landlady had had a change of mind and was no longer willing to part with the bird. Hudson was desolate. He spent days trying to persuade her once again to let it go, even offering money. But her mind was set. 'I do not think I have ever spoken sharper things to any person about the hideous cruelty of keeping birds in unnatural conditions than to this woman: but it was all useless,' he fumed.[49]

The effect on Hudson was to be shattering. He described himself as 'miserable and angrier than I have felt for many a long day'.[50] He left the White Horse and went outside though the rain was pouring, befitting his mood. He was so discombobulated that he got lost in the woods near Midhurst as darkness was falling. When he finally found his way out, he was drenched but still heated with indignation, continuing to seethe about what he called 'the good people who torture birds out of love for them'.[51]

While Hudson was doing his fieldwork, his colleagues were busy in the metropolis. The Bird Society had been instrumental in a Bill being introduced in Parliament seeking to change the system in which birds, to be protected, had to be added by local councils to a protected list. What was sought was a system where all birds were automatically protected, unless added to a list indicating otherwise. This would be a kind of 'presumed protected' system: innocent unless proven guilty of misdemeanours against humanity. A similar Bill would be brought again in 1900.

In early July 1899, Bird Society leader Etta Lemon gave an address on 'Our Duties to Wild Animals' at the International Congress of Women in Westminster, London.[52] Two of the society's vice-presidents, the MPs Sir Herbert Maxwell and Sir Edward Grey, were also on the platform. 'What an impoverished nature and earth future generations will inherit from us!' she declared passionately from the lectern. 'God's footstool, yes: but with all the shining golden threads picked out of its embroidery.'[53]

Lemon later reflected on Hudson's guiding influence and his early reassurances:

> He had unfailing sympathy with young and inexperi-
> enced workers [and gave them every encouragement]:
> When I was young [she was around 28 when they
> met, and he 19 years her senior] he helped me over
> many difficult places connected with my position as
> Honorary Secretary of the Society for the Protection
> of Birds. On one occasion, when I was terribly

nervous at having to speak at an important meeting,
he encouraged me by saying 'It does not in the least
matter what you say, for you look all right in that nice
frock and becoming hat'. It struck me as so funny that
all nervousness vanished.[54]

'I don't know if *Downland* is selling well', Hudson reported to Don
Roberto soon after publication, while noting that 'it has been very
favourably reviewed. If it is not a success I must lay down my pen and
take up the shovel and the hoe-o-o.'[55] Although not a bestseller, *Nature in
Downland* was attracting favourable comment from his friends, including
Don Roberto and Sir Edward Grey. Life had seldom looked better for
Hudson than in 1900, the final summer of the nineteenth century.

Halcyon days – a cottage by the river

I am sure Sir Edward Grey was instrumental in planting the idea in Hudson's
mind that he should write next about Hampshire. Grey was also influential
in first suggesting and then preparing some of the ground for Hudson's
application for a civil list pension, with his 60th birthday approaching
in 1901. Hudson would first need to become a British citizen, and in
spring 1900 he was busy with the paperwork this required: seven printed
documents to be signed in front of the Commissioner of Oaths, including
five requiring the signatures of his sponsors. 'It is no easy matter to become
a citizen of this small country', he groaned to Eliza Phillips.[56] According
to Grey biographer G.M. Trevelyan, 'it was on Grey's advice that Arthur
Balfour as Prime Minister gave a pension from the Civil List to Hudson.'[57]

Hudson also had the satisfaction of knowing he and his colleagues had
done all he could on the campaign front for Kew Gardens, the Old Deer
Park and the birds more widely, and progress was tangible. His application
for British citizenship was rubber-stamped on 5 June. He would now be
eligible for the pension, but by no means guaranteed to get it. Demand,
inevitably, was high.

In spring 1900, Hudson had a front row seat at the unveiling of Thomas
Henry Huxley's statue at the British Museum of Natural History and was
within touching distance of Prince Edward who as heir to the throne did
the honours. Hudson was struck by the future king's charm, while being
much less complimentary about other dignitaries present, whose identities
are lost in his ensuing letter to Eliza Phillips as their names were made
illegible by someone who evidently wished to spare the feelings of those

named. Did someone – perhaps Phillips herself I wonder – erase the names, while not wishing to destroy the whole letter? Or did she give Hudson his letters back before she passed away in 1916, which may have enabled him to doctor them in this way? Or it could have been a later custodian of the letter. Hudson's letters have in many cases had an interesting after-life.

Best of all, Hudson had the use of the Greys' cottage on the River Itchen in Hampshire for the summer. Sir Edward and Lady Dorothy so deeply valued the privacy and anonymity they found there that they loaned it to almost no one else. It was their 'sacred place', that Sir Edward had built as a hideaway from the stresses and strains of London and parliamentary life, where he could spend time immersed in nature and where it was forbidden to talk about work. Hudson would become one of the chosen few.

A letter from Edward to Dorothy in August 1901 gives some sense of the bond between them, and their beloved cottage:

> It's one o'clock and I have just got here and I feel as if
> my heart was too full and might burst; the place is so
> sacred … I feel as if I must keep coming in every half
> hour to write to you … It is very strange that you aren't
> here; stranger than I thought, but I suppose it wouldn't
> be so strange to you, as I am so often away. What
> wonderful days you must have here without me![58]

One of Hudson's letters indicates that Grey had shown him to a kingfisher's nest on the riverbank nearby, presumably on his induction visit. Hudson was especially pleased by the survival of these birds, having been assured locally that they had been wiped out by keepers of the nearby fish hatchery. And now that Hampshire had belatedly got a bird protection order, thanks in no small part to Hudson's lobbying, he could dare to hope that 'the barbarians may be restrained in their zeal for destroying this lovely bird'.[59]

If Hudson followed the pattern set by Edward and Dorothy, getting to the cottage would have involved rising at dawn on a Saturday to catch the 6 am train at Waterloo. The train would take him south to the tiny village of Itchen Abbas, with its wayside station on a single branch line. From there it was a short walk, leaving the road and on to a track through an avenue of lime trees. Itchen Abbas could also be reached by a faster train to Winchester, half an hour on a one-horse fly, or around an hour by bicycle, or a walk of several miles following the riverbank.

It was at the cottage that Hudson witnessed the particular depth of Lady Dorothy's connection with nature. 'When her husband could not come down, she would dismiss the one servant [Susan Drover from the village looked after the place for the Greys] and spend the night out of sight and sound of any other human habitation until the following morning.'[60] Dorothy had what Hudson called Henry David Thoreau's 'sense of an un-accountable friendliness in nature, like an atmosphere sustaining him'.[61]

Hudson brought his wife Emily to the cottage in late July 1900, in sultry summer weather. He placed a table and chair under a lime tree behind the cottage and spent hours sitting there in the shade with his books and notes, soothed by the hum of the bees on the lime flowers, and the tinkling of goldfinches overhead. Breezes riffled the miles of tall plants on the water meadow, from which a medley of bird sounds issued. Crooning turtle doves were 'numberless'. Sir Edward had accidentally flushed one from its nest while casting his fishing line in the river. 'Even here in the cool of the big lime trees I sit under most of the day it is hot – what must it be like in loathsome London!' Hudson mused.[62]

Some of the bird species present in the vicinity seem extraordinary now. Hudson described cirl buntings nesting beside the cottage, and wrynecks, and there were red-backed shrikes, the latter two being migratory species that today are almost unknown in Britain, but for occasional passage birds and the odd nesting attempt. Grey had shown him the cirl bunting nest, a species lost today from all but small pockets in the West Country where it is subject to intensive conservation efforts. Arriving with Emily and keen to show her, Hudson found the nest now empty. Happily, the male bunting was still singing nearby. Two weeks later he noticed fledged cirl buntings being fed by adult birds, indicating another productive nest nearby. In addition, there were nightingales, cuckoos and grasshopper warblers.

This Hampshire hideaway was called Fishing Cottage, but Lady Dorothy fondly described it as 'the tin hut'. Built ten years earlier, it was now wreathed in climbing plants in full flower – honeysuckle, clematis and climbing rose – and thronging with nesting birds, and bees and butterflies. Only the brick chimney, rising from the ruck, a large window framed in twisted stems, and a red corrugated metal roof might give it away at close range.

Inside, it had a small sitting room with blue linen on the walls, a matching carpet and a blue and white chintz sofa positioned so occupants could enjoy the view over a small lawn sloping down to the meadow beyond, and beyond that the river. There was a stove, a basket for firewood and basic cooking utensils. The living room had two long shelves of books, including some of Hudson's of course. The *Cottage Diary* lay on a table by the sofa, in which the Greys recorded their nature notes – of which this one by Dorothy gives a further taste: 'June 11th [1900]. The nightingale left off its night singing, but still sings a little in the day.'[63]

In his first letter from the cottage, Hudson describes the meditative pattern of their day – rising early, him going to fetch milk from a farm at the village, otherwise seeing almost no one else and being untroubled by newspapers. 'It is a sweet little hermitage and will I trust do us both good',[64] he wrote to Eliza Phillips. His surviving letters to Phillips and Emma Hubbard give a picturesque sense of halcyon days. 'It is a most lovely spot, greener in July than it is possible to imagine',[65] he rhapsodised.

They attended the quaint little church for Sunday services, with prayers said for the vicar, who was dying. As ever, Hudson's plumage radar was on, and he reported with some satisfaction that 'the few worshippers had no feathers on their heads'.[66] Perhaps he could for once relax.

Emma Hubbard sent him the review from *Nature* of his *Downland* book, along with some peaches. 'Like giving champagne to a yokel', he joked. He wrote again at the end of July. 'We are quite alone … and when we go lock up the cottage and leave it to the wild creatures. The little old village of Itchen Abbas is near, tho' out of sight, but next to nothing is to be had in the one small general shop.'[67] Luckily he had brought his bicycle, and could make runs to Winchester to replenish their supplies.

The Hudsons were soothed by the sound of gentle wind in the limes, willows and poplars and, on the rare occasions it rained, the sound of it drumming on the roof at night. It was the perfect spot for writing his book about Hampshire. 'With green shade of old limes about us, and the river flowing past close by … We are our own servants, and sometimes see no person all day; and only once during the last week have we seen a newspaper.'[68] He wrote to Don Roberto that they hoped to stay all summer, until the end of August. 'The Greys want us to stay as long as we can.'[69]

From the cottage they could see across the river to the Avington woods, which he had read were the last inland breeding place in southern Britain of the raven, before it disappeared in 1885. The land had been owned by the brother of romantic poet Percy Bysshe Shelley. Hudson's local rovings took him there, and one day in August he encountered the son of the owner and, as usual disarming his no doubt surprised host, soon got to talking about birds in general and the absent ravens in particular. Hudson shared with him the idea of bringing the birds back. They had another chance meeting a few days later. Young Shelley reported that his father wanted to see Hudson, to discuss the raven idea further. Returning to the cottage, Hudson wrote to Eliza Phillips and asked her to send him a copy of their *Lost British Birds* pamphlet, to pass on.

It wasn't just the work of the Bird Society that Hudson continued to promote at every opportunity. He had brought copies of Henry Salt's new *The Humane Review*, and passed one on to a Miss Williams in Winchester, a 'great worker for the animals',[70] who would now join the Humanitarian League, an animal welfare body that he had long supported.

Late in the summer, to the Hudsons' great pleasure, the cirl buntings built yet another nest and new eggs were being incubated. But the autumn was approaching and, sadly, after a cold night, he found that they had abandoned the nest. The pretty eggs were icy cold when he touched them. Some late summer storms, complete with hailstones, may have contributed to the abandonment of this late breeding attempt. He consoled himself and Emily that it might be for the best that the birds gave it up before the hatchlings had to face the shorter days and colder weather.

In the end, the contented couple extended their stay until well into September, enjoying clear blue skies, although this meant cold nights. Owls now hooted before daybreak, and robins found their sombre voices again. The mists of early autumn greeted him now for his dawn walks to collect the milk. They now breakfasted later – at 6.45 am – on their coffee and bacon, and the flocks of goldfinches began to disperse. He now counted just nine but was able to get within a yard of them by sitting still among the tall plants where they gathered seeds.

'Every day brings so many beautiful sights to us', he wrote again to Eliza Phillips on 13 September. 'And the weather is so fine and the long holiday is doing my wife so much good, that we are determined to say on as long as possible.'[71]

There is the first hint of Emily's advancing years and diminishing mobility: 'she bicycles not, and cannot walk much.'[72] Hudson also mentions in a letter that he can't leave her alone there, and that their rambles are confined to afternoon strolls. Their walks had often been a bit of a mismatch; him striding ahead or veering off-road to investigate something of interest. Emily was finding it increasingly difficult to keep up with him, while never wanting to hold him back. But they were still able to stroll each afternoon and go mushrooming on the nearby downs.

Absorbed as he was in such diverting surroundings, Hudson could hear duty calling: 'I dare say I shall be back before our next committee meeting',[73] he wrote to Phillips. The swallows were gathering for autumn departure south by the time they finally, with reluctance, ended their long stay at the cottage. He had collected a 'great mass of material' that he hoped might be shaped into a book and see the light of day before long.[74] It would form the basis of *Hampshire Days*, published in 1903. In a break with habit (he had an avowed dislike of book dedications), Hudson dedicated the book to Sir Edward and Lady Grey: 'Northumbrians with Hampshire written in their hearts'.

It had been the summer of a lifetime for the Hudsons, and a highly unusual one. It is never explained exactly how Emily had been able to spend so many weeks with her husband on this occasion, but the fact that the accommodation was free, and they had their own space, made it both possible and desirable for Emily to prolong her stay. Up to now work commitments and costs had meant she only had a few days off here and there, usually on public holidays. Was she now resting on doctor's orders? Whatever the pretext, it had been a rare chance for Mr and Mrs Hudson to take stock of their lives.[75]

I wonder if it marked a turning point in their relationship, as Will approached his 60th year, and Emily perhaps her 75th? Did they find time to properly discuss and assess the pattern of their lives and contemplate a new arrangement? By now, with British citizenship imminent, Hudson could anticipate a potential civil list pension of £150 a year, which in 12 months' time would give him a new-found financial security, and increased freedom. Given the backing of Grey and other influential friends, he could allow himself a degree of confidence in securing this.[76]

There is a strong clue that while ensconced there the Hudsons discussed him getting a place of his own, while he sampled what that might be like (at least in warmer weather). Writing to Hubbard on 19 September, he clarifies something he has spoken of before and of which she has reminded him, namely that a cottage in Hampshire would be the most suitable place for him. 'I did not mean that Mrs Hudson would live in it', he stresses, 'she will be in London I daresay.'[77]

He betrays some uncertainty about being able to choose the right place, and cutting ties with London, before concluding: 'I must I suppose have a *pied-a-terre* in London, and go there when I must.'[78] A small place, for occasional use in the city, and a cottage in the country … was he at last daring to dream of an escape from London? And would he chase it?

One of his remarks underlines that Emily Hudson knew when to make clear that it would be good for both of them if Hudson wasn't under her feet for a time: 'My wife has been hinting – the hints get plainer each day – that she would be much better off if I would go away for two or three days.'[79] Four years on from that idyllic Hampshire summer, Hudson still hadn't taken the plunge on an alternative or second home.

Through the Greys, Hudson befriended Bishop of London Mandell Creighton and his wife Louise. The Creightons were among the privileged few who were also invited to use Fishing Cottage. In August 1903 Sir Edward wrote to Mrs Creighton after she asked if they could pay something for the privilege: 'It is to be paid to the birds in the form of bread,' he advised her, 'which is to be put on the grass in front of the glass door of the sitting room'.[80] He also offered the following touching glimpse of the cottage interior: 'Cottage book is put in the shelves next to *Hampshire Days*. There is very likely a bird's nest in the stove chimney.'[81]

The late 1900 general election was won by the Conservatives, with their allies in the Liberal Party. Edward Grey of course retained his seat: there would be no escape from the political front line. It was dubbed the 'Khaki Election', the Conservative campaign boosted by the general belief that the South African (aka Boer) War was coming to an end and had been won. In fact, the war would grind on for another two years. 'It is 10 o'clock now and guns are going off and the population of Westbourne Park is getting drunk all for joy that Pretoria has fallen', Hudson must have sighed, perhaps turning away from the top window at Tower House to return to the letter he was writing. 'Well, I'm about tired of the war, and want to be back among the birds, beetles and snakes of the Forest.'[82] He must have wished he was back in the cottage; for him, the 'most delightful spot to one who looks on the wars of kites and crows as of more importance than our petty human affairs.'[83]

Feathers flying

The military may have been persuaded to give up their plumes following Queen Victoria's order in 1899, but even this lofty example made little difference to the fashion industry. 'The dealers declare that the demand for birds of every description will this year be greater than ever,' the *Fashion Paper* crowed in 1900. Their words were promptly seized upon and quoted by the Bird Society in a pamphlet distributed at tuppence a dozen. In the Edwardian decade, over 14 million pounds (weight) of exotic feathers would be imported into Britain with a value of nearly £20 million.[84] There is something rather David and Goliath about the ongoing contest between campaigners and industry.

Given the scale of the multi-million-dollar trade and the worldwide range of its sources, hoping to end the importation of bird-skins and plumage might by now have seemed futile. But there was another strategy: preventing the export of plumage from source countries (India would show the way, in 1902). The limitation with this, however, was that the British Empire, extensive as it was, only went so far. It would be impossible to effect any export bans elsewhere. International, joined-up agreement was needed. The British government made enquiries, starting with some neighbouring European countries. The response was predictable.

There were other practical difficulties: if you couldn't reasonably ban the trade in plumage of *all* birds, how would the average customs official know what was legal or not, being unqualified to tell one type of plume from another. Legal traffic has always been used as a cover for illegal. Similar thorny questions had obstructed legislation against eggers. Who could tell a reeve's egg from that of a redshank?

And there were some in the scientific and museum communities who were nervous about the possible collateral impact of any bans on their sourcing of specimens. In the first decade of the twentieth century, the Zoological Society of London, the Selborne Society and the Linnaean Society were at odds with the Bird Society over the detail of proposed law. These other organisations were minded to seek a compromise with commercial interests and an 'economic preservation of birds' approach – what some might today call 'sustainable' or 'controlled' trade.

The Bird Society stood firm, refusing to have anything to do with an industry that would casually issue lies about how plumes were procured, question whether slaughtered egret colonies and the like were actually threatened, or suffered, and tell customers that the plumes weren't real. Without an outright ban, the Bird Society reasoned, any residual legal trade would disguise and continue to enable the trafficking of protected, threatened species by interests who had shown themselves to be unscrupulous at best.

Hudson raged against more lies being told by the plumage industry – which was also claiming that the plumes they purveyed were naturally moulted, and that the adult plumage of birds of paradise were, as Hudson paraphrased, a 'token of senility, and the bird should therefore be shot (like a rogue elephant) for the good of the race and incidentally to defeat the plumage legislation'.[85] He had a conspiracy theory too that some leading zoologists might even have accepted bribes in return for complicity in this deceit. This would lead him into trouble.

As the new century loomed, the words were now pouring faster than ever from Hudson's pen.

PART 2

TWENTIETH CENTURY

CHAPTER 7

Early Edwardians

T he first year of the twentieth century might have been momentous anyway, but it marked a major threshold in British empire and world history with the death of Queen Victoria. While the queen had belatedly confirmed the order that the use of plumage by the military had to cease, it would be another year before regiments such as the Royal Horse Artillery officially stopped wearing bird plumes in their military finery. It was taking time to find alternatives.[1]

The 64-year-long Victorian era came to an end on 22 January 1901, when the queen passed away at her beloved Osborne House on the Isle of Wight. As news spread to London, and as Hudson sat writing in his belfry, people left their houses and poured onto the streets. Queues ten deep were reported at Whiteley's department store, with its restaurant on the top floor under a glass dome (which would become a regular haunt for Hudson), shoppers anxious to buy jet-black mourning clothes, including bird plumage specially dyed for the occasion.

Hudson wrote that same day to Missy Bontine, Don Roberto's mother, with brief reference to the historic moment: 'It is hard to realise that the Queen is really dead, and that our head is a King!'[2] Missy Bontine's younger son Charlie would skipper the boat bringing her back to the mainland for the funeral.

Hudson's main purpose in writing was to thank Missy Bontine for her interesting letter, and a donation she had sent him to help with the work of the Bird Society. He let her know that the society had just moved in with the Zoological Society to make use of rooms within their headquarters at Hanover Square. It meant that Hudson, always a regular browser of the Zoological Library there, would be sharing the premises with his old collaborator Dr Philip Sclater, who enjoyed, as part of his employment package, a luxurious flat on the top floor, above the offices. It was by now more than 30 years since Hudson had first begun to correspond with Dr Sclater from Argentina, and followed many years of collaboration between the two men in which the creative tensions were heartfelt. It was either a very small world and this was some curious coincidental re-convergence of the two men's lives, a coming together again of groups that didn't actually have an awful lot in common, besides an interest in non-human life forms, or else the link between Hudson and Sclater was instrumental in the move. I think we can take it that it was Hudson's connection with Sclater and the Zoological Society of London that catalysed this arrangement.

Missy Bontine, meanwhile, had been gathering intelligence on rookeries in the city, including one established by the birds near the Prince's Gate. Hudson had been doing some investigations of his own. He mentions having worked out why another rookery, at Connaught Square, had been abandoned by the birds: they had been deliberately driven away by the keeper of the square, who had 'even trapped and killed some of the birds'. Hudson reported a satisfactory ending: 'when this was found out the miscreant was hauled over the coals. Then the forgiving birds came back, and last summer had about twenty nests.'

Considering his radicalism, Don Roberto reflected generously on the Victorian era now passed: 'sixty years of progress' and an 'era brought about by steam and electricity ... wages at least thrice higher ... England's dominions more than thrice extended; arts, sciences ... a thousand times advanced'.[3] He even joined the crowds to witness the queen's funeral procession pass:

> The silent crowds stood reverently all dressed in black.
> At length, when the last soldier had ridden out of
> sight, the torrent of humanity broke into myriad waves,
> leaving upon the grass of the down-trodden park its
> scum of sandwich papers, which, like the foam of some
> great ocean, clung to the railings, around the roots
> of trees, was driven firmly before the wind over the
> boot-stained grass, or trodden deep into the mud ... At
> length they all dispersed, and a well-bred and well-fed
> dog or two roamed to and fro, sniffing disdainfully at
> the remains of the rejected food ...
> Lastly, a man grown old in the long reign of
> the much-mourned ruler, whose funeral procession
> had just passed, stumbled about, slipping upon the
> muddy grass, and taking up a paper from the mud fed
> ravenously ... then, whistling a snatch from a forgotten
> opera, slouched slowly onward and was swallowed by
> the gloom.[4]

On that murky day of the royal funeral in February 1901, the Duke of Portland rode behind the Prince of Wales and Kaiser Wilhelm of Germany as Queen Victoria's coffin on its gun carriage was taken from Victoria Station through the streets of London to Hyde Park. Bands played Chopin's *Marche Funèbre* and the watching crowds stood hatless and silent.

The duke seems to have a gift of second sight. Shortly before the funeral he had a dream that the state coach got stuck in the archway at Horse Guards Parade on its way to Westminster Abbey, and that the top had to be sawn off to free it and allow it to pass. He relayed the dream to the Crown Equerry, and suggested they measure the coach. Sure enough, it was 2 feet too high to clear the archway, the level of the road having been raised since its last outing on a state occasion. Adjustments were promptly made, saving the day.[5]

Winifred, Duchess of Portland had a formal role at Edward's coronation, as what is called a canopy duchess, and acted as mistress of the robes to Queen Alexandra for a time. With Edwardian decadence in full swing, the Portlands now held smaller and more intimate parties at Welbeck 'more or less all the year round', as the duke fondly recalled in his memoir. They had guests for the shooting season at their Langwell estate in north Scotland 'all through the autumn', and in winter the Welbeck shooting parties were 'continual'.

'The spring was cold and backward', Hudson recorded in his notes early in 1901, the climate in keeping with the mood of mourning in London. His thoughts and sympathies as always were primarily with the birds: 'East winds which prevailed in March and April probably proved fatal to large numbers of the more delicate migrants.'[6] At the same time, relations between the Bird Society and other ornithological bodies were becoming increasingly frosty. There are mentions in Hudson's letters of his attendance at meetings of the BOU in the late 1890s that hint at the trouble brewing.

He would eventually withdraw his membership in 1908,[7] and in 1900 wrote to both the BOU and the British Ornithologists' Club setting out his determination to see private collecting criminalised. He sent a copy of this letter to Eliza Phillips.[8] It appears that the BOU had written to the Bird Society with a quibble about its unflinching position against collectors and taxidermy, and this brought things to a head. Despite the progress that had been made in strengthening bird protection laws, Hudson continued to argue for more effective legislation, and to let the BOU and the BOC be in no doubt that he would not be deflected from his aim of securing an act of Parliament to curb the excesses of the bird-killers and stuffers, 'to forbid the making of collections of British birds by private persons'.[9] It is worth noting that he was writing on this emotive topic from Sir Edward Grey's Fishing Cottage. I sometimes wonder if Hudson was ever actually able to switch off from bird protection business and relax.

There are further hints of the escalating tension between the Bird Society and other organisations, as Hudson enlisted the backing of leading ornithologist Henry Dresser, who attended the April meeting to advise them. Dresser had previously had similar 'fights' with leading men of other associations. Hudson had another motive in approaching Dresser: the society needed a new treasurer, and Dresser was a contender, one of two 'good business men of standing' being considered. The other was author and activist Ernest Bell, who would later be an executor of Hudson's will. Hudson and Eliza Phillips were also trying to persuade current chairman Montagu Sharpe to stay on a while longer, at least until their Bird Protection Bill – to be led in Parliament by Mr Bigwood – passed through one or both Houses. By July 1901 the Bigwood Bill would be in the charge of London MP Mr Stephens.

Another Hudson letter to Phillips gives a flavour of the dynamic within the society by this time, and his role as chief agitator. The new bill was being drafted by chairman Montagu Sharpe, and was the subject of a meeting in March 1901, at which Hudson raised objections, pointing out its flaws. Sir Edward Grey made the same points by letter, no doubt after conferring with his friend. Sharpe was somewhat dismayed to find the room down on his proposals but agreed to recast them, and to convene the group again after Easter.[10]

Hudson ruffled the feathers of the prominent politician Sydney Buxton by arguing that whatever it might achieve, the law as drafted would do nothing on the crucial matter, for him, of preventing collectors from destroying rare birds. Buxton, it should be noted, had a collection of his own. The discussion around this was so prolonged there was no time to discuss any other society business. The Bird Society by now had six 'watchers' in place to guard rare nesting birds at key locations across Britain, and this growing network would become a distinct department.

The Bird Society was continuing to attract new friends in high places. Eliza Phillips sought Hudson's advice on behalf of a wealthy, titled acquaintance on the care of captive lovebirds. In so doing she forgot the limits on his expertise. 'Alas I am the last person in all this populous land to advise you about the ailments of love birds, or any other feathered creature in a cage', he replied, perhaps a little more wearily than she at least was used to. He reminded her that, where birds are concerned, he preferred 'to look on them in their own vast sun-lit forests, tumbling down, a shiny green rain out of a hot blue sky, making the air ring with their shrill glad multitudinous cries ... I know that caged birds scratch and pull each other's feathers out ... poor caged birds, infested with mites and other parasites ... it makes me shiver to see them ... I should prefer to see them dead and at peace – poor little man-tortured beings. I mean if I must see them here in this strange land.'[11] It is not recorded whether Phillips passed on his points to her friend.

The *faux pas* over Alice Keppel was put firmly behind the Duchess of Portland and her husband, and normal service resumed with the royals in the new Edwardian dawn. In August 1902 Winifred was one of four duchesses holding a pole at each corner of the canopy under which Queen Alexandra was anointed. Winnie served as mistress of the robes to the queen on other occasions and would perform the same canopy-bearing role for Queen Mary in later years. She remained unshakeably well connected.

The Portlands learned to be adaptable in changing times. 'Despite being a committed horse man, and recognising the danger that the new fad in car ownership might pose to horse breeding, the duke gave orders for some of the best types of cars to be supplied to him.' The duchess had doubted that their friends would 'forsake the horse in favour of mechanical locomotion. That time, however, came about, and now the Duchess is claimed as a patroness of the car, which if prosy, compared with the delights of horsemanship, is, nevertheless, useful for accomplishing distances which horses are not expected to cover.'[12]

The duchess presided over many of the society's annual meetings, unless detained elsewhere. But for all her apparent dedication, if Hudson's few surviving remarks about her can be relied upon as a guide, he remained unconvinced. 'The Duchess of Portland has not much to say when on her legs', he observed in a letter of February 1905. He was even-handed in his critique of the stage performers: 'and our poet [Laureate] Alfred Austin was most disappointing'. Hudson could be a difficult man to please, and could always be relied on for an unvarnished review.[13]

A few years ago, an RSPB librarian showed me telegrams exchanged between Mary Russell, Duchess of Bedford, and Bird Society president Winifred Cavendish-Bentinck, Duchess of Portland, in which they discuss how they might influence the new Queen Alexandra to support the plumage campaign. The matter was also high on the agenda of Bird

Society meetings. The Duke of Bedford would sometimes preside over the annual meeting, if the Duchess of Portland was otherwise engaged.

Hudson continued to pull some strings in the background. 'I have had to suspend my own work for a couple of days to do the work, or help with it, of our Birds' Society to get the Annual Report etc. ready for a Committee Meeting on Tuesday next, and arrange for the General Meeting on Feb. 10', he wrote to Morley Roberts in January 1903. 'The Duke of Bedford is going to preside, but we can't find any grand eloquent speakers to support him. All the orators are away.'[14]

The Dukedom of Bedford's Woburn estate is well known to this day for its impressive collections of exotic fauna, and has been the launch-pad for colonisation of large parts of the country by escaping muntjac and water deer from China, and grey squirrels from North America. Having made acquaintance with the Bedfords at the bird meetings, Hudson was invited by the duchess to their estate at Woburn in late April 1903, 'to go and see the beasts there'.[15] But did he actually take up her invitation and make the short journey north to Bedfordshire? I made an enquiry to Woburn to see if they had any record of it, but I haven't yet been able to confirm if they do. I would like to think Hudson followed through with it, and if he did it would be fascinating to know how it went. It would be a pity if he didn't take the opportunity to befriend Duchess Mary Russell properly, as she was an expert aviator and ornithologist. Her support for the Bird Society and her keen interest in bird migration would have given her and Hudson much to talk about. And while he might have been ambivalent about the aviaries at Woburn, he continued to make use of the facilities in London's Zoological Gardens, although tensions were mounting in the offices.[16]

Tensions at the zoo

It's not clear why it took ten years after the publication of *Argentine Ornithology*, his joint work with Dr Sclater, but Hudson was finally elected a fellow of the Zoological Society in 1898. Among the perks of this were free tickets to London Zoo, which he shared freely with friends. Hudson's dislike of pets and caged things in general did not prevent him from recognising the interest in animals that zoos, whatever their faults – and of course they were many in these earlier years of their development – could foster in the public's mind.

Hudson's preference was for visiting on Saturday afternoons when he could also indulge his love of music and the band that played on the lawn.[17] In 1910, he encouraged one budding ornithologist friend, John

Harding, to visit and see a spectacular South American addition to the zoo's bird collection – the colourfully named and brilliantly plumaged cock-of-the-rock.[18]

Harding must have enjoyed his visit. On a later occasion Hudson sent him more tickets, but Harding had to return them as they were unsigned. Hudson duly signed them, adding three more besides. He was obviously becoming more absent-minded in older age and told Harding that he'd made the same oversight when giving tickets to a female friend of his, 'a suffragist too, though a non-militant, and so [she] boldly put my name to the cards and found it all right'. Harding, whom Hudson would later persuade to succeed him on the RSPB council, was obviously more rule-bound than Hudson's radical friends.

With his zoo associations Hudson was starting to find his advice being sought on bird-keeping matters more often than he wished. In June 1901, Mrs Close of Kirtlington Park, whom Hudson had visited at her immense Palladian palace near Oxford, asked if he knew where she might source a Patagonian eagle owl in London to pair with one that her friend owned. 'Such a sight would be fitting to "draw iron tears down Pento's cheek,"' Hudson groaned, 'and I am not as hard as Pento.'[19] It's another of Hudson's occasional and sometimes obscure Biblical references, learned as a child.

On captive animals Hudson was clearly conflicted. For all his tirades against the incarceration of birds and other animals, he was of course also fascinated by the glimpses of their behaviour that it allowed. He was often entranced by animal behaviour at a time when it was beginning to be investigated more closely. One incident he witnessed at the zoo, when two vultures seemed to be playing with a moulted wing feather, exemplifies his inner conflict. 'It was an amusing exhibition but it saddened me at the same time to see these two wretched prisoners for life trying to get a little happiness out of the fallen feather', he wrote.[20] Hudson was certainly less militant on animal welfare than his friends Henry Salt and, later, John Galsworthy. Morley Roberts recalled Galsworthy getting into trouble for publicly calling London Zoo a 'ghastly jail'.[21]

Not long after the Bird Society ensconced itself in Hanover Square as tenant of the Zoological Society of London and its secretary, that reunion of unlikely bedfellows, Dr Sclater's long tenure as Zoological Society secretary was finally drawing to a close. He had held the post for 42 years. His son William was in the running to succeed him, but first there would be an election, in which Hudson as a Fellow was entitled to vote. In the opposite corner was a man called Dr Peter Chalmers Mitchell. Given Hudson's supposed ambivalence towards Sclater senior, we might suppose that he would be in the Mitchell camp; but whatever Hudson's personal and professional misalignment with Sclater senior, he was clearly throwing his weight behind his ornithologist son. In April 1903 he wrote to Eliza Phillips: 'I must go back for the great meeting of the Zoological [Society] and give my vote for Mr Sclater.'[22] But was this simply a pro-Sclater junior position, or did Hudson have misgivings about the other candidate, Mitchell?

The meeting took place on 29 April, and the following day Hudson wrote to Emma Hubbard with the news that Mitchell had 'gotten the victory', adding 'I am sorry; but most persons I know sound glad.'[23] Sclater junior seems to have been based in South Africa at the time of this 1903 contest for the leadership, so that might not have helped his election campaign. That same year, Sclater senior cofounded the Society for the Preservation of the Wild Fauna of the Empire, later Fauna and Flora International.

I wonder if Hudson had particular reason to fear Mitchell's agenda. I also wonder if the role would have been both antipathetic to animal welfarists and to any threats they might pose to the practicalities of running a zoo in Edwardian Britain – whatever the ideals of the wider fellowship. Was an industry with infrastructure to supply plumage and skins globally also tied up with the supply of live animals?

Hudson resigned from the British Ornithologists' Union in 1908. The following incident from 1910 helps explain why. Hudson noticed an advert for a pair of African hooded cranes, which a London dealer was selling – alive – for £80. 'These birds are now on the point of extinction,' the advert announced. Hudson promptly wrote to *The Times* in protest. He also cited Walter Rothschild when he described the attitude of these sellers and collectors, who were so determined to secure the last surviving wild specimens for their collections. 'It appears a damnable proceeding,' he asserted. *The Times* ran an editorial supportive of Hudson's position.

The dealer responded. He had, he said, procured the cranes 'at the request of the greatest ornithologists of the day, numbering more than 40 persons'. Rothschild also wrote to *The Times*, arguing that 'it would be better for our successors if these last specimens were preserved in museums'.

Hudson's ire was hardly likely to be soothed by this. 'The recent final capture of the British Ornithological Union by the private collectors and exterminators of rare birds has made it pretty plain to everyone that ornithologists and collectors are not distinguishable in this matter,' he raged. 'They are engaged in the same business.'

Several BOU members wrote to *The Times* in an attempt to reassure the public that they were committed to bird protection. Clearly, not all members were apparently blasé about species loss. The British Ornithologists' Club was having to manage internal conflicts between those of a collector mentality and those concerned to conserve wild birds in the wild.[24]

When the Zoological Society of London moved offices in 1909, as the Hanover Square building was about to be demolished, the Bird Society didn't go with it to its new base at Regent's Park. Instead, it relocated to top-floor offices at 23 Queen Anne's Gate, Westminster. Sclater junior also followed a different career path, and in that same year became curator of the Bird Room at the Natural History Museum.

The close working arrangement that may have papered over some of the cracks in the relationship between the two organisations now over, it was not long before a divergence of views became more apparent, as the plumage campaign and its counter-campaign grew in intensity.

Of course, there were many other battles being fought. Of all the discoveries made in this quest to animate the life of W.H. Hudson, one of the most eye-opening has been that he was campaigning to save albatrosses at the turn of the twentieth century, and rallying to the cause some more very famous names.

Saving the albatross

Supporters of global conservation today will be well aware that the RSPB and its partners in BirdLife International have been running an Albatross Task Force for more than two decades, to save these majestic and symbolic birds from needless destruction by long-line fisheries, which operate from many countries in the southern oceans. I have enjoyed letting my marine colleagues know that an early prototype version of this work was taking place over a century ago, led by the birdman from the Pampas. I had hoped that Hudson might have first become acquainted with albatrosses on his emigrant journey across the south Atlantic from Argentina, and these later words made me wonder if he might have done: 'The mere sight of this noblest pelagic fowl, the great Wandering Albatross, is a moving event in the life of any person, even as is that of the soaring condor among native mountains; and, in a less degree, that of the golden eagle, the one great bird which happily still survives in the northernmost parts of our country [Britain].'[25]

I have imagined him witnessing a fellow passenger engaged in the nefarious practice of 'fishing' for these birds. It was this 'sport' that Hudson despised and set out to have banned. Hudson first alerted the world to the abuse of albatrosses as early as 1893, when he 'made my little protest against this modern sport ... English gentlemen amusing themselves by taking albatrosses with hook and line'.[26] That he had been able to weave such a topic into his *Birds in a Village* illustrates the 'local to global' scope of his thinking and vision.

It seems that he was reminded of the practice by Emma Hubbard, who had noticed letters on the subject in the eminent science journal *Nature*, in which she was sometimes published. She tried to add her voice to the dissent, having, in Hudson's words, 'sent an indignant protest to the paper which the Editor politely refused to insert, altho' Mrs H. has been an occasional contributor to that journal ever since it came into existence'.[27]

The authors of letters in *Nature* that had caught Hubbard's eye were not concerned about the welfare of the birds; nor were they expressing distaste at their capture. On the contrary, what had piqued the interest of these correspondents was the fact that captured birds had somehow stayed alive when stored below deck, half-frozen. In fact, the ship's staff were minded to conduct experiments on a future voyage to further test the resilience of albatrosses to this torture. The ship's owner, Sir William Corry, MP, apparently displayed 'a lively interest in these investigations'.[28]

Reminded of this revolting business, Hudson's fury was renewed, and his displeasure was shared by his old amigo Don Roberto Cunninghame Graham. They lost no time in taking their campaign to the pages of the *Saturday Review*, a prominent London weekly newspaper. A long letter from Hudson appeared on 24 November 1900: 'The senseless destruction of this noble bird is still being carried on for the amusement of officers and passengers on our ocean-going steamers', he fumed. 'That any man who has any poetry in him, any reverence for life, any sense of mystery and glory of this visible world, an instinct of humanity, or of truth, who is not a ruffian at heart as well as a Philistine, can find pleasure in torturing and killing such birds, is a thing to wonder at.'[29] In signing off, he called on the leading naturalists' societies and unions to add their voice to that of the Bird Society, which of course he had already brought on board.

When he received a copy of the newspaper carrying his letter, Hudson hurried to Kew to show Hubbard, and wrote to Eliza Phillips enclosing another copy. His attention now turned to bringing the matter to the attention of Parliament. With Don Roberto's help he enlisted the support of Labour founder and soon-to-be leader Keir Hardie, who agreed to raise

Don Roberto Cunninghame Graham and his beloved stallion Pampa at Rotten Row, Hyde Park. 'His wild antics might disturb a devotional frame of mind', Hudson once said of his kindred gaucho spirit and alter ego.[30] Image courtesy of the National Portrait Gallery.

the matter on their behalf in the House of Commons. When this was delayed, Hardie wrote to Hudson on 13 December 1900 to explain why his questions could not be asked that day. Hardie assured him they would be asked the next.[31]

Hudson attended a Bird Society committee meeting that same day. 'I am to write a leaflet on the Albatross', he told Hubbard afterwards, indicating that this had been discussed and agreed with his colleagues.[32]

'Keir Hardie asked his question', Hudson was soon able to report, along with the response of Mr Ritchie, who spoke on behalf of the government, without adding much: 'He knew nothing about the subject',[33] Hudson added.

Undeterred, Hudson was still campaigning hard on the last day of the nineteenth century, gathering as many facts as he could muster. Don Roberto's knowledge and connections were proving invaluable, as usual. He even had his own nickname for the birds, which seems to have puzzled Hudson. 'In spite of your explanation, my thick skull has not taken in the reason of your calling the albatross "Gritte Poule", he wrote to his friend. 'It may dawn on me some day.'[34] Anyone who might think Hudson only 'shot from the hip' and might be inattentive to detail should note his meticulousness on this occasion, as he wrote to Don Roberto:

> About the albatross, Louis Becke's testimony would be most valuable to me, and what I should like to know from him is whether it is not a fact that these birds of the Southern seas are not taken with hook and line from steamships as well as sailing vessels.
>
> A stay-at-home ornithologist here, who is considered a great authority, pooh-poohs the idea that these birds are being exterminated, and affirms that they are not caught so much now as they were formerly, simply because they can't be taken from a steamer, as it moves too fast. Becke must know about this better than men who spend their lives in the British Museum. If you know him, please ask him.[35]

Hudson requested and received a copy of the parliamentary report with Keir Hardie's question, loaning it to Eliza Phillips. He lobbied other editors to ensure there would be more on the albatross issue in forthcoming magazines. The *Saturday Review* had run a leader column that took up the issue again. Don Roberto now had a letter published there, and it prompted a response 'to these attacks' from the implicated ship owner, Sir William Corry, which the editor – obviously on the campaigners' side – showed to Hudson before it was published.

Now it was Hudson's turn to respond. 'If I can put in something in reply to your remarks as to Keir Hardie's questions in the House I shall do so,' he wrote to Don Roberto on 7 January. 'I am trying to get some reliable up-to-date information about the "sport" before writing a pamphlet,'[36] he re-emphasised. He had written to someone called Bullen, who might have more insights, but Bullen was ill with bronchitis. Impatient for information, Hudson seems to have been a little unsympathetic. 'Bullen being a religious man should be on the side of the angels', he remarked, with the caveat, 'but it doesn't always follow.'[37] Hudson's pamphlet would remain a work in progress while he awaited responses from these other authorities.

Novelist Joseph Conrad now joined the fray, being an old seaman himself, keen to back up his literary friends. He met Hudson and reported that he had also written to the *Saturday Review*, though his letter wasn't printed. 'I am not surprised,' Hudson wrote afterwards to Eliza Phillips. 'I told him he had better not take up subjects before finding out something about them, as otherwise he only discredits himself.'[38] Let's hope Hudson expressed his caution more gently than this suggests. Poor Conrad was only trying to help...

Hudson returned to his draughty Tower House belfry to work on the pamphlet and compose his next missive. 'It is very cold today up in this sky-room,' he wrote to Phillips. 'The windows do not keep the wind out, and tho' I am as near the fire as it can sit my fingers are so still that I can hardly hold the pen. I hope you do not feel it as much in your large rooms.' He added, without hinting at his feelings on this other matter (although we might guess), 'I suppose you saw Phillip Crowley's death: he had the largest private collection of birds' eggs in England.'[39]

This was all happening during a difficult period for Don Roberto and Gabriela Cunninghame Graham, who had been clearing the great house at Gartmore, finally forced to sell by rising estate debts. The house had been bought by 'new money' – a wealthy industrialist. One evening, after a day of supervising removals, Don Roberto slumped in a chair at his desk, exhausted, and shared his thoughts in a letter to friend and publishing guru Edward Garnett: 'This has been a long and heart-rending day. To see everything one put up (16 years ago, when young), taken down, & the blank look of all, you can imagine how it affects one ... It is bright moonlight now and the familiar trees look spiritual and perhaps reproachful. I wish it would rain, and look ugly.'[40]

Hudson flanked by Morley Roberts (on his right) and Lancelot Cranmer-Byng, probably in Soho, near the Mont Blanc restaurant on Gerrard Street, frequented by an illustrious circle of literati in the Edwardian era.

Don Roberto and Gabriela were moving to a smaller though still grand and imposing house at Ardoch, by the River Clyde, which they filled with what family treasures had not been auctioned off. Losing Gartmore was of course a terrible wrench for Don Roberto, after it had been for so many generations in his family, but he was now free at least from the many years of headaches and tiresome administrative burden and could perhaps approach the new century with his trademark zest for life renewed.

Despite his own personal trials, Don Roberto was undoubtedly instrumental in bringing Hudson to some extent out of his shell. Back in 1893, at the outset of their friendship, Hudson, despite his recent publishing successes and albeit in a phase of depression after a bitter winter, had written to William Canton: 'I scarcely know one literary man in London – what have I to do with men who can work all day, and meet one another in the evening at their clubs, read papers and talk and laugh, and enjoy it, and then go home in the small hours, glad to think that another day is so near!'[41] Nearly ten years on, this was all about to change.

Edward Garnett and the literati

'I'm not one of you damned writers: I'm a naturalist from La Plata.' For me, this spontaneous outburst defines Hudson, in his own words. Ford Madox Hueffer/Ford remembered Hudson once proclaiming this, perhaps in the heat of verbal battle at a literary lunch, the normally measured old sage provoked by one of his playful young associates.

Hudson had many admirers, some of whom helped him at vital moments in his writing career in different ways. But perhaps his most important ally in publishing was the man regarded as one of the most influential editors in English literature – Edward Garnett. He was also described as 'the first Englishman who did not care a damn for the Victorian social conventions and very little about money, or success'.[42] This was the verdict of his son David 'Bunny' Garnett. The chemistry between Garnett senior and the normally guarded Hudson was always going to be fascinating.

As was often the case, Don Roberto may have been a catalyst in bringing them together, recommending Garnett to Hudson and vice versa. Don Roberto would insist that he was an 'essayist and an impressionist, and secondly a story-teller', in his animated debates with Garnett.[43] In 1900, Don Roberto had told Hudson about Joseph Conrad and urged him to read his novels. By the time Hudson met him, Conrad's *Heart of Darkness* was

John and Ada Galsworthy. 'People hold their breath when they hear his name',
Hudson said of his later Nobel Prize-winning friend, and champion.
Image courtesy of the National Portrait Gallery.

recently published and his route was clear to further fame, if not fortune. A little earlier, Conrad had brought his friend John Galsworthy along to Garnett's Kent home. Galsworthy would pick up the baton and, after a faltering start with Garnett, never look back.

Garnett had identified in Don Roberto's provocative and satirical essays exactly the sort of challenging prose he wanted to nurture and encourage. Garnett would sometimes share with Hudson letters received from Don Roberto. That said, Don Roberto seems to have been something of a satellite to this core literary group, perhaps leading too hectic and far-flung a lifestyle to ever be a regular lunch participant. There is one letter from Hudson in 1902 inviting Don Roberto to the Mont Blanc restaurant in Soho's Gerrard Street, with the incentive that Edward Garnett would probably be there, so he is likely to have been at least an occasional attendee.

There is a neat symmetry of the Edwardian era and Hudson's personal golden age, as he became naturalised as a British citizen, and was awarded the civil list pension for his contribution to literature. Although notoriously reticent, he seems to have found a comfort zone with these informal and moderately boozy gatherings.

Garnett was a 'star-maker' for several decades in the early twentieth century. His biographer Helen Smith sums up the qualities Garnett searched for and admired in a writer: 'the ability to suggest the intangible from the palpable; a willingness to shake the reader out of his or her settled perceptions, and the facility to make a small, apparently insignificant detail reveal the depths of a situation'.[44]

Garnett mentored other luminaries, for instance Liam O'Flaherty, E.M. Forster, and D.H. Lawrence, all of whom openly acknowledged the part he played in harnessing their talent and honing their work. Hudson's name might seem out of place mentioned in such company, but in fact Garnett placed Hudson on an altar of his own: 'I have known several men of genius, remarkable minds, but no man's personality has ever fascinated me like Hudson's.'[45] He also spoke of how others 'succumbed quickly to the spell of his personality'.[46] It's quite an accolade, considering the roster of Garnett's protégés, and the charisma and stardust of most of them. There was definitely 'something about Hudson'.

The manner of their meeting is wonderfully cinematic, and timely. It was September 1901; Hudson had not long secured his civil list pension and his official British identity. Perhaps he was walking London's streets with a new-found self-assurance, beginning to shed the impostor syndrome by which he had been burdened. If this *were* a movie, Garnett might narrate the words he wrote about the occasion of their meeting: 'It was my last day as Heinemann's "reader" and I was clearing up my work when a lad announced "Mr Hudson!" and looking through the window I saw a tall dark man standing on the leads outside my little room.'[47]

Hudson had sent his manuscript of *El Ombú* – short stories about life on the old Pampa (to give it its proper name), written from the point of view of an elderly man – to this publisher, and Garnett, having loved it, had been trying to persuade the owner to take it on as a book.

"But we shan't sell it!" the boss insisted, irrespective of its literary merits. You can picture Mr Heinemann pushing the wad of papers back across a leather-topped oak desk towards the exasperated young editor. "What the hell is an *El Ombú* anyway?" he might have groused.

Disagreements of this kind with his employer precipitated Garnett's departure from the firm. He was a publisher's reader who cared passionately about writers he believed in, and less so about the commercial success of his firm. He was ready to part ways. It was at this crossroads, filmic moment that Hudson had come to his door to learn the verdict.

Amid the chaos of his office, Garnett greeted Hudson, they shook hands, confirmed their names and regarded each other in silence for a few moments. I can hear Garnett speaking the words that he later wrote. '[Sir,] you have written a masterpiece. Its grave beauty, its tragic sweetness, indeed … swept me off my feet.'

I can also picture Hudson's level stare, his disbelief, interrogating Garnett's face, searching for the sincerity of this utterance. He needn't have doubted. Garnett's candour was well established. 'It's my last day here,' Garnett, might have said, breaking the tension, gesturing at the piles of books, papers and boxes around him. 'Where can I meet you?'

I can picture them then walking briskly through the crowds and past market stalls to the Mont Blanc restaurant, a focal point for the loose alliance of literary figures of which Garnett is a pivot point. So begins a new chapter for Hudson and a close working friendship that will last till the end of his life. Not all of Garnett's relationships were so enduring.

Although a generation separated them, the two men would soon have found they had much in common, including Irish heritage on their mother's side, and that they had married within three years of each other. They had published their first novels around the same time too, with similarly underwhelming impacts. Both could also be described as frustrated poets. They had a shared affection for the nature writing of Richard Jefferies. Garnett's first reading job was with Fisher Unwin, which had published Hudson's A Crystal Age (if Garnett had rated this niche book, it's not recorded). Garnett had left school early and like Hudson was an autodidact – self-taught through voracious reading. And Garnett's father was keeper of books at Hudson's home from home, the British Museum Library.

Both men liked to talk, and to read, but the literary range and energy of Garnett dazzled Hudson, as it did most people. He was in awe of Garnett's ability to devour manuscripts and embellish them, reviewing texts with rapier incisiveness. Garnett also lavished books on Hudson, encouraging him to read Russian literature, some of it translated by his wife Constance. Anna Karenina by Tolstoy was one of the first books Garnett gave him, sent as a Christmas present.[48]

The literary radicalism of Garnett was eye-opening for Hudson too. He was at the forefront of a new wave in British literature, on a mission to shake-up a reactionary establishment. He had a particular love of anyone who could shine a light on the stuffy, intransigent world of contemporary society. 'Write about the English',[49] he implored Galsworthy. He spoke of Conrad 'visualising for us aspects of life we are constitutionally unable to perceive'.[50] And if this core group of authors centred on the Mont Blanc had one theme in common it was the outsider's perspective that Garnett quickly recognised and valued. It is intriguing to think of Hudson – sometimes simplistically labelled as a conservative – in the thick of all this cultural subversion and upheaval.

A few weeks after his first meeting with Hudson, Garnett was now ensconced in his new job at the offices of Duckworth. He would spend two days a week there, meeting his new associate when possible on Tuesdays at the Mont Blanc, the rest of the time reading and commenting on manuscripts and penning advice to authors from a book-lined room at his home in rural Kent – The Cearne – which Galsworthy's wife Ada described as 'rather primeval – its owner too'.[51]

The first surviving letter from Hudson comes soon after. Formalities are already being shed. 'Dear Garnett, Will you kindly allow me to drop the Mr?' Hudson requested. Garnett had asked to see the manuscript of El Ombú again, intent on championing it to Duckworth. 'The MS is at home and I will gladly send it to you on my return', Hudson promised. It turns out that he had been offered some kind of deal by Heinemann after all, but knew he could do better: 'Heinemann's terms did not suit me; so I asked for it back.'[52]

On the occasions Hudson couldn't make it to the weekly lunches, he would write instead, which gives a clear idea of the kind of topics they liked to discuss. A June 1902 letter gives a sense of Hudson's life as it appeared to Garnett, the older man visiting Sir Edward and Lady Dorothy Grey at their cottage in Hampshire, discussing the books that Garnett had given him, being absorbed in nature, living with local people rather than seeking the comfort of hotels, and being 'out-of-the-world' in terms of news.[53] It was, all in all, a far cry from the fashionable eateries of Soho and its teeming, noisy streets.

Garnett was and would remain intrigued by Hudson, so often out of town somewhere stumbling around in the underwood, communing with nature and – despite his precarious health – getting 'killed' by stinging flies. 'That I like, but I find there are people who don't like it', Hudson warned, perhaps uncertain if Garnett would actually enjoy joining him on one of his forest explorations. 'The flies in the King's Copse were almost not to be borne.'[54]

At the first hint of interest shown by Garnett in nature, Hudson tried to get him along not to the outdoor world of biting insects but to the Bird Society's Westminster general meeting in early 1902. He failed.

Garnett therefore missed being part of a large crowd enjoying a platform busy with speakers. The newspapers had been 'liberal' in their coverage, with '*The Thunderer*' (aka *The Times*) devoting a leader column to commenting on it.

Garnett or no Garnett, Hudson was soon back in the land of the biting insects.

More Hampshire Days

Almost 30 years had passed since William Henry Hudson's life in England began in Hampshire on that bright, early May morning in 1874, when he stepped off *The Ebro* at Southampton harbour. He retained an affinity for the county thereafter, and it was perhaps inevitable that he would write a book about the place where he first fell in love with this 'land of morning' as he went straight into the woods and fields to see and hear British birds – even if he could name almost none of them at that stage.

His classic 1903 book *Hampshire Days* is a love letter to the county, its wildlife, people and places, conceived at Sir Edward Grey's fishing cottage on the River Itchen. Another favourite Hampshire base for Hudson was Roydon House (or Manor) at Brockenhurst. Much as he loved the solitude at Itchen Abbas, he also liked to experience Hampshire – and his other favoured destinations – through the lives of the locals and 'to live with people, eating at their table, being a member of the family so to speak'.[55]

He described Roydon House in lyrical terms:

> The ideal beautiful house – a wonderful gem of red brick in its green and flowery setting … And the birds are wonderful. No cat or dog to frighten them: the shyest ones have become tame. In a yew close to the front door a bullfinch has a nest full of young; and a couple of yards from the bullfinch a bottle-tit [long-tailed tit, today] has a round nest as big as a coconut hanging from a yew twig. You can look into it and see the mother sitting on her young. Close by a robin is sitting on a cuckoo's egg; and as for thrushes, blackbirds and starlings one could fill half a bushel with the young birds in the small garden. In the evening you hear owls hooting, nightjars reeling and woodcock grunting and whistling.[56]

The pattern of Hudson's life at Roydon in November 1902 involved breakfast with the family at 7 am, writing till lunchtime, a walk in the afternoon, then nothing much in the evening, until supper of bread and milk before bed at ten.[57] It all sounds simple and relaxing, but there was one snag with Roydon, as Hudson warned Ford Madox Hueffer (who later changed his last name to the more English Ford): 'you cannot smoke a pipe with Mr Hooker as he is against all the things which the unregenerate man loves'.[58]

Perhaps for this reason, Hudson also frequented the otherwise less salubrious and rulebound Rollstone, where the farmer Mr Snell met him off the train at Beaulieu and took him in his horse cart to the house, where 'you have to put up with very coarse fare', Mrs Snell being too 'absorbed in her proper work' to 'trouble much about the stranger in her gates'.[59]

Hudson's relentless in-person campaigning for his beloved birds inevitably included efforts to create 'a general bird protection scheme in Hampshire',[60] and he was particularly keen to see Wolmer Forest preserved from the persecutors. 'I am doing what I can in my small way by writing to the Hampshire people I know on the subject.'[61] He had enlisted local support – his close friend the vicar of Milton was on board: 'Mr Kelsall will do all in his power',[62] Hudson reported.

Hudson's other favourite base in Hampshire was a house called The Pines at Silchester. During an untypically warm spell in May, he described getting up at 6.30 in the morning to 'get a little walk in the cool time and hear the birds'.[63] His campaigning focus included the common there:

> which I am trying to save from complete disfigurement
> by the wholesale removal of gravel ... the lord of the
> manor (the Duke of Wellington) and the commoners
> have been tamely submitting to this vandalism: but
> something can be done to prevent the utter ruin of
> a beautiful spot, and I hope to be not less successful
> about this place than I was some time ago about the
> Old Deer Park [at Richmond].[64]

'The Common is much abused,' Hudson reported, 'the lord of the manor's agent and the commoners spiting each other by doing all the damage they can to the poor ground, but the birds stick to it. I see and hear nightingales, nightjars, magpies, jays, owls, butcher-birds ... water voles, wood mice.' He even missed a Bird Society meeting in London while he campaigned to end the plundering of the common for its gravel by road builders and menders, which left it 'disgracefully disfigured'.[65]

When, in 1906, asked by 'a young Scotch naturalist' to suggest the best place in southern England to find birds he wouldn't encounter in Scotland,[66] Hudson wrote back to recommend Selborne and the Hampshire landscape around it. The young Scot later sent a note on the exotic species he found, which included the Dartford warbler. 'I am always grateful to be told of a new locality of the bird,' Hudson responded, 'and always express the hope that it will be told only to those who are anxious to preserve it from the collectors.'[67]

Some of his explorations of Hampshire and the New Forest reveal Hudson as an all-round naturalist, with his forensic eye and endless patience, and – ahead of his time – a belief in the value of studying his subjects in the field, not just in the laboratory. He described finding crickets in the forest, and 'going there day after day to spend long hours in pursuit of my small quarry. Not to kill and preserve their diminutive corpses in a cabinet, but solely to witness the comedy of their brilliant little lives.'[68]

He called the New Forest 'one hundred and fifty square miles of level country which contains the most beautiful forest scenery in England'. The rise in its popularity with an increasingly mobile, vehicle-owning urban public (of which, of course, he was one) made him fear for its future. 'The Forest has been known and loved by a limited number of persons always; the general public have only discovered it in recent years. For one visitor twenty years ago there are scores, probably hundreds, today … there will be thousands in five years' time.'[69] He spent long hours deep in the woods, quietly watching, listening and communing with nature, returning to one spot in particular where he felt at home:

> I came to prefer one spot for my midday rest in the
> central part of the wood, where a stone cross had been
> erected some seventy or eighty years before. It was
> placed there to mark the spot known from of old as
> Dead Man's Plack. According to tradition, it was just
> here that King Edgar slew his friend and favourite Earl
> Athelwold.[70]

Hudson's short book about this legend was published late in his life: Hampshire was ever in his thoughts. Although he never moved from his London home in the end, had he done so it seems likely he would have chosen a place in Hampshire. It seems apt to allow him space to describe another magical moment he experienced there:

> In the New Forest; when walking there one day, the
> loveliness of that green leafy world, its silence and its
> melody and the divine sunlight, so wrought on me
> that for a few precious moments it produced a mystical
> state, that rare condition of beautiful illusions when
> the feet are off the ground, when, on some occasions,
> we appear to be one with nature, unbodied like the
> poet's bird, floating, diffused in it. There are also other
> occasions when this transfigured aspect of nature
> produces the idea that we are in communion with or in
> the presence of unearthly entities.[71]

Hudson could lose himself in the forest. 'We are out of the world as regards news', he wrote in summer 1902, 'and on Wednesday passed the day in the belief that the King had died.'[72] The coronation of Edward VII and Alexandra had been set for 26 June but was postponed when Edward developed abdominal pains and had to have surgery. Plans for celebration events such as providing 500,000 square meals for Londoners went ahead without the royal couple, and Hudson. The king was finally crowned on 9 August 1902 with no rare bird plumes on display.

The success of *Hampshire Days* brought renewed and increased demand for Hudson's work from publishers. Hudson never minced his words, and older age, success and a growing reputation, citizenship and civil list pensions would not make him any less brave and occasionally scathing in his opinions, especially of those who continued to threaten the survival of rare and beautiful birds. When *Hampshire Days* was reviewed in *The Times* he commented: 'It censures me for speaking disrespectfully of naturalists. When I lash out at collectors I expect to be hit back.'[73]

Behind the lens

The earliest trace of Hudson dabbling in photography is from July 1901. He was about to turn 60, and learned around this time that he was to receive a pension for his contribution to literature. Perhaps he treated himself to one of the new Box Brownie cameras that had been introduced in early 1900, bringing snapshot photography to the masses. His first photographs

'I couldn't endure to have that dreary and repelling portrait stuck in,' Hudson wrote, so he cycled 24 miles in June 1903 to have this picture taken, in the New Forest. This image was later the template for the painting above the fireplace.

were of a little red dog, so that a friend could illustrate it for him for an article he was writing.[74]

He took photographs of subjects on which he wanted Emma Hubbard to base her illustrations, including for *Hampshire Days*. He mentions a bridge being made of stone, as it isn't clear to her from the photograph he has supplied. This quote gives some sense of what it must have been like to live at a time when photographs were a novelty, a completely new perspective on what the world looks like: 'The camera sees what the eye does not see, and seeing ignores,' as he put it in a letter to Hubbard.[75]

Hudson also purchased prints to give to Hubbard. On one occasion he bought six from a Mr Green for a shilling each: Itchen scenes to be reproduced in pen and ink.[76] The new technology of course didn't always cooperate, which may have been teething trouble or operator inexperience. In November 1902, he was trying to take a photo of Roydon House when his camera jammed – it 'does not click,' as he put it. He intended to take it to the photographer in Lymington.[77]

Quite possibly at the amateur photographer Hudson's prompting, the Bird Society discussed the newly emerging hobby of photography at its 1904 general meeting as offering the outdoor enthusiast an alternative to shooting wildlife with a gun. Sadly, none of Hudson's own photographs seems to have survived his late life bonfire of keepsakes, notes and letters.

A long-forgotten letter reveals the story behind the photograph that provided a template for the oil painting that hangs above the fireplace at RSPB headquarters. Writing to Emma Hubbard in June 1903, Hudson reports that his publisher Longman has sent him an urgent request for a portrait photograph, needed by a magazine set to publish a review of *Hampshire Days*. Hudson has obviously insisted they don't use an existing shot of him. 'I couldn't endure to have that dreary and repelling portrait stuck in,' he groans. 'I had to get down here, to Lyndhurst of all places ... I went on Sunday evening.' It would involve a journey of around 24 miles but – typically – Hudson cycled, although 'the sky was one black cloud, and the wind blew hard and cold against me the whole way'.[78]

He struck a deal with a local photographer, 'a fellow named Short', who had picked up on coverage of *Hampshire Days* in the local press. Obviously at the cutting-edge of Edwardian portraiture, Short suggested that the picture be taken out of doors, 'in character, binoculars in hand'. Hudson noted 'the good advertisement he gets for himself' when photographing celebrities.

The following morning, they went out from the studio and into the forest, and 'found a spot where big trees gave a suitable shadow, and a spot with bracken to sit down on'. Five shots were taken, and it would be another day before the prints were seen by Hudson. In the meantime, he received a further letter from the editor of *The World's Work*, anxious for the prints, deadline looming. The one chosen is the photograph on which the painted portrait of Hudson by Frank Brooks was later based, and that has hung on the walls of successive RSPB headquarters for a century.[79]

A Suffolk pilgrimage

In September 1904, Hudson wrote a long letter to Eliza Phillips from East Anglia after a weekend of exploring.[80] He was keen to share with her what he had discovered on this pilgrimage, and a rare visit to this part of England. He had been drawn to Suffolk on the trail of Robert Bloomfield, a poet who had lived there a century before and who had been an early inspiration. 'I had never been at school, and lived in the open air with the birds and beasts',[81] Hudson would tell people, although he was literate from a young age, as he sifted his parents' books for information about nature. He often found this in snatches and snippets of poetry, including that of Bloomfield, whose subject was 'people's toilsome life in a remote agricultural district'.[82]

Bloomfield is recalled by some for his work *The Farmer's Boy* of 1780. The eager young Hudson found the book in a second-hand shop in south Buenos Aires. The poem was pure gold to the young naturalist, providing him with his first view of England. When Hudson left Argentina, Bloomfield's poem came with him on the ship. Like Gilbert White's *Natural History of Selborne* and Darwin's *Origin of Species*, it was one of Hudson's inspirations for longing to visit, and chase the dream of becoming a professional naturalist.

John Clare, no less, called Bloomfield the first of the rural bards and he had, according to Hudson, been known as '*the* Suffolk poet' in the early nineteenth century. Hudson's letter to Phillips also gave me a route-map to follow in retracing his Bloomfield pilgrimage. I know the Breckland landscape quite well, with Thetford Forest today at its heart, although the villages linked with Bloomfield are a little way off the beaten track. Bringing his bicycle, Hudson took a train from London to Brandon on the Thursday, noting to Eliza Phillips that at this small town 'they still strike flints'. What defines this landscape is its thin, stony soil, sparing it some of the more intense arable agriculture in the more fertile lands that surround it.

The autumn day of Hudson's visit was so warm and sunny that he decided without delay to explore further afield, and cycled the 7 miles or so south-east to Thetford, across the border into Norfolk and an 'open country of bracken and pine without a house in sight'. Thetford Forest is today the largest conifer plantation in lowland England, but it didn't exist when Hudson cycled there. It was planted in the wake of the First World War, to compensate for the loss of so much timber to the war effort and reduce future reliance on imports. Hudson reported seeing a stone-curlew, even then an uncommon bird of open, rough ground – which flew up from the roadside as he pedalled past. Isolated populations survive today, carefully protected.

Thetford, he found, was too big for his tastes. Preferring even smaller settlements, he set off again for Honington, Bloomfield's birthplace, a further 7 miles south-east. He was assured that he would find accommodation at the Fox Inn, but when he arrived there was no room for him. (I was surprised and pleased to find that not only does the public house continue to trade, but it is also still, 120 years later, called the Fox Inn.) The locals directed him to Troston, an even smaller village 2 miles to the south-west. Alas there was no room there either. It was shooting season, and a visiting shooter had taken the last room.

In typical Hudson style, he knocked on some doors, and found a room to rent in a carter's cottage. I am sure they felt honoured to play host to this soft-spoken, well-mannered and genial naturalist, with his genuine interest in their lives, their thoughts, their opinions.

On the Friday he cycled back to Honington, to get properly acquainted with the village. A further half-mile along a country lane brought him to Sapiston – where Bloomfield had worked as a plough boy. The landlady of The Fox had suggested he could get a room at the rectory there. He discovered upon enquiry that, even a century on, Bloomfield was still well recognised by most of the locals, but not all. 'I went and visited the Rector, a learned man, and expected to find him greatly interested in my poor little poet. But he was not! The learned man was in his garden, in his old clothes and very bad boots. He said he had heard of Bloomfield but took no interest in him.'[83]

'Are you an archaeologist or what?' the rector asked, scrutinising him, perhaps struggling to place this stranger with the inquiring mind. 'I'm supposed to be a naturalist but take an intelligent interest in churches and most things' Hudson replied.[84] His host gave Hudson a tour of the local church, then his library, no doubt proudly. Hudson thought it 'a very poor one for a scholar'. When chat turned to birds, the rector told him of a nightingale that had sung at the rectory for two or three summers until one day, as his wife sat writing, the bird dashed itself against a window. He found it dying from this collision, and believed that the death of his wife that same year might somehow be connected.

As they bade each other goodbye, the rector asked him his name. Upon hearing the reply 'Hudson', he stopped short. 'No relation to the naturalist in La Plata, I suppose?' And finding that before him did indeed stand the author of exotic origin, the rector 'sat down on the grass and started talking afresh'. After this extended dialogue, Hudson 'rambled about and went in to a good number of cottages. They all seemed pleased to have Bloomfield talked about.' He was told by the locals that other pilgrims still came, most of them from North America.[85]

Hudson returned to look for Troston Hall. It was the squire of that estate, a man called Capel Lofft, who had 'discovered' the talent of Bloomfield, and published and promoted the lowly farmworker. Bloomfield described jackdaws in the belfry, sparrows by thousands descending on crops. He also recounted his ongoing efforts to discourage crows by decorating fields, fences and branches with their corpses, soon to be ignored by the living birds, as they ignored the straw men and their straw shotguns.

Bloomfield described himself in the third person:

> His life was constant, cheerful servitude...
> Strange to the world, he wore a bashful look
> The fields his study, Nature was his book

Hudson described 'strolling about in the shade of the venerable trees in Troston Park'.[86] Squire Capel Lofft had given the trees individual names – so there were oak, elm, ash and chestnut called Homer, Sophocles, Virgil and Milton. Hudson found the squire's Elizabethan mansion house now empty, with just a gardener and his wife and children in residence, owls living inside, and many rows of house martin mud nests plastered on the eves outside. They told him that the latest squire in the Lofft family had gone bankrupt, and the house contents had been sold off to pay debts.

Hudson spent several hours exploring Troston's park and gardens. I looked, as he did, for a kingfisher by the stream, without success; but I was pleased to note the tinkling song of a goldfinch. He described finding goldfinches here, pointing them out to the caretaker and his family and showing them to the gardener's 'highly delighted' children through his 'binocular'. They told him their name for this bird was King Harry, of which Hudson approved.

There are still sheep here too, rare breeds in one small paddock at Sapiston and a large flock gleaning on arable fields at Troston. Hudson had long related to the shepherding parts and descriptions of lambs in *The Farmer's Boy*, including this pleasing comparison of gambolling lambs and rose blossom in Bloomfield's description of spring.

> Or, if a gale with strength unusual blow,
> Scattering the wild briar-roses into snow,
> Their little limbs increasing efforts try,
> Like a torn rose the fair assemblage fly.

Bloomfield was no poetic idealist. His descriptions of the treatment of horses make the reader wince. And Hudson, true to form, pounces on the opportunity to prick contemporary consciences. He argued that horses might have suffered the same cruelties as in Bloomfield's day had not motor vehicles been recently invented, and the public would be as little affected as we are 'at the horrors enacted behind the closed doors of the physiological laboratories, the atrocity of the steel trap, the continual murdering by our big game hunters of all the noblest animals left on the globe, and finally the annual massacre of millions of beautiful birds in their breeding time to provide ornaments for the hats of our women'.[87]

By the time he had finished writing his letter to Phillips on the Monday, Hudson had moved on to a different Suffolk village, Euston. Here he found men engaged in rebuilding Euston Hall, which had been devastated by fire two years earlier. He sat by the river and watched several herons, dozens of coots, moorhens and wild duck, and more than a hundred pewits, as he calls them – more commonly called lapwings today.

Like Bloomfield, Hudson was not a sentimentalist about rural life, although he was often dismissed as such by the abusers of animals against whom he fought so vigorously. He could see that animal cruelty perpetrated for profit or 'sport' would not be ended without intervention. Hudson recognised that Bloomfield had long since fallen out of fashion, and that even in his heyday his popularity was so extreme that it was scoffed at by people of higher taste; for example, 'Byron laughed at it'.[88] But Hudson cared little for high tastes, for airs and graces and literary pretensions. He acknowledged but made no apology for paying this tribute to Bloomfield, who most literary types would now think a 'small verse-maker'. But his 'tender love and compassion for the lower animals' was sincere,[89] even if he lived in the shadow of his contemporaries, Coleridge, Shelley and Wordsworth.

'I fancy', Hudson wrote in conclusion, 'I shall sleep better tonight having discharged this ancient debt which has been long on my conscience.'[90] The following morning, his time now up, he cycled to Bury St Edmunds to catch his train back to London.

Linda Gardiner

An important appointment was made by the growing Bird Society in 1900, with the arrival of someone billed as a clerical assistant to Etta Lemon. 'Miss Gardiner came to us with recommendations from Mrs Suckling, and the Secretary of the RSPCA,' Lemon later recalled, 'in whose office in Jermyn Street she had become well acquainted with office routine and duties.'[91] This was the very office in which the northern and southern England Bird Societies had met and merged nearly a decade earlier.

It's obvious that Linda Gardiner immediately offered much more than administrative support. She was a qualified and experienced journalist, who had worked on regional newspapers for over 20 years. She had also written several novels.[92] Her parents were both senior journalists, and her older sister Marian an artist.

Gardiner's arrival coincided with the society lurching into the modern era and a further surge of impact, and she was a primary reason for it. Within a few years, she led on the writing and producing of the society's new mouthpiece, *Bird Notes and News*, and helping to conceive, launch and grow its education programme, focused on the newly devised Bird and Tree Competition for schools.

We get the first glimpse of Gardiner in Hudson's letters in January 1901, with a hint at how things were shaping up for the small team in their new Hanover Square office, c/o the Zoological Society. Hudson tells Don Roberto that 'the Hon Secretary [Etta Lemon] is ill, out of town, while the assistant secretary is on duty every day'.[93] I can picture Gardiner bashing resolutely on a heavy iron typewriter, surrounded by piles of correspondence and books, the engine room of the operation. She was totally committed and, despite the workload, enjoying a degree of autonomy and freedom to bring together her journalistic skills, creative talents and love of nature for this cause.

She would have the added fulfilment of seeing the organisation continue to grow in stature and influence day by day, at the forefront of something significant and life changing. It was a far cry from the hectic, noisy, often stressful regional newspaper offices in provincial towns that she had known, much as she might have thrived on those in earlier life. It had been a nomadic childhood, the Gardiner family having moved from Darlington to Norfolk, then to Buckinghamshire and Hampshire, as her parents rose in the newspaper business.

Gardiner now found herself rubbing shoulders with some influential people – including Dr Sclater in the larger office along the corridor, as well as the great and the good who patronised the society and who made occasional appearances at Hanover Square. She would have seen Mr Hudson almost as often as she did Mrs Lemon. He had quickly grown very attached to this intelligent, dedicated, independent and worldly spinster, who shared his love of books and the outdoors, and knew well some of his favourite parts of the country – East Anglia, where she spent her childhood, and Hampshire, where she last lived and worked with her parents, who were by now at retirement age. Hudson could see that Gardiner was devoted to her artist sister, older but similar in age. The Gardiner sisters still lived and holidayed together, enjoying excursions in the outdoors; painting, reading and writing. Neither had ever felt much incentive or inclination to marry.

Gardiner was 20 years younger than Hudson, and he enjoyed having someone to take under his wing, although she had as much to teach him, at least about the workings of the media, as he could impart to her. In March 1901, Hudson mentioned to Eliza Phillips that he intended to drop in at the office to see what Miss Gardiner had found in the newspapers, or if she had any correspondence with which he could help.[94]

Hudson sent Gardiner a number of books in April, on the pretext of freeing up some space while he had some redecorating done – suggesting it was a large consignment. He recommended she read the work of his friend George Gissing, who was and remains best known for his novel *New Grub Street*, about the life of a newspaperman in late Victorian London whose marriage is failing as, like his other writer friends, he struggles in 'the valley of the shadow of books'.[95] One way or another, Hudson and Gardiner had a lot in common and much to discuss.[96]

She was also kind enough to type his manuscripts for him, and she trusted him enough to later confide in him her frustration at not being able to write books of her own anymore. He offered encouragement, even suggesting he envied her the role she had editing the society's newsletter. There are also hints of her exasperation with others interfering with her editorial decisions. When she had her feet firmly under the table, she declared that in future she intended to please herself and nobody else. Hudson did not discourage this, at least by letter.[97]

Impatient for change, Hudson had frustrations of his own, and could sometimes find the 'bird meetings' stodgy.[98] However, he was satisfied with the publicity the organisation was continuing to attract, boosted by Gardiner's expertise, with *The Times* an ever-reliable outlet. He was buttonholed by the operational boss after the 1903 meeting. 'Mrs Lemon detained me to speak of artificial plumes,' he told Eliza Phillips, 'about which she wants me to write to the papers.' He duly drew up a letter that Lemon signed and sent to the press, 'to put a stop to that fraud of selling real as imitation feathers for aigrettes'.[99] When the letter was published on 14 April it roused the by now very elderly Professor Alfred Newton to take up his quill pen and write to them both.

The society was pressing the government for a committee of enquiry to be set up to look at ways of strengthening bird protection, though it would be another decade before the Home Office finally granted this, in 1913.

With so much activity and a growing wage bill, the society ended the year 1903 in debt. It was time to petition for a royal charter.

Hudson did something unusual after the bird meeting in early January 1904: he wrote to Emma Hubbard as soon as it was over. 'Nearly 5 O'Clock, so I'm rather in a hurry',[100] he scrawled. He had been at Hubbard's house the evening before, and hoped that 'I did not 'bore you to death ... by staying so late'. It's not clear what it was about the bird meeting that had made him anxious or that needed to be reported so quickly, but Montagu Sharpe seems to have blocked a proposal to have fewer regular meetings – reducing them to quarterly. Had there been some issues of falling attendance or declining momentum? Sharpe proposed instead to increase the number of branch members eligible to join. Hudson argued that the further-flung representatives would only turn up once 'in a blue moon'.[101]

The question of 'incorporating the society' had also come to the fore – of lesser interest to Hudson than the progress of a bill to ban pole traps, which had earlier failed at first reading.[102] On 4 March 1904, he reported that the Pole Trap Bill had just been read in the Lords for a second time. 'We may be sure that it will be law very soon. There was no opposition.'[103] The Bird Society pamphlet on barn owls was cited in the debate.

At long last the practice of setting snap traps on top of poles, where birds of all kinds would fall victim and die lingering and painful deaths, caught by the legs in the iron jaws of the traps, was criminalised. Hudson had described meeting a gamekeeper who told him that he had no problem with owls, but had wiped them out in any case as collateral damage because they were as likely to land in his pole traps as the crows and other birds he was targeting. Pole traps are no longer a routine feature of rural life, but this law would continue to be flouted and some people still use these repugnant contraptions to this day.

The Home Secretary, meanwhile, had written to advise the society that it would be pointless to introduce the consolidated Bird Protection Bill in that session of Parliament.[104] That same month Hudson dined with Sir Edward Grey and Sydney Buxton, who both admitted to him that the existing laws were not adequate in protecting rare species from the collectors. Hudson was already hatching his plan to produce a full-colour book version of his *Lost British Birds* pamphlet, to up the ante on this.[105]

In May 1904, the Bird Society held what would become its biggest meeting yet, at St James's Hall, as part of a programme of activities around receiving its royal charter. It was so crowded that Hudson, even with his height advantage, couldn't find Catherine Hall, hostess and sponsor of those first meetings back in 1889 and obviously still in the thick of it, as he later reported to Eliza Phillips, now retired but obviously keen to be kept informed. The event was covered in the *Daily Chronicle*. Canon Wilberforce of Westminster Abbey was among the speakers. The society was officially Royal from 3 November, requiring Etta's husband Frank Lemon to assume the mantle of honorary secretary.

To mark the occasion, Manchester founder Emily Williamson was invited to reflect on the first 15 years of the organisation in an address to the annual meeting. She recounted two early incidents in the life of her Society that had stayed with her. One was receiving a letter from a working man who lived in the slums of Manchester, expressing his support, the other a helpful notice in *Punch*, the satirical magazine that routinely lampooned the wearers of dead birds. Williamson also congratulated her colleagues on three triumphs: getting so many well-known names on board as backers, their wisdom and moderation in avoiding the pitfalls of 'faddism and fanaticism', and their zeal and energy.

There was even a name-check for the Rajah of Sarawak, a vast kingdom in Borneo under the administrative control of an Englishman, Charles Brooke, in her parting words, as exemplifying the range of those within the society's ambit. And there were hearty thanks to Her Grace the Duchess of Portland, president since the organisation was little more than a shared idea, and duly elected once more at the meeting to continue as the figurehead.[106]

The surface glamour of these Bird Society occasions and the lofty patrons arrayed there belied the furiously paddling going on beneath the surface. In January 1906, Hudson reported to Algernon Gissing that he was again preoccupied with his part in producing the society's Annual Report, 'over which I am just now tearing my soul to pieces',[107] as the meeting was imminent. After this one, Hudson helped to furnish Her Grace with the form of words needed to approach Queen Alexandra, to request her help to save white herons and birds of paradise. 'If it were known that your Majesty disapproved of a fashion so indefensibly cruel, and involving such bad consequences, that fashion would, we are convinced, speedily die out.'[108]

The response from the Palace was music to their ears:

> The Queen desires me to say in answer to your letter
> that she gives you (Winifred Portland) as President full
> permission to use her name in any way you think best

to conduce to the protection of all living creatures,
and I am desired to add that Her Majesty never wears
osprey feathers herself, and will certainly do all in her
power to discourage the cruelty practiced on these
beautiful birds.[109]

'What the royal veto may do', Hudson reflected in private, 'where naturalists and humanitarians have long spoken in vain, remains to be seen.'[110] While all were waiting to see, the hard yards of campaigning carried on apace. Hudson's own brand of direct intervention and doorstepping extended beyond local authorities and could sometimes include schools.

Children and education

One day while walking in Hampshire Hudson encountered a group of schoolboys who were persecuting birds. Enraged, and having established which educational establishment they belonged to, Hudson found the place and 'burst into the school in a passion',[111] to protest to the head teacher about the boys' behaviour.

He later returned to the school to give the schoolmistress – Miss Linda Thatcher – the relevant Bird Society literature and a copy of animal rights campaigner Henry Salt's poetry anthology. Miss Thatcher – 'the most lovable person I have met in Silchester',[112] he called her – told him that, following his earlier visit and intervention, the offending boys had expressed contrition, and all was now well. She had been reading to them from the pamphlets that Hudson had supplied, and he had been assured that 'the children have a very different feeling about the birds now', as he reported to Eliza Phillips.[113]

He was also heartened by a story he had heard from a Mrs Visger about a boys' school at Ealing. Blue tits had been found nesting in a hole in the school gate post, which had become a matter of great pride to all 60 boys in the school, who of course had to pass by the nest-hole several times each day, and had become very protective of their birds.[114]

Nearing her 80th year, Eliza Phillips gave her final speech at the Bird Society's 1902 annual meeting. She used the occasion to talk about the Bird and Tree Competition for schools, launched the previous year. 'Promoting the study of Nature among all classes will do noble work,' she declared, 'for it must tend to spread abroad in the hearts of both teachers and pupils the best and surest, nay, the *only* efficient motive power for good – the Love of God.'[115]

Although Hudson could never overcome his aversion to addressing an audience of adults, late in 1902 he gave a lantern slide talk to children at Silchester, one of the very few such talks he ever gave. Linda Gardiner provided him with 80 images for his hour and a half, and the vicar Mr Sweetapple provided the lantern to illuminate them.[116] Another of these rare occasions arose in spring 1906 when Hudson agreed to speak at a Band of Mercy meeting in Penzance, feeling obliged, having 'urged its formation on some friends when I was last there'.[117] Bands of Mercy were local organisations that encouraged kindness to animals.

It's fair to say Hudson wasn't a fan of formal education, missing out on which had never done him much harm (he may have reasoned). 'Schools are factories,' he would growl.[118] 'No boy should be shut up in school till he is twelve years old. For the rural districts now particularly. The Schoolmaster … will not then find it so easy to rub out the boy's soul.'[119]

He described an incident of the kind that confirmed his lack of faith in the ability of formal school lessons to foster a proper love of nature. On a trip to Norfolk in 1912 he befriended the son of his landlady, and was assured that the boy cared for nothing but nature. After walking with the boy along the coast, Hudson discovered that in fact his knowledge was sadly lacking, and a damning indictment of Norwich High School. Hudson of course was candid. 'I told him that his school teaching was all rot, that if I had ever been in a school where they teach you that you may know everything by reading about it in books, but do not teach you to think and observe for yourself, I should have known no more than he does.'[120]

Perhaps inevitably, given the strong wills involved, Hudson didn't always see eye to eye with Etta Lemon on the content of the society's education materials. She was not in favour of his A Linnet for Sixpence story being produced as a leaflet for schools, for example, writing to tell him that 'what was wanted was a leaflet dealing with the facts about the linnet and its destruction'. This may reflect a Bird Society evolving in her mind from the anecdotal, emotional approaches of the early years into something more 'fact-based'.

'That', as far as Hudson was concerned, 'was another matter.' But – perhaps fortunately – it was all academic. Treasurer Ernest Bell had advised in any case that they couldn't afford to print it, but was keen to see it published in his welfare publication The Humane Review.

Hudson also pioneered some of what we might now call market research. On another occasion, his Bird Society peers deemed his story A Thrush That Never Lived unsuitably complex to reproduce as a leaflet for primary schools. Unconvinced, Hudson sent it anonymously to a schoolteacher he knew and then visited to see if she had used it, and if so what reaction it had met with among the pupils. He was pleased to report to Eliza Phillips that it had been 'pounced upon' by the children.[121] It was duly produced as a Bird Society leaflet.[122]

It is noticeable that girls' schools were usually the winners of the Bird and Tree essay competition, and of course this was remarked upon by the organisers. Hudson would remain a judge and custodian of the Bird and Tree Competition for the rest of his life. He later described reading the Hampshire children's essays for 1916. 'I came upon one by a little boy which ends as follows: "One of our schoolboys had a heron given him, so his mother cooked it and when it was done it was tough and had a nasty taste."'[123]

Hudson had just one venture into writing books for children. 'Just now I have been doing a child or boy's book – which tells the adventures of "A little boy lost" in the wilderness.'[124] Hudson was pleased to learn that the artwork for the book would follow the style of those illustrating Grimm's Fairy Tales. When it was published, he got word that the book was selling, and due to be reviewed in *The Speaker*. He also had a letter from the editor of *The Spectator*, keen that Hudson write for them. Soon after this, a favourable review of his 'book for babes' appeared in that title.[125]

This excerpt from a letter he sent along with a copy of his book may be the only surviving example of Hudson writing to a child:

> My dear Elsa,
> I have great pleasure in sending you as a Xmas card a story of mine about a little boy lost and hope you won't find it perfect nonsense. I think you said you don't believe in Fairies ... but we needn't argue at Xmas time ...
> I am going to put in the parcel along with my book a *real* fairy-tale, told in rhyme by a woman poet ... many people think Christina Rosetti [sic] was the best lady poet we have ever had in England. And *I* think that *Goblin Market* is the loveliest thing she ever wrote.[126]

Hudson was unfailingly generous with presents for his friends' children, and at Christmas 1902 he sent sweets to the Gissings, recalling his own tastes as a child. 'Little children like sweeties better than "food for the mind" however we sugar it. That's how it was with me, I know, when I was little. I didn't have a mind, and as far as I remember did not begin to suspect I would ever have such a thing till I was 14 or 15.'[127]

Although he didn't care much for cards of any kind, much preferring letters, Hudson had encouraged poet Alfred Austin to provide the words and artist Archibald Thorburn a painting of a tern for the Bird Society to use in creating a Christmas card, titled 'Peace and goodwill to the birds,' which raised more funds for the cause.

As the new century progressed, and with success and some savings now in the bank, Hudson could begin to dream of visiting the United States, with which bird protection alliances had been formed, and where the stakes were getting higher.

American dreams and nightmares

In 1901, SPB chairman Montagu Sharpe had visited the Audubon Society, the Bird Society's counterpart there. There was a growing bond of support between the two countries, with lantern slides donated from the United States to enhance presentations, and news and lessons of respective plumage campaigns shared.

In 1905, conflict over the plumage trade became deadly as it emerged from Florida that the warden Guy Bradley had been killed while trying to protect nesting heron and egret colonies. He had been courageous enough to confront three men he found hunting birds for plumage in the swamps. He was shot and left for dead, his body later found by his brother, 10 miles downriver. A man called Walter Smith gave himself up to the authorities the next day. He claimed that he had been shot at first by Bradley, although there was no evidence that Bradley's gun had been fired. It was said that if he had fired, he would not have missed. A jury later concluded that there was insufficient evidence to convict Smith, causing a national outcry. Funds were raised by the public to rehome Bradley's widow and two young children, their home having been burned down by Smith's associates while he was being held for trial.[128]

Before law changes in 1900, Bradley had been employed by the plumage industry to source nesting egrets. He then became one of the first paid wardens in the United States. An obituary was published in *Bird-Lore*, with Bradley described as:

> A faithful and devoted warden, who was a young
> and sturdy man, cut off in a moment, for what? That

a few more plume birds might be secured to adorn
heartless women's bonnets. Heretofore the price has
been the life of the birds, now is added human blood.
Every great movement must have its martyrs, and Guy
M. Bradley is the first martyr in bird protection.[129]

In June 1905, Hudson had lunch at the invitation of Sir Edward and
Lady Dorothy Grey, and they must have discussed this tragic news from
the United States. 'If I can get up enough energy I may go to America for
a few weeks,' Hudson mentioned in passing to Don Roberto, in August.
'The Leland steamer will take you direct to Boston [from Liverpool], so
that I could visit some relations in New Hampshire and Maine without
going to New York, which would be an advantage to me.'[130]

A year on and he was still hankering. 'I hope to go before long to New
England. Maine and New Hampshire and Vermont where my mother's
relations are. I've never seen any of them nor her native place and have a
sort of desire – a kind of pious or superstitious feeling – to pay it a visit.'[131]
It's not clear how seriously he pursued this notion. In any case he had more
on his plate than ever, closer to home. A further measure of Hudson's
growing fame in the wake of his early twentieth-century publishing success
was the invitation he received from portrait artist William Rothenstein to
come and sit for him at his house in a prestigious district of north London.

CHAPTER 8

Modernists

Rothenstein and the Hampstead art scene

William Rothenstein was an admirer of Hudson's work, and Hudson's response to an invitation to sit indicates that he had already met the painter, and his wife Alice. They had both warmed to him, and Hudson would become and remain a close friend and correspondent of both.[1]

Hudson's reputation for evasiveness and self-consciousness about his appearance might have led Rothenstein to expect a polite but firm refusal, but one way or another Hudson was prepared to consent, although the fates intervened and there would be a delay: Hudson developed an untimely, late-onset spot. He wrote again on 29 December 1903, coming quickly to the point:

> Dear Rothenstein,
> I cannot go to you just yet as I have a painful pimple
> on the bridge of my nose which must go away before
> I sit [at] yours. I hope it won't last more than a day or
> so longer.[2]

When the blemish had run its course, Hudson sat for Rothenstein at his home and studio at 26 Church Row. It was an address frequented by many illustrious figures, and a fruitful place to network. It was there that Hudson had met one of the most celebrated names in literature of the time – Thomas Hardy. Their chat confirmed a shared interest in the West Country, and Hudson was promptly invited to visit Hardy at his Dorchester home later in the summer of 1903.[3]

In late summer, Hudson tried to overcome a low mood, and fulfil his promise to visit Hardy at his Dorset home. In the event he spent quite a long time loitering outside Hardy's gate, overcome with a combination of shyness and self-doubt, perhaps hoping Hardy might appear at the door and offer some further encouragement. 'I walked by Hardy's house two or

three times and though I'm quite sure he would have received me kindly
I went not into his gate,' he reported to Garnett.[4] 'I have not been to
see Hardy only because I have not felt well enough,' he told Hubbard. 'I
may go tomorrow: but I will not be sure.'[5] There is no record of Hudson
following through with the visit, and I think we can be fairly certain it
never happened.

Rothenstein was also keen for Hudson to meet and dine with botanist
Frank Darwin – son of Charles – who lectured in botany at Cambridge. He
was by this time a widower with a daughter of 17, called Frances, whom
Hudson also befriended. When Rothenstein's painting of Frank Darwin
was hung in the National Portrait Gallery with five others, Hudson went
to view it, and duly congratulated the artist 'most heartily' on the work.[6]
Now it would be Hudson's turn, and further arrangements were made for
him to sit.

In April 1904, Alice Rothenstein invited Hudson to one of their
by-now regular lunch meetings, and in reply he mentions that he is looking
forward to seeing 'that picture your husband is doing' when he returns to
town for his next bird meeting. Alice obviously enjoyed his company, and
invited him to join her on a trip to Yorkshire in July, and in November
to an art viewing.[7] He later invited her to a music evening at his Tower
House, mentioning the musicians he had also invited.[8]

Hudson was sitting for Rothenstein again in November 1906 – he
even delayed his winter migration to Cornwall to accommodate his friend.
'He's doing it for his own pleasure and as it may be worth something to
him I agreed to put off my journey',[9] Hudson reported. He had hoped to
be out of London from November to Christmas but postponed till the new
year.[10] Just before Christmas Hudson was again returning to sit – it seems
that the light had not been right for the artist. On 23 December he wrote:
'Rothenstein has not wholly done with me yet. You see we have been
having foggy and very dull weather for some time past. I sat on Saturday
and shall have to give him one more sitting.'[11]

When Hudson was finally able to view the results in the National
Portrait Gallery he went along with Emily, but couldn't pretend to
like what he saw. Don Roberto (whom Rothenstein had painted like a
Velasquez) tried to reassure his friend that the portraits were flattering.
Hudson would not be persuaded. 'Of course with your head and Conrad
too – at a distance – [you] can afford to look with equanimity on the Rothensteins with their pencils, but when it comes to a man with distinctly
Palaeolithic lineaments the case is slightly different. He does not feel so
pleased at the idea of his portrait being held up before a smiling public.'[12]

The friendship endured. In February 1908, Hudson promised Alice that he would 'drop in one afternoon while Miss Darwin is with you'.[13] In 1910, Miss Frances Darwin – by now Mrs Cornford – had published her first book of poetry, a copy of which Rothenstein gave to Hudson as a gift.[14] It was in fact Hudson who introduced Rothenstein to Alfred Russel Wallace, by letter – a Darwin for a Wallace. Such was networking in Hampstead in the Edwardian period.

Last days with Emma Hubbard

Having saved Kew Gardens from the late Victorian equivalent of the bulldozers, Hudson could return to enjoying the gardens and green spaces there with renewed peace of mind. In early spring 1901 he watched pied wagtails as they gathered to roost in a patch of bamboo near the pond, counting 78. His ornithologist friend Henry Dresser thought the numbers unprecedented in England.[15]

As an acknowledgement of their lasting gratitude, the Kew authorities invited Hudson to write a report on the birdlife of the Gardens, and in doing so he listed around 80 species. 'Even in a perfectly rural district it would not be easy to find so great a variety in the same space,' he noted, 'and it is, indeed, this variety and abundance of bird-music which to the lover of Nature give to Kew Gardens their principal charm.'[16]

Among the species he had recorded are a sobering number scarce or absent from the locality today: tree pipit, redstart, lesser spotted woodpecker, nightingale, red-backed shrike, spotted flycatcher, swallow, cuckoo, wryneck, nightjar, barn owl, turtle dove, grey partridge and hooded crow. Although already known to be declining, the corncrake could still be heard on a spring day in the meadows of Syon House nearby. House sparrows were so numerous and seemingly impossible to discourage that they were considered a pest. Hudson's pied wagtail roost had now grown to 150 birds. Sadly, the wood-wrens that Hudson and Hubbard had been so keen to hear in 1894 stopped coming back to breed in the oak and beech wood near the Temperate House five years later.

It is notable that the great spotted woodpecker didn't make Hudson's Kew list, while he often heard (if not saw) its smaller relative. This would change within a decade, and in 1913 he reported great spotted woodpeckers adopting one of the nesting boxes installed in Richmond Park, a then innovative measure brought in to help London's birdlife. It was a milestone moment, and the first reference by Hudson I can find to birds using one of these new contraptions.

Hudson returned to the subject of woodpeckers in 1921, and was able to report that they 'are safer in London than they would be in any wood or forest in the country where keepers shoot them to give a pretty bird to some friend or to their children'.[17] London was an unlikely sanctuary for birds as it had been for Hudson himself. A century on, great spotted woodpeckers are widespread and abundant in Britain, while lesser spotted are now extremely rare and localised.

Hudson also took a close paternal interest in the welfare of the great crested grebes that were trying to breed on Pen Ponds.[18] This species was one of the most coveted by collectors and plumassiers, and had been widely decimated. 'Yesterday I had a day in Richmond Park and saw the Gt. crested grebes just arrived for the summer at the Pen Ponds,' he told Roberts.[19] In 1904, he described to Hubbard watching the male grebe – 'the cock or drake or gander, or whatever he calls himself' – relieving the female of incubation duties at their nest. 'Two young grebes have been brought out,' he told her, excited. 'It has even made the press.'[20]

Emma Hubbard died at her Kew home in early June 1905, aged 77, after a long period of deteriorating health, throughout which Hudson had been a frequent visitor and comfort. 'I saw her at 5 o'clock on Thursday last when I came back,' he wrote to Algernon Gissing, 'and she conversed very brightly and said she had a sense of life which she had not felt for a long time. Only eight hours later the end came.'[21] The *Times* carried a long obituary, of which this is a part:

> Mrs Hubbard was a not infrequent correspondent
> of *Nature* and other scientific periodicals, more par-
> ticularly on the subject of birds and their ways, and
> several of her poems, either with or without her name,
> have appeared in *Longman's Magazine* and elsewhere.
> She was also an artist of considerable power, and her
> delineations of some of our rarer insects, drawn from
> the life, are remarkable for the knowledge they indicate
> of insect habits and attitudes, as well as for their
> scrupulous anatomical accuracy.
>
> Her principal pictorial work, however, consists
> of a series of watercolour sketches in Kew Gardens,
> for many years one of her favourite haunts. These
> drawings, remarkable not only for their artistic merit,
> but as an historical record of the garden woodlands
> at the beginning of the 20th century, may, it is hoped,

like those of Miss [Marianne] North, find a permanent
and appropriate home in one of the museums in the
gardens themselves. In the more laborious and far less
generally appreciated art of indexing scientific works
Mrs Hubbard was exceptionally successful.[22]

'It was a great relief to me,' Hudson wrote to Roberts of the recognition
Hubbard received, 'as she was a great friend and helped me with my work.
She was a very remarkable woman: it is wonderful to think that a big
circle of friends and all of her own family … all regarded her as guide,
philosopher and friend'.[23] 'Her life of love for everybody and everything
was very beautiful,' he told Eliza Phillips.[24]

I found a local newspaper cutting in the RSPB archive, from the
granddaughter of Emma Hubbard. She was writing to express her pleasure
at reading about local schoolgirls looking after the Hudsons' grave in
Worthing:

> I am now a senior citizen but in my youth I met
> Mr Hudson and have never forgotten the strong
> impression he made upon me. After [Emma Hubbard's]
> death in 1905, her daughter, my aunt Mrs Wynard
> Hooper, kept up the friendship, and it was at her
> house that I met Mr Hudson. I remember him as a tall,
> gaunt old man, with the face of a hawk, and looking
> completely out of place in my aunt's elegant drawing
> room, among her smartly dressed friends. He sat in a
> corner, completely silent, looking like a caged eagle.[25]

Hudson retained his affection for Kew for as long as he lived, above
other London parks, finding it 'more restful to the eye and mind, more
shady, more picturesque. It is really a most wonderful place, and strange it
should almost surprise me with its charm seeing that I have been familiar
with it quite forty years.'[26] He recalled having written 'more appreciatively
about it than any one before (or after) me', having sung its praises in his
1898 book *Birds in London*.

Hudson's love for Kew – and for Emma Hubbard too – are perhaps
captured perhaps best in this short sentence: 'There nature herself came
to me in the flesh and conversed with me in words,' to which he added,
'the best thing I ever wrote. O vain silly scribbler that I be.'[27] He wrote
this in a 1918 letter to Ranee of Sarawak Margaret Brooke, who would

in due course help to fill the gap in his life that followed Hubbard's passing.

Hudson would remain proud of the Kew campaign for as long as he lived. He recalled in his final year of life how Lord Crawford had been a key ally in saving the Old Deer Park from desecration by the Physical Laboratory. 'By now it would have been covered with buildings',[28] Hudson could now reflect, with some satisfaction. Lord Crawford was made chair of the government committee overseeing the creation of bird sanctuaries in London's parks. Hudson's friend John Harding was appointed to this committee, as well as succeeding Hudson on the RSPB council.

Cowboys and angels

In November 1902, while Hudson was away from home, Emily contacted her husband to let him know that a letter had arrived from Don Roberto. Hudson's reply is telling about the dynamic of their friendship: 'As for Graham's letter I fancy it is in Spanish. But I am not going to try to read it until this afternoon or after dinner. I must go to church and his wild antics might disturb a devotional frame of mind.'[29]

While researching in the RSPB archive I found a rare surviving and neatly written letter from Missy Bontine Cunninghame Graham to Hudson, penned on 28 June 1903.

> Dear Mr Hudson
> I cannot tell you how much I was disappointed at hearing from my son that I had missed seeing you in the park this morning.
> If I had only know you were to be there I would have gone out earlier; as it was I seem to have got to the park half an hour too late.
> I only go there occasionally to see Robert on Pampa, and to give the old horse sugar. Beside the pleasure of meeting you.[30]

Happily, there would be other opportunities. Hudson later recalled another spring day when he was joined by Missy Bontine to watch Don Roberto on Pampa, 'when the picazo behaved very proudly and seemed 'proud of his pride', tossing his mane and pawing the earth', an occasion which was 'still very vivid in my mind'.[31]

Did Hudson ever take the opportunity to ride Don Roberto's horse at Rotten Row? A curious thing is the scarcity of specific mentions of horse-riding during all Hudson's years in Britain. I have found just two traces of him travelling by this means, or riding purely for recreation. In 1899, there is a reference to *not* doing it, for economic reasons, in a letter to Emma Hubbard from Sussex: 'I can enjoy everything on horseback, but alas! The possessor of a horse, even a broken down old screw, wants to make a fortune out of him, and so I may not ride.'[32] His apparent unwillingness is especially peculiar as the instinct for horse-manship seemed to live on. Morley Roberts remembered Hudson in London on an idling 'knife-board' or open bus absent-mindedly hitting the side of it with his umbrella, forgetting where he was and perhaps transported momentarily back to days in the saddle or bareback on the wide Pampas.

In September 1903 Hudson tried to visit Don Roberto and Missy Bontine while they were holidaying at Kings Worthy rectory in Hampshire. He had been delayed by the passing through Winchester and Basingstoke of many thousands of troops and cavalry, who had returned from South Africa.[33] The occasion had kicked up so much dust he decided to take his bicycle on the train instead of cycling, and by the time he reached the rectory his friends had left. He cycled on to revisit Fishing Cottage at Itchen Abbas, expecting to find Mrs Creighton there, but the place was empty and locked so he sat for an hour on a chair reacquainting himself with the view across the marsh and river.

Hudson was a genuine admirer of Don Roberto's prose, and offered warm encouragement. Praise from Hudson was all the more valuable because it was sparingly dispensed: his friends could always count on its sincerity. The fact that Hudson praised Don Roberto's work and was pleased to see glowing reviews for his latest book *Success* in a letter to Eliza Phillips confirms the authenticity of his regard.[34] Don Roberto knew he could also rely on 'Huddy' for a candid view of his writing, and his impassioned pronouncements. If he didn't approve of something, he would say so. 'I have a colder mind, and sometimes think that the you-be-downness comes out a little too much,'[35] he told him, on another occasion. Hudson sometimes thought his friend's style 'almost too brutal in its realism'.[36] For all his effervescence in speech and demeanour, Don Roberto could be bleak on the page – too bleak even for Hudson, on occasion. 'I wish you had omitted some of your side-thrusts at our times ... the human animal has become softer and more human during the last three centuries, in spite of Russia and perhaps ourselves in S. Africa and elsewhere.'[37] Joseph Conrad, meanwhile, once described Don Roberto's more withering stories as 'the philosophy of unutterable scorn'.[38]

Don Roberto dedicated to Hudson his book about the Spanish conquistador Hernando de Soto, late in 1903, sending him a copy as a seasonal gift. Hudson was moved, calling it 'the noblest Christmas box I have ever received'. It brought much-needed cheer. 'Winter is dark and dreary in London and one pines for March and April,' he added. 'With all good wishes for 1904.'[39]

'It was only fit that a man who has spent so many days on horseback on lonely plains and mountains should be the biographer of such a one as de Soto,' Hudson wrote, when he had read the book.[40] It prompted a discussion between them about war and civilisation, and atrocities committed in the name of religion, and otherwise. 'The strong crush the weak and seem to revel in cruelty whether they wear the religious mask or not,'[41] Hudson lamented.

With his pen now in full flow, Don Roberto declined exhortations from his political acolytes to stand as a Labour candidate in the 1905 general election. Hudson thought his reasons sound.[42] But dark clouds were gathering. On one rendezvous with Don Roberto and Pampa the stallion in Hyde Park, Hudson learned of his friend's concern about the health of Gabriela.[43] In late summer 1906, she was on another literary research trip in France while Don Roberto remained at home at Ardoch, on the banks of the River Clyde, working on his book about the Jesuit missionaries in Paraguay, A Vanished Arcadia. A telegram arrived for him. Gabriela had now taken seriously ill. He left immediately for the long journey to southern France to be by her side.

Tragically, Gabriela never recovered, and passed away in September 1906 at the age of just 45. She was brought home to be buried in the Grahams' ancestral burial site within the ruined walls of the priory on Inchmahome Island in the Lake of Menteith. Don Roberto dug the grave himself.

A painting of Gabriela was hung in the Portrait Gallery. Hudson wrote to Roberts in late November 1906, having been there solely to view 'poor Mrs. Cunninghame Graham's portrait by Lavery'. He wasn't impressed. 'She had an infinite sadness in her face which that dull artist, or painter of portraits, has not caught – and could not. C.G. is in London just now but I have not seen him.'[44]

'I have been going through an immense mass of letters before destroying them',[45] Hudson reported in 1906. These were letters that had been returned to him from executors for a friend who had recently passed away. I wonder if it was Gabriela, to whom no Hudson letters survive, or Emma Hubbard, to whom many letters do. If the latter, then it is further confirmation that he was selective in his burning. He had also been given 14 years' worth of correspondence from a woman who 'does not wish to leave this labour to others at her death'.[46] This timescale suggests a Bird Society colleague, perhaps Eliza Phillips. Again, quite a number of selected letters to her did evade the flames.

In the summer of 1907, Don Roberto knew that his beloved Argentine stallion Pampa was now too old for the London life, and he put him out to grass on paddocks near Weybridge in Surrey. Hearing the news, Hudson wrote: 'I feel grief and pleasure at the same time … Grieved that his days of active life with you on his back are ended and glad that you have been able to find him such a haven of rest.'[47] Hudson describes this horse in his *Naturalist in La Plata*. It prompted Hudson to briefly contemplate the phase of inertia he was experiencing. 'Perhaps, like Pampa, I've done with activities,' he added. 'I'm doing little now.'[48] Don Roberto soon acquired another mustang from the Pampas. Hudson was pleased at this news. 'I'm glad you've got another Argentine horse', he told him, 'and hope to see him in the Park one day.'[49]

While on a visit to family in the west of Scotland, I had the good fortune of being able to visit Don Roberto's former home Ardoch House, on the kind invitation of current owner Professor Tommy McKay. Tommy showed me the library in which Don Roberto wrote many of his books, and took down from the shelf his copy of Hudson's *El Ombú* with its dedication to the friend he called 'El singularísimo escritor Inglese … who has lived and knows (even to the marrow as they themselves say) the horsemen of the Pampas, and who alone of European writers has rendered something of the vanishing colour of that remote life.'[50]

Green Mansions

Edward Garnett was instrumental in Hudson finally completing his novel *Green Mansions*, which had been a work in progress for over a decade.[51] It had been gestating in his mind since the early 1890s. 'Mainly I am working on my first British bird book, and something about an explorer in the forests of Venezuela', he wrote at that time, obviously distracted by his obligations to produce the ornithology textbook. As he was out of town,

Hudson asked his wife Emily to send a copy of his manuscript to Garnett, and confided his misgivings about it:

> [I] was again struck painfully by the cumbersomeness of the form. Perhaps some little alteration might be made here… the introductory chapters seem too slow: the story doesn't move at all, it simply sits still and stews contentedly in its own juice; and it doesn't even stew, or boil, *a barbolloner*, but simmers placidly away, like a saucepot of cocoa-nibs that has all the day before it. This too might be remedied to some extent.
>
> There are, I daresay, some good points in the book, especially the hero's feeling for nature; and he being a Venezuelan some might say that it is all wrong. But of course it is a delusion that this feeling is confined to our race and that it is a thing of to-day. It is as strong in some of the old Spanish poets as in some of the modern English poets which show it most, and I have known S. Americans with that passion as strong in them as any Englishman…
>
> But you can see all this and much besides better than I can, and I will no further seek its merits to disclose or draw its frailties from their typewritten abode.[52]

It is fascinating to note the doubts Hudson was having over a book that was about to catapult him to whole new levels of fame and some fortune, with a little help from his friends – and especially his reader Garnett.

We can piece together the day when *Green Mansions* was accepted for publication: it was 17 November 1903. In the morning, Hudson dropped in to the Bird Society office at Hanover Square where it had been agreed that his pamphlet *The Yellow Bird* was going to print, with 5,000 copies being ordered. He then collected Garnett from the office at Duckworth's, who had just agreed – after editorial input from Garnett – to publish the novel.

In celebratory mood, the pair headed to Gerrard Street for lunch and were joined at the Mont Blanc by Joseph Conrad, G.K. Chesterton and Hilaire Belloc. Hudson recalled the dynamic at the lunch-table: 'It was very amusing to listen to Belloc's tattle; even G.C. [G.K. Chesterton], the brilliant one, had his lights rather put out. I suppose Belloc looked on me as a sort of outsider – a naturalist who had no right to eat his shilling lunch in such company.'[53] In fact Belloc had told one of the group that he liked and had read *A Crystal Age* more than any other book. 'He was amazed to hear that I had written it,' Hudson commented with glee. Things were certainly looking up for the humble field naturalist.[54]

Although I have found no description of it by Hudson, nor of any other of the scenarios at lunch (what a pity – and mystery – that there are no photographs taken at the dining table) the Mont Blanc, as its name might suggest, offered a French-themed menu at affordable prices attractive to cultured types. The regularly booked literary table was upstairs. It was long, with Garnett at its head, a rough cloth under the carafes of plonk, plates and books side by side, and the gabble of excited, opinionated fellows competing for air-time. Painted alpine scenes adorned the walls, a far cry from the spectacle and tumult of the lanes and market squares nearby, where the rich aromas of crushed and mouldering fruit and vegetables were infused with fresh horse dung and smoke. Hudson occasionally dined elsewhere, but Garnett's correspondence confirms his general reluctance to branch out in this way.

Not long after, Hudson was invited to the home of E.V. Lucas, who joined the staff of *Punch* that year.[55] Also present on that occasion were Conrad, J.M. Barrie (author of *Peter Pan*, which was first published later that year) and Maurice Hewlett. I wonder if they discussed their respective children's books: Hudson's one venture into writing for children – *A Little Boy Lost* – appeared in 1905. Hudson mentions that Barrie wanted him to send his first published book *The Purple Land* to a different publisher, for a revised edition, sure that it would now achieve the sales and recognition it deserved, but had failed to get first time round.[56]

When *Green Mansions* was published in 1904, Joseph Conrad expressed his admiration for the story, telling Don Roberto he thought it 'very fine … remarkably harmonious, nothing too much, the right note of humanity, the right tone of expression, a sort of earnest quietness absolutely fascinating to one's mind in the din of this age of blatant expression'.[57] Hudson would exchange books with Conrad, who gave him a copy of his *Nostromo*, in November 1904, not long after receiving *Hampshire Days*.[58]

Described as 'a romance of the tropical forest', *Green Mansions* tells the story of a man who flees revolution in Caracas and ends up deep in the 'jungles' of Guyana, where he finds and falls in love with a mysterious 'bird-girl' called Rima. This strange meditation on the beauty and fragility of nature and love captured the hearts of thousands of readers.

Hudson mentioned to Garnett the response of Gabriela Cunninghame Graham, who had returned from Italy having just read it and written him a 'remarkable letter'. She was going to review it for the *Saturday*, and she didn't hold back in her praise.[59]

The ethereal singing of his 'bird-girl' character Rima may be a nod to Hudson's own early attraction to his wife Emily's singing voice, and of course his love for the voices of so many songbirds. An encounter with a shepherdess high on the Sussex Downs might have been an additional inspiration. 'I constantly heard her oft-repeated calls and long piercing cries sounding wonderfully loud and distinct even at a distance,' he wrote in *Nature in Downland*, which described a number of encounters with shepherds in the course of his rambles. 'It was like the shrill echoing cries of some clear-voiced big bird – some great forest owl, or eagle, or giant ibis, or rail, or courlan, in some far land where great birds with glorious voices have not all been extirpated.'[60]

Battling the censors

Besides their pushing of literary boundaries, the Mont Blanc circle of writers of which Hudson was now part were soon organising themselves against the censorship of subjects deemed risqué by the government. The subject must have occupied much of the air-time in the Mont Blanc by the middle of the Edwardian decade, Hudson perhaps more in listening mode now, as Garnett and co. fulminated against their oppressors. Inspired by the works of Henrik Ibsen, and keen to help English drama free itself from the shackles of the drawing room and add a dash of European modernism, Garnett's own ambitions as a playwright were being curtailed by the censor, but also to a degree by Galsworthy, who would urge rewrites of some of Garnett's raunchier passages.

Conservative values died hard. The literati were up against archaic legislation designed to protect 'good manners and decorum, or the public peace', essentially vetoing discussion of sex, politics and religion.[61] When one of his plays was banned, Garnett made public both the script and his correspondence with the censor, and called for organised protest. In October 1907, Galsworthy used his influence to bring other big names on board, including Don Roberto, George Bernard Shaw, H.G. Wells and others to petition the prime minister, Campbell Bannerman. Galsworthy drafted their letter to *The Times* and enlisted 71 signatories, with J.M. Barrie, Conrad, Thomas Hardy, Henry James, Somerset Maugham, Edward Thomas and W.B. Yeats adding their names.

The campaign came to a head with a demonstration that gathered in Trafalgar Square (Don Roberto was in the thick of it again, though not beaten up by police on this occasion, nor arrested) before marching on Downing Street, with George Bernard Shaw at its head. Garnett and Galsworthy walked behind him, pole-bearers of a red banner declaring 'Down with the Censor', of whom W.B. Yeats helped to carry an effigy.[62]

For a time, I wondered if Hudson was ready to be on board with all this liberal radicalism, but I needn't have doubted it. In October 1907, he asked Morley Roberts to send him one of two copies of Garnett's censored play, asking to know his opinion of it. 'I fear G. won't be very pleased with my judgment,' Hudson confessed, candid as ever, 'but his letter to the Censor should have good results.'[63] A week later, he was urging Roberts to sign the petition, as authors as well as dramatists were now adding their names. 'There are so many who would like to put their names', he urged. 'We want all arts to be free … We all have the possibilities in us: and even if we haven't (I haven't).'[64]

By December, all the signatories were sitting for portrait photographs as part of the publicity campaign. In a break with tradition, or perhaps just caught up in the positivity of it all, Hudson was actually pleased with the results of his sitting, and showered praise on the photographer, Marie Leon.

On New Year's Day 1908, he wrote to Garnett with 'just a line to say I'm not ill, though not robust', adding, 'I sign, as you wish', indicating that he had added his name. He also enclosed for Garnett a box of cigars, and for Constance Garnett a sprig of lavender.[65]

The petitioners were granted an audience with Home Secretary Gladstone in February 1908. 'Are you I wonder going to be present at the Deputation?' Hudson asked Garnett, adding, 'well I suppose so and if I have the time I will be there too.'[66] It is recorded that Garnett didn't attend, for some reason,[67] and it would be untypical of him if Hudson had.

The campaigners sent letters to The Times and petitions to the king. After all that, did the freedom-loving artists and writers win? Eventually, the government appointed a Joint Committee on Censorship in the summer of 1909, bringing the parties together to see what might be updated. The petitioners achieved some concessions, and the censor lost his job three years later. But the law remained largely intact for another 60 years. Garnett's play, appropriately titled The Feud, had some airings in minor venues, but was mostly panned. He will have taken some solace from the fact that Hudson was kind about it, writing to Garnett in April 1909 to congratulate him on 'so fine a piece of work … wonderfully vivid and powerful. One wonders that so peaceful a son of civilisation as yourself could find it in you to make these bloody-minded barbarians so fearfully real.' Galsworthy's less controversial play *The Silver Box*, meanwhile, fared much better than *The Feud*.

Hudson assured his friend that he hoped to find a way of seeing *The Feud* performed, but in June gave his apologies as he couldn't get back to London in time, while hoping to persuade his wife Emily and her friends to go along.

Garnett continued to write plays, and to encourage Hudson to come and see them. On another occasion, Hudson had to politely decline owing to other commitments.[68] After this exchange about the play, a gap of 15 months follows till the next surviving letter to Garnett, in October 1910.

'You have to see him and hear him',[69] poet Edward Thomas said of Garnett, in explaining to a friend where his mentor's strengths lay, which were not on the page. The frustration of this for Garnett, whatever his genius, would increasingly bubble to the surface. Thomas, whom he had invited to The Cearne, where they rowed furiously about Walt Whitman, and became fast friends, was one of his later Edwardian protégés. E.M. Forster was another. The dramatic device of the incomer illuminating the inhumanity of the ruling classes recurs in the work of Forster. It is old Mr Emerson in *A Room With a View*, of whom a clergyman says: 'he has the merit – if it is one – of saying exactly what he means'.[70] There may be echoes of Hudson here. But it was D.H. Lawrence in particular, nurtured by the now wounded Garnett, who was about to take polite society by storm.

The Gissings

One name that doesn't seem to figure in the literary circle of this period is that of Hudson's old friend George Gissing. In a letter written at New Year 1903/4 to William Rothenstein, Hudson relayed the sad news of Gissing's death. 'In the eighties and early nineties Gissing, Morley Roberts and I were three very poor Bohemians living in London and very much together,'[71] he wrote to Don Roberto. Morley Roberts and H.G. Wells had both gone to Gissing's aid in southern France when news came that he was dying.

Although Gissing's life was turbulent, the news still came as a shock. Gissing's brother Algernon had had no news on his health when Hudson enquired at Christmas. Hudson was one of Gissing's six closest friends (by his own calculation). H.G. Wells had gone to France, where Gissing was living in exile, to check on his state of health, and it was Wells who had found him dying. Wells immediately sent a wire to Morley Roberts to come at once. Roberts left that same day. Gissing died on 28 December.

A week later, Hudson was writing to the Royal Literary Fund to request support for Algernon Gissing. 'I have not known a more deserving case than his', he urged. His novels were 'considerably above the average', and he may have been 'overshadowed by the greater reputation of his brother'. 'I have been in a similar sad position,' he added. 'Had I not been helped by a grant ... I should probably not have lived to write other books and see better days.'[72] Hudson had received his support from the Royal Society while engaged in the lengthy and otherwise profitless business of co-authoring *Argentine Ornithology* with Dr Sclater. Belatedly, some of Hudson's other early and profitless titles were now poised for a new lease of life.

A Crystal Age revisited

In 1893, six years after *A Crystal Age* was first published, the publisher Fisher Unwin had come calling, keen to try again with the book. Hudson believed that if revised and improved, and with his name on the cover this time, it would be 'sure to command a sale'. He also wanted to secure a copyright in America.[73] Negotiations broke down, no doubt because Unwin didn't want to bear the additional production costs of a new version.

Ten years later, with Hudson's fame now on another plane, Unwin returned to the table once more. Needless to say, Hudson still would not consent to it being reissued without major renovation. He reminded the publisher that he 'paid me nothing' in 1887, and that in any case he couldn't understand this renewed eagerness. 'All the books since written, issued in my name, though they have been greatly praised, have a very limited sale,' he told Unwin, which is typically and painfully honest, if hardly calculated to strengthen his hand in a negotiation.[74]

'I made a last attempt to get Fisher Unwin to resign his copyright in *A Crystal Age*,' he told Roberts, 'but he would not do so for love or money.'[75] Hudson explained to Edward Garnett that he had offered £15 to Unwin for the rights, and been turned down. 'So I've sold it him and told him to publish – and be damned.'[76]

A Crystal Age was finally republished in 1906, with Hudson's name now on the cover, the publisher hoping it would ride on the coat-tails of his now escalating fame and the success of *Green Mansions*. It did at least have a fresh appearance and a new preface. 'I hope the cover won't frighten you', he warned Roberts when sending him a copy, adding – a little peevishly – 'I care precious little about the book as a fact and don't care whether it succeeds or not.'[77]

Hudson's journalist friend William Canton had been a loyal champion of *A Crystal Age*. He had written a long piece for *The Bookman* in 1895 in a series about 'neglected books', describing it as 'not only a naturalist's day-dream filled with the poetry of nature, an evolutionist's reverie in which the ape and tiger vanish from our blood, but a charmingly naïve humorous, and withal tragic love-story'.[78]

The new edition met with some favourable critical reaction, with *The Times* carrying a long and complimentary review. Hilaire Belloc was by now literary editor of the *Morning Post*, and wrote a review 'all about Belloc with very little about the book',[79] as Hudson grumpily observed. But despite this second, enhanced iteration, and for all its apparent influence on better-remembered authors, *A Crystal Age* would never join the ranks of the celebrated futuristic fantasies, or earn Hudson the fame of authors such as H.G. Wells, Arthur Conan Doyle and Rider Haggard.

Hudson must have met Doyle, who was a supporter of the Bird Society and even spoke at one of the annual meetings. Meanwhile, although they had mutual friends, if Hudson and H.G. Wells ever met I have found just one hint of it. He seems to have had mixed views of Wells's blockbusters. *War of the Worlds* he wasn't keen on, but he acknowledged that 'one remembers *The Time Machine* because it has a human interest',[80] in a letter to Garnett in 1905.

I wonder if privately he could see that the power of Wells's short story is in how it relates back to the known world, with the time traveller present in his London environment, with his gentlemen friends, as well as in the recognisable past and mind-boggling future. Hudson's *Crystal Age* protagonist Smith lacks this relatability and anchoring, and while the author may not have cared to structure his narratives to please the reader, and seems to have been resistant to interference (even, in this case, from Edward Garnett), perhaps deep down he knew that narrative fiction was not his forte, and never would be. With the publication of *Green Mansions*, his novel-writing days were now over. He would stick to what he was best at – the observation and description of nature.

I've never forgotten how in the movie made of *The Time Machine* in 1960 the fashion boutique shop-window dummy opposite the time traveller's house marks the passage of time by its rapid changes of outfit, including feathered hats – an arresting special effect by Ealing Studios. It might have served as a lesson to other late Victorian novelists on how to accelerate a narrative…

Although the realisation came late, Hudson's experience of dabbling in different genres and finding his strong suit would prove valuable to share with a young man of letters with whom he was about to form a very close bond.

Edward Thomas

Of all Hudson's friends, Edward Thomas may be the most fondly remembered literary figure of all.[81] 'War poets' have always held a special place in the affections of the public, for articulating the tragedy of the conflict that turned the world upside down from August 1914, and shattered so many lives, and families. The rapport between the two men – one young, aspiring and troubled, the other ageing and often a little world weary – tells us a lot about both.

They first met in 1906, introduced by Edward Garnett at one of the regular literary lunches in Soho. It seems they hit if off straight away. Thomas was in his early 20s, tall and lean, like Hudson, but fair-haired and thinning on top, unlike Hudson. That he didn't wear a hat marked him apart at a time when almost all men did. He carried an ash-wood walking stick. In photographs he looks angsty, brooding, offended even. He shared

Edward Thomas and Helen Thomas, both age 21 around the turn of the century. Edward revered Hudson. 'Except William Morris, there is no other man I would sometimes like to have been, no other writing man'. Images courtesy of the Estate of Helen Thomas.

Hudson's intense love of nature and his enquiring mind, and his yearning to explore the landscape on foot and bicycle, especially that of south-west England.

He was already an admirer. 'The writing gives us a sense of the actual bodily presence of a man,'[82] wrote Thomas, of Hudson's prose style. He was said by his wife Helen to have revered Hudson. She recalled of Hudson that 'he and Edward constantly wrote to each other and met often in London. Edward responded to this somewhat austere, solitary man, and honoured him above all writers of the time.'[83]

Her statement is backed up by what Edward himself once said of Hudson: 'except William Morris, there is no other man I would sometimes like to have been, no other writing man'.[84] When Thomas began work on a biography of nature writer Richard Jefferies the year after they met, he asked Hudson to accept the dedication. Hudson was not a fan of dedications in books, but made an exception in this case. It was published in 1909.

The difference in their backgrounds was not lost on Hudson. When Thomas was applying to become a lecturer at Oxford, he asked Hudson for a reference. 'How funny!' he told Margaret Brooke. 'A wild man of the woods giving a testimonial to an Oxford graduate known by his literary books to the world!'[85] References to Thomas in Hudson's letters show that he always retained an active interest in and concern for the younger writer's well-being and fortunes. Thomas invited Hudson to comment on his early poems, although was not easily persuaded to edit his work. Hudson believed he should 'follow his own genius',[86] and that way lay poetry.

Notwithstanding Thomas's sensitivity, Hudson was typically candid about his friend's strengths and weaknesses as a writer. Hudson told Edward Garnett that Thomas's only novel, *The Happy-go-lucky Morgans*, 'interested me greatly, but I don't think it will interest the reading public one bit … I believe he has taken the wrong path and is wandering lost in the vast wilderness.'[87] With views like these, earnestly expressed as always, Hudson may have been instrumental in persuading Thomas to concentrate on poetry. 'He is essentially a poet, one would say of the Celtic variety.'[88]

While the other laurelled war poets Rupert Brooke and Siegfried Sassoon were skinny-dipping in Grantchester meadows and fox-hunting in rural Kent, Thomas's soul was restless and his family life turbulent. In November 1911, Hudson received word from Garnett that Thomas was having a nervous breakdown. Two years later, in March 1913, Hudson was once again 'much distressed',[89] about his friend. Helen Thomas reported that her husband was exhausted and in despair. 'Why will he work so incessantly and so furiously?' Hudson pleaded.[90] In October 1913, Thomas received a bad review in *The Times*, but still wouldn't admit to Hudson that in his critical study of author and critic Walter Pater he had chosen the wrong subject to write about.[91]

American poet Robert Frost was the other great influence on Thomas's decision to concentrate on poetry. Shortly before the outbreak of war, Hudson had a lunch appointment with Thomas, who was to bring Frost, but an excursion the previous day had taken its toll on Hudson and he cried off. Thomas instead came round to visit him, and sat with him for half an hour at Tower House. It may have been Hudson's only chance to meet Frost.[92]

I am sure Thomas must have described to Hudson a turning point incident in which he and Frost had been confronted and threatened by an armed gamekeeper while trespassing on a shooting estate. Thomas's loss of nerve and Frost's relative bravery in the face of threats from this armed man had unsettled him, and is thought by some to have been a factor in his eventual decision to enlist.

Along with Joseph Conrad, of Hudson's admirers Edward Thomas is probably the most enduring name in literature. 'I love the man,' Thomas wrote.[93] But it was another respected literary figure, even more successful, who would exert probably the greatest influence on the further success that awaited Hudson.

John Galsworthy

Edward Garnett was always looking for talented new authors to cultivate. Inevitably, with the strong characters involved, there would be signs of wear on some of his earlier alliances, some sooner rather than later. Conrad, for example, would weary of Garnett's repeated positioning of him in reviews and essays as a 'Slavic outsider'. Simmering tensions over Russia also spilled over. The Polish Conrads had history, and he did not share Garnett's unconditional love of the 'Russian soul'.[94] Garnett and Galsworthy would also exhaust each other after a running battle of wills over the consecutive books that would become the five-volume *Forsyte Saga*. Garnett would even travel to Galsworthy's Devon farmhouse retreat and spend days toughing it out over details of characterisation and plot.

Of Galsworthy's portrait of the aristocracy in novel form, *The Patrician*, Garnett screeched: 'You don't know these people well enough.'[95]

But Galsworthy had data to back his corner, and duly produced a list of 130 aristocrats whom he had met, even if briefly. He was sticking to his guns more forcefully than usual, normally open-minded enough to bow to Garnett's clear-eyed objectivity and street-wisdom on publishing matters. *The Patrician* marked a parting of the ways. Galsworthy would now fly unassisted, and America beckoned. Significantly, Garnett sought Hudson's view of Galsworthy's play soon after.

'You wanted my opinion,' Hudson replied in March 1911, having to put his critique in writing as he couldn't attend the next Mont Blanc gathering. His verdict was both glowing and qualified. 'It is not so real as *The Man of Property*'. He essentially agreed with Garnett's view that the characters were not drawn from life, but from types. 'As you know I'm extremely reluctant to praise anything, good or not', he caveated, and without pause or line-break he mentioned two of his own friends in high places: 'I saw [Wilfrid] Blunt on Thursday ... C. Graham [Don Roberto] writes to me from Rome where he went on a visit to his mother.'[96]

Galsworthy and Conrad eventually agreed that Garnett wasn't interested in them now owing to their 'beastly success'.[97] On the other hand, Hudson's spats with Garnett were tame by comparison and they remained close, attending evening events together, including a Fabian Society reception, at which George Bernard Shaw gave a speech. 'Did Shaw ever neglect an opportunity of blowing his own brass trumpet?'[98] Hudson wondered afterwards. Both he and Garnett much preferred the speech given by author Anatole France. In Garnett's estimation, 'the insularity and lack of cosmopolitan culture shown by the English speakers contrasted painfully with the urbanity and wide European vision of the great French writer'.[99]

While Garnett helped Hudson on points of detail, Galsworthy changed the landscape for him through his wider influence on opinion. Hudson could scarcely have hoped for more heavyweight men of letters to back his corner. Galsworthy was literary royalty, his greatness recognised when he was awarded the Nobel Prize for Literature in 1932, a year before he died at the age of 65. This highest of accolades was awarded in recognition of 'his distinguished art of narration, which takes its highest form in *The Forsyte Saga*'.

From this lofty pinnacle, Galsworthy remained a household name for at least another 50 years – he was certainly a well-known name in our house. His epic Forsyte family yarn was adapted for television in the 1960s, and reshown in the 1970s and later, when I was old enough to wonder what all the fuss was about while my mum and big sisters were eagerly tuning in. He was also awarded a knighthood, as early as 1918, but was bold enough to turn it down. He later accepted an Order of Merit. Galsworthy's credentials were never in question, and that he held W.H. Hudson in such high esteem is of considerable significance. He loved not only Hudson's writing, but the man himself. Alongside Joseph Conrad and Edward Thomas, Galsworthy is probably the most enduring name among Hudson's literary fans. Unlike Conrad and Thomas, Galsworthy also devoted a lot of time and energy to elevating Hudson to the position of renown that he occupied by the end of his life. I've been intrigued by what it was that drew him to adopt and support Hudson in this way, of all the people that he knew.

They seem to have first become acquainted at the turn of the twentieth century, when Hudson began to lunch regularly in this literary circle. Galsworthy was much in demand elsewhere and not the most regular attendee, but he took his place at the table when he could. An undated letter from Hudson thanking Galsworthy for his book *Man of Devon* seems likely to have been sent in 1901, the year the book came out. The next clear trace of their relationship comes in September 1903. Hudson wrote to Emma Hubbard that he had lunched with Galsworthy, as well as Conrad, Hilaire Belloc, G.K. Chesterton and Edward Garnett.

Galsworthy obviously warmed quickly to Hudson at these lunch meetings. In January 1904, Hudson wrote thanking him for the gift of a copy of his new novel *The Island Pharisees*, the first that Galsworthy published under his real name. It tells the story of a young man educated at Eton and Oxford who meets a foreign visitor and, getting to know him, begins to see upper-class British society from an outsider's perspective. It isn't hard to imagine lunch-table chats with Hudson and Conrad helping Galsworthy's characters to form in his mind.

In a later critique, David 'Bunny' Garnett thought the book 'inspired by hatred' and wondered why Galsworthy had such antipathy to his own class.[100] He concluded that it was frustration, borne of having fallen in love with his cousin's wife, Ada, against all the rules of polite society.[101] But I think too, as for Hudson, it stemmed from his loathing of the brutal treatment of animals by many privileged types who ought to have behaved better. Galsworthy was a committed campaigner for animal welfare.

Hudson gave a typically even and candid critique of *The Island Pharisees*, and promised in return to send Galsworthy his own new novel. 'Please don't trouble to take any particular pleasure from *Green Mansions*,' he begged. 'I've never taken myself seriously as a fictionist, and am not going to try to begin now.'[102]

A Hudson mention of 'always expecting to see him on Wednesday afternoons' suggests an open invitation and that Galsworthy had lately been too busy to drop by at Tower House.[103] He was a man in great demand. Galsworthy's diaries indicate a hectic lifestyle of international travel and literary fine dining, at a range of venues including the Savoy – which puts the relative narrowness of Hudson's social horizons in perspective. Despite the limits on Galsworthy's time, the friendship grew in strength, such was his admiration for Hudson, and perhaps because he could see the relative lack of reward he had received for his literary endeavours. The respect was mutual, and Hudson was an admirer of Galsworthy's character. 'He's a good fellow, and that's better than being merely a successful author,' he told Morley Roberts.[104]

The two men had a shared Devon lineage, on their father's side, which may have contributed to a sense of kinship. 'He was intensely interested in his own origins and descent through a long line of Devon farmers,'[105] Hudson noted. That said, I don't detect in Hudson a similarly strong affinity for the place – Clyst Hydon – from which his own paternal grandfather is said to have originated. When Hudson made an excursion to South Devon in autumn 1912, it was to look for wild ravens on the sea cliffs there, the odd pair having managed to evade the guns and traps. When Galsworthy also went to Devon that autumn, it was on an ancestral pilgrimage to Wembury, near Plymouth.

Hudson had more luck than he had found for Garnett in being able to attend Galsworthy's plays, on more than one occasion with the Ranee of Sarawak Margaret Brooke (see Chapter 9). In autumn 1906, Hudson was in the audience with his wife Emily for a performance of *The Silver Box*, writing afterwards to his friend: 'we were going to congratulate you after the play but saw you shut in by such a vast throng of eager friends that we had to retire'.[106] I can picture Hudson anxious to side-step the gaggle of hangers-on. Galsworthy was sorry to miss them in that foyer, and needed some added reassurance that the Hudsons had actually enjoyed the show, prompting Hudson to write again: 'you feel that I don't like it … as a fact I did like it very well',[107] he insisted. His approval obviously meant a lot to the celebrated playwright.

Not long afterwards, Galsworthy dedicated his latest book, *A Country House*, to Hudson, and sent him a copy early in 1907.[108] Hudson responded politely, saying he was very proud, and proceeded to provide another detailed and frank critique. I quote the following as I think it says a lot about Hudson – agnostic, but sympathetic to most honest men of the cloth.[109]

> It is possible that the feeling against country
> clergymen, so common among intellectual people, is
> the reason of my very strong feeling in their favour.
> That would seem to show a contradicting or mulish
> spirit in me; but perhaps I am not so bad as that; I must
> try at all events to think that I like country parsons not
> because others dislike them but because they really are
> very good people.[110]

Hudson wasn't a habitual theatregoer, but he was also present for a performance of Galsworthy's play *Strife* in 1909, again in the company of Ranee Margaret Brooke. He afterwards congratulated the playwright on having put 'reality on the stage'.[111] Hudson had spotted Mrs Asquith (the prime minister's wife) in the audience, and even consented to go backstage after the show with Ranee Margaret to meet her friend Mrs Granville Barker, also known as actress Lillah McCarthy.

All in all, it seems that most of these wealthy, intellectual and cultured men that Hudson knew in most cases liked him more than they liked each other. It is Hudson for whom the warmest affection is reserved and expressed. It had remained a matter of regret to Bunny Garnett that by a matter of days Hudson didn't live long enough to read his first published book. 'I should have liked his opinion of it more than most people's,'[112] he rather touchingly reflected.

When Bunny was old enough to have a glass of wine and hold his own in cultured, grown-up company his father started bringing him to lunch with the literati. Among this stellar cast, he liked most of all to sit between Hudson and Edward Thomas. Bunny noted that in contrast to Hudson's more adversarial rapport with Garnett senior, 'his heart went out to Edward Thomas without reservation'.[113] From these gatherings Bunny could also see that his father and Joseph Conrad 'both loved Hudson'.[114]

In Bunny Garnett's recollections of the other diners, perhaps surprisingly, he isn't complimentary about Galsworthy. It is worth noting, however, that Galsworthy later refused to back Bunny's application for exemption when he refused as a pacifist to join the war in 1916 (Bunny had other high-profile options and was backed by Maynard Keynes instead). Bunny thought the character of Bosinney in Galsworthy's *Forsyte Saga* was modelled on his father Edward. Bunny also recalled that both his father and Conrad disapproved of the first biography of Hudson, which was published in 1924, just over a year after his death. They 'disliked Morley Roberts' book' for daring to publicly dissect Hudson's unhappy years of poverty and rejection.[115] The younger Garnett's further impressions of Hudson deserve a section of their own.

Bunny Garnett remembers

1981. A boutique house in a small, sleepy town in southern France called Montcuq, off the beaten, chalky track. Edward Garnett's son David – or 'Bunny', as he is still known – is by now nearing his 90th year, sitting at his table, reflecting on an eventful life as he composes his memoir, which he will call *Great Friends*. It is a collection of portraits of the eminent literary figures he has had the privilege of knowing. Hudson is one, along with some of his associates – Joseph Conrad, Edward Thomas, John Galsworthy, Ford Madox Ford. It's like the line-up of a dream dinner party, or the coming back to life of an Edwardian period lunch gathering at the Mont Blanc. I was also delighted to find a chapter on T.H. White, the troubled author Bunny befriended in later life.

I can see Garnett as he writes, not least because he is shown doing so on the cover of the book, hair snow white and with his carafe of claret, but also because by a quirk of fate I can see in my mind's eye Montcuq, having stayed there for a memorable fortnight in June 1991, in a gite with friends, communing with hoopoes, orioles, butterflies and reptiles by day, frogs and nightingales after dark. I can also imagine sitting with the now old man Garnett, enjoying his reminiscences on a café terrasse. If only I had got there ten years earlier.

Bunny Garnett was a boy when he first knew Hudson, who was an occasional visitor to the Garnett family home in Kent, The Cearne. Bunny (he got the name from the rabbit-skin cloak that he liked to wear) explored the woods with their guest. He recalled several of Hudson's visits, and being treated as an equal from the first. It confirms Hudson's usual easy rapport with children.

Hudson would become something of a mentor, accompanying Bunny to explore the dense woods cloaking the steep hills behind the house. On one occasion Bunny led Hudson hoping to show him an unusual frog he had found in a pond on a private, neighbouring estate. The boy was nervous that they might be caught trespassing by the gamekeeper. Typically, Hudson scoffed at such fears, putting him at ease. He had yet to meet a keeper he couldn't disarm with his force of personality.

Arriving at the pond, Bunny recalled Hudson treading carefully and then waiting and watching silently, patiently, until eventually Hudson pointed, gently, and sure enough there was the mystery amphibian. Hudson was able to identify it as a natterjack toad, indeed an unusual find, even back then. Today, this species is limited to just a few localities in the UK.

Bunny recalled Hudson's curious attire – 'even to my nine-year-old eyes it looked strange' – of rough tweed coat with pockets in the tails, waistcoat and trousers that matched, starched high collar and cuffs and laced boots, even as they stepped quietly through the woods, looking and listening.[116] Hudson looked like a man from another place and time, and of course he was both. It is easy to understand Hudson's air of invincibility enhanced by this disarming dress: besides his physical stature and bearing he must have looked a bit like an apparition. He had many an encounter with a gamekeeper and trapper in the course of his rovings, but rarely a confrontation. Somehow, they usually found plenty to discuss.

A letter from Hudson to Bunny's father Edward Garnett links neatly with Bunny's recollection. In October 1904, Hudson was invited to visit Bird Society founder, sponsor, host and treasurer Catherine Hall at her Little Squerryes home at Westerham, Kent, 'to help her entertain her rough boys' for an hour.[117] At first I thought this must be a reference to her pack of rescue dogs (she had six, and a cat, according to Hudson's letter to Emma Hubbard in November 1904),[118] but it seems she had a role supervising some occasionally unruly children and Hudson was going to give them his bird talk. On another occasion he mentions having given a talk to the boys about 'England under the regime of the wolf'.[119] It is another slightly freakish coincidence that Catherine Hall, host of the first ever meetings of the Bird Society in Notting Hill and close friend of Eliza Phillips, also had a house so close to the Garnetts' in the Kent countryside. It would stretch plausibility in a work of fiction.

It was on Hudson's suggestion that Bunny come over to Hall's house, to ensure he could find his way back to The Cearne. 'His woodcraft and weapons would be guidance and protection in that wilderness dark',[120] Hudson explained, only half-joking. After lunch he went with 'the boy' to visit Miss Hall for tea, and to meet the dogs and cats. It was dark when they made the return journey, and there was no path. Bunny, in his own imagination a backwoodsman and pioneer, knew every turning and quite possibly every tree, ensuring they didn't get lost.[121] Hudson later reported to Emma Hubbard on their adventure, saying 'it was beautiful down there in the woods'.[122]

Young Garnett had other illustrious figures as playmates. When John Galsworthy visited, Bunny was chief of a tribe of Indians, and named his new recruit Running Elk. Old sea dog Joseph Conrad was similarly enlisted, in his case to captain a ship (a large wicker basket) when he popped round from his house, which was also nearby.

Bunny also recalled evening campfires at The Cearne, the adults sipping the Spanish wine that Hudson had brought – a Spanish Valdepenas that he had taken to, partly because of its low cost – it also helped Garnett senior to overcome his sleeplessness. It was superior to the Australian plonk on sale in Westerham's grocery store and, no doubt fuelled by this, their debates were sometimes animated. Bunny recollected a disagreement between the two men over Hudson's strongest suit as an author. Garnett had told him he was better at character description than writing about snakes and birds. If this was intended as bait, Hudson rose to it: 'I am a naturalist', he fumed, in the firelight, eyes blazing. 'I care nothing about people. I write about what I observe. I don't care for made-up stories that amuse people like you!'[123]

Was Edward Garnett serious? I don't think we can be sure. But perhaps he did at least know that any passable fiction would be a more lucrative line for Hudson now that he was established. And a line from a Hudson letter to Garnett indicates that while Garnett had some passing interest in wildlife and admired Hudson's communion with it, he was, in Hudson's verdict, 'more interested in people than birds'.[124]

Sometimes, as the embers of the campfire dimmed, the grown-ups would venture into the warm woodland night to look for fire-flies and nightjars. Bunny would creep along behind them, despite having been pointed towards bed by his parents. Leading the way, Hudson would imitate the long-drawn-out churring call of the nightjars, until the birds came close and displayed, clapping their wings and swooping around them in the darkness.

As a young man, Bunny for a time aspired to be a botanist. Hudson corresponded with his father on this subject at the time, and his letter is worth quoting in detail:

> I hope David is not going to confine himself to
> mushrooms and toadstools, or even to the entire
> vegetable kingdom, lest his fate should be that of the
> pale small boy full of wonder at the world and life
> described in Lord Lytton's *Botanist's Grave*. He sets
> out to find the secret of life and ends by being a 'great'

botanist, a writer of monographs which nobody reads
and are not intended to be read. That bitter poem of
Lytton's is one of my favourites.[125]

Hudson sent Bunny his children's story – *A Little Boy Lost* – even before
it was submitted for publication. Bunny recalls not liking it much, but
whether he said so at the time, and it had any bearing on the finished
article, is unclear.

Bunny's memoir also provides a rare eyewitness account of the
Hudsons at home. His mother had become friends with Emily Hudson,
and Bunny would accompany his mother to visit her at Tower House. He
remembered an occasion in 1906, by which time Emily was already 'a frail,
little old woman, with dark eyes, dressed in black'.[126] The visits were fairly
regular, and on one of them he recalled the genuine affection and concern
Hudson showed towards his wife, which contrasted with the distance he
maintained from others. 'It was clear that he loved her and wanted to
make her happy.'[127] Bunny and his mother also visited John Galsworthy
around this time, when he was living in secret with Ada, who was not yet
divorced.

I had another coincidental intersection with Bunny Garnett's homes
when in November 2019 I found myself in the small Cambridgeshire
village of Hilton, in a sound studio, recording a programme about Rachel
Carson's *Silent Spring* for BBC Radio 4. I discovered that just around the
corner was Bunny's former home, Hilton Hall, which has been described
as 'the most beautiful of all the Bloomsbury houses'.[128] Bunny bought
it in 1924 and it became a hub for Bloomsbury Group gatherings, with
attendees including Lawrence of Arabia, Dorothy Brett, Aldous Huxley
and Virginia Woolf. I was delighted to discover that it was also here that
Garnett entertained T.H. White, in a later life. I had been unaware of the
house's location while writing about White in *Looking for the Goshawk*.

In spring 2022, I gave a talk about Hudson to the RSPB group for East
Surrey. The next morning I set off early to look for Etta Lemon's former
house set in the wooded common at Redhill, before heading east into
Kent to see if I could find The Cearne. I walked through the woods for
a time, and hadn't anticipated what a spectacular setting this would be,
the wooded slopes commanding immense views to the south. Following
the map I had come across by chance in a Westerham charity shop,
I found a small sign marked 'The Cearne'. Two men clearing up a fallen
tree responded helpfully to my enquiry, and one led me straight to the
owner of the house next door, who in turn took me to meet the owners of

the Garnetts' former home. I could now picture those summer evenings around the campfire, Hudson perhaps recalling his former gaucho life as stories were shared. I resolved to return one day and retrace Hudson and Bunny's path (or one quite like it) through the woods to find Catherine Hall's former country home, back in the direction of Westerham.

While the traces of Catherine Hall are ghostly, and tantalising, and an impression is left of a quiet, self-effacing and determined philanthropist, Hudson was about to meet someone larger than life and very present in the historical records. Of Hudson's friendships, the one he subsequently developed with Margaret Brooke is the most improbable – and colourful – of all.

CHAPTER 9

Later Edwardians

Margaret Brooke, the Ranee of Sarawak

Even by the standards of the era, Ranee of Sarawak Margaret Brooke was a flamboyant figure, with an extraordinary back-story. She and her family provided a rich and steady source of headlines for the newspapers and comic material for the music halls. We may wonder today what a Ranee of Sarawak actually is – it sounds like an instalment of Harry Potter – but long before the Edwardian period the ranee and her estranged husband the rajah had become household names.

Brooke was desperately keen to meet Hudson. She had been in thrall to him ever since reading *Green Mansions*. Hudson and Brooke became an unlikely pairing, but when she set her mind to adding another artist, writer or musician to her friend collection, Ranee Margaret usually got her way. 'I was always ready to pounce on the possessors of any virtues admired by me in literature or any of the arts,' she proclaimed in her second of three memoirs. 'Please let me know you, for in so doing you would lift me nearer the stars' might have been her rallying cry, as she made her advances.[1]

Her description of meeting the socialist writer and artist William Morris at the house of the Pre-Raphaelite artist Edward Burne-Jones gives a snapshot of her modus operandi, but also her essential modesty:

> I longed to know him, having revelled in his *Earthly Paradise*. But he rather frightened me. I don't know why. I did try to talk to him, but our conversations never came to much. I felt that he was far removed from my simple remarks, in reply to which he used to look down at me, scratch his head and say 'oh'. I think I rather bored him.[2]

Luckily for the readily star-struck ranee, not all the celebrities of the day were such hard work. She could count among her very close friends Oscar

The fabled Ranee of Sarawak Margaret Brooke, society hostess. Hudson's early impression was of 'a great lady who pours herself out in her own extravagant way … a very fine woman'. Image courtesy of the Brooke Trust.

and Constance Wilde, novelist Henry James and composer Edward Elgar. Oscar Wilde had dedicated his fairy tale *The Young King* to her. But when she first read and then plotted to meet and ensnare W.H. Hudson she was about to enmesh her life with that of the cultural figure she would adore most of all.

Some biographers have it that Hudson was befriended by the ranee in 1911, but it was actually very much earlier, in June 1904.[3] She describes how her heart became set on meeting the mysterious author, although assured by those in society who knew of him that she was wasting her time, that he was reclusive and wouldn't meet anyone, and certainly not someone like her. 'I came to know him in a somewhat unconventional way', she reflected, after being 'enthralled' by Gabriela Cunninghame Graham's critique of *Green Mansions* in the *Saturday Review*.

> What the gifted lady had to say about it simply set me
> agog with excitement. I immediately telegraphed for
> it to Bickers, my bookseller, who despatched it to me
> forthwith. I at once began it, and can never describe
> my feeling on reading it except to say that I thought
> it one of the loveliest things I had met with in current
> English literature. So tender, so beautiful, so close
> to Nature!

Her years among the lush tropical forests of Sarawak were evoked for her once more, and she was sent in to rhapsodies:

> You could almost scent the fragrance of tropical
> vegetation as you went through its pages. Here
> and there the cruelty and ugliness of flesh-eating
> individuals; and then the purity, simplicity and divinity
> of Rima – half bird, half woman! I had to stay awhile
> over the poignancy of its last pages.
>   I could think of nothing but *Green Mansions* for
> days. It obsessed me. I read it over and over again and
> determined that I must somehow or other meet face to
> face the author of this wonderful book. I inquired of all
> my literary friends, 'Do you know the author of *Green
> Mansions*?' No, I found no one who did.[4]

She was staying with a friend, when fate intervened:

On the last evening of my visit I was rather bored to
find that a lady whom I did not know had come to
dine. When we went in to dinner she and I sat on
either side of our host. I was just about to dip my spoon
into the soup when something made me aware that the
newcomer was a friend of Hudson. My spoon still in
the air, I caught and fixed her eye.

'You know W H Hudson,' I said, 'the author of
Green Mansions.'

'Yes, I do,' she replied in rather a surprised tone,
'but how did you know I did? He is such a retiring man
and holds himself aloof from everyone.'

'It came to my mind that you did. And now that
you do know him, do please arrange that I may know
him too.'

'That would be impossible,' she replied, 'he will not
meet anybody.'

I thought it better not to wrangle over the dinner-
table, so said no more; but later, when the ladies were
alone in the drawing-room, I returned to the subject.

'Look here,' said I to the lady, with my best smile,
'you really must manage somehow that I get to know
Hudson.'

'Well, if you really mean it,' she replied (somewhat
sillily, I thought), 'you would have to come to tea at my
flat and pretend that you were staying with me. I could
send you a wire next time I am expecting him to come
and see me.'

She was a kind lady and I gladly fell in with
her plan. Next day she went back to London and
I returned to Ascot. About a week later I received a
telegram and hied myself off to London. Arriving at
the lady's flat about half-past three, I took off my hat
and tried to look as much at home and as innocent
as possible. As the clock on the mantel-piece chimed
the half-hour after four, Hudson arrived. On seeing me
he frowned slightly until it was explained to him that

> I was staying with my hostess, having come to London
> to consult a doctor.
>
> 'You are Rajah Brooke's wife,' he said. 'Have you
> lived out in Sarawak?'
>
> 'Yes,' I said.
>
> And after that he and I got on like a house on fire.
> We actually made plans for future meetings![5]

Her exotic credentials helped her overcome Hudson's fabled reserve, as did her avowed love of birds, and perhaps her enthusiasm for the atmosphere of his romance of the tropical forest. But it is evident that Margaret Brooke had so much effortless warmth, vivacity and charisma that few could resist her charms – not even Hudson at his gruffest. His early impression was of 'a great lady who pours herself out in her own extravagant way'.[6] 'A very fine woman … She said all kinds of wild extravagant things in praise of *Green Mansions* and is a great lover of nature and wild creatures,'[7] as he told Emma Hubbard. But he was soon sufficiently at ease in Brooke's vibrant company to set aside his customary reticence. It was the start of a friendship that would last till the end of Hudson's life.

Ranee Margaret had overseas credentials to more than match his own. 'Until I married, my real name was Ghita, culled by my French friends out of Marguerite,' she must have told him, and as she wrote in her memoir. 'I hate the name of Margaret – it is ugly.' I haven't seen Hudson use this nickname for her, but he often called her 'your highness', and referred to her as 'the Ranee'. She was born in Paris in 1849, in suitably melodramatic fashion, when her mother, caught up in the excitement of Rossini's *William Tell* at the Opera House, went suddenly into labour.

The drama of her aristocratic family began earlier, when they had to flee their chateau during the upheavals of the French Revolution. Her great-grandmother was an infant at that time and was smuggled out of the country to Holland, returning years later to reclaim the property:

> My parents reigned supreme over their lands and
> chateau at Epinay … During the first years of their
> married life my father and mother lived with the old
> lady … France has always remained for me my most
> beloved land … not that I do not realise that England
> is the greatest country in the world and that English

men and women are in their character, their loyalty,
their truth, above all others. But to me the French are
so lovable in this sense, that if they take to you they
will talk to you from their hearts. They are so entirely
free from that English reserve which almost amounts to
a bugbear – the fear of *giving themselves away*.[8]

Ranee Margaret had no such fear, and I think she relished Hudson's
sometimes brute frankness. After that first meeting they reconvened
just two days later, at the Bird Society's offices at Hanover Square,
where Hudson showed her bird books, and pictures of birds. 'Thus
began another lovely friendship', she recalled, with evident fondness.
She thought him very handsome, and 'an outstanding representative of
English literature'.[9]

Tall and spare, he was one of the most active men
I have ever met. His face was long and thin with high
cheek-bones, his nose was aquiline, his beard and
moustache short, while his piercing eyes, dark and
beautiful, reminded me of an eagle. Like so many other
really great authors, he was very modest, and appeared
to be unaware of the excellency of his writings.[10]

I love Ranee Margaret's colourful account of plotting to meet him,
although a Hudson letter to Eliza Phillips that I came across later, from
June 1904, indicates that he had heard that Lady Thompson wanted to
introduce him to the ranee (or 'Rawnee Brooks', as he labelled her at this
point).[11]

Hudson, always drawn to those with lived experience of the wider
world, will have been eager to know more about the fabled kingdom of
Sarawak, a vast chunk of Borneo gifted by the Sultan of Brunei to James
Brooke, an explorer and diplomat who had helped the sultan to quell a
local uprising. As reward for this he became the first Rajah of Sarawak,
ruler of a territory the size of a large English county, with 500 miles of
coastline. It would be extended over time and grow to the size of England
itself. James was the uncle of Charles Brooke, Ranee Margaret's husband,
who had inherited the kingdom.

Ranee Margaret no doubt related to Hudson many of the less auspicious
details about life in Sarawak that couldn't be printed in the newspapers.
Her now estranged husband had inherited Sarawak because his uncle was

sonless. Now in his forties, he was all too aware that if he wanted an heir of his own he had better do something about it. Noticing that his 40-year-old, land-owning cousin had been widowed, he came to England with the intention of proposing to her.

While staying with his cousin in Wiltshire, he decided that her 18-year-old daughter Margaret might provide more robust breeding stock and stand a better chance of adapting to life in the tropics. Despite having barely uttered a word to young Margaret in all the days of his visit, he proposed to her by scribbling a brief note. Ranee Margaret was under no illusions: 'he had married me, firstly and lastly, because I was young and very healthy', she later reflected, 'and what he wanted above everything was an heir'.[12] Bored with her life she was prepared to go against the wishes of her mother – who heartily disapproved – and take the leap.

Her honeymoon and subsequent married life took lovelessness to their logical extreme, but Ranee Margaret remained philosophical throughout. She crossed the globe to a new life in the Far East, bore three children in rapid succession and then tragically lost all three to cholera while taking them on a visit home to England. She promptly provided three more male children – an heir, a spare and a thespian.

Her principal duty completed, and the children in school, Ranee Margaret left the rajah, whom she thought was bored with her, and was once again living in England. She could now enjoy the high life of a society hostess, with homes in London, Ascot and Cornwall, and a chauffeur-driven car to take Hudson out in. 'It is to me a great delight to travel with you in your car', he told her, while regretting on occasion the wider impacts of motoring on the well-being of society. He liked that it gave 'a better sight of the long road than one has in a railway carriage'.[13]

The pair enjoyed their own golden age in the latter half of the Edwardian decade, with many eventful days spent exploring Cornwall from her Badger's Holt cottage base in Lelant, the 'violet village'. Her love of birds extended to keeping parrots and macaws among her menagerie of pets, and Hudson had little difficulty cajoling her into getting involved with the Bird Society. He would invite her to the annual meetings: 'as you know the Duchess [of Portland] you would be obliged I suppose to ascend to that eminence [the platform]', he mentions to her in one letter.[14] He also lost no time in asking her to enlist the support of the rajah in the plumage campaign, no doubt to stem the flow of spectacular plumes and bird-skins from Borneo. Hudson drafted one letter for the rajah that he shared with Eliza Phillips in August 1904.[15]

Ranee Margaret may have added spice and a dash of decadence to Hudson's life, but it was the next generation of her family that would be the real eye-opener.

The Grey Friars Orchestra. Ranee of Sarawak Margaret Brooke is at the piano, flanked by the seated Brett sisters, Dorothy (left) and Sylvia front and centre, both holding drumsticks. 'She could dominate a room with her personality and her magnetic eyes, and enchant everyone in it,' Sylvia said of Ranee Margaret.
Image courtesy of the Brooke Trust.

Fun and games with the Brett sisters

In the 1890s, when her marriage to the rajah had run its course, her Sarawak adventures behind her and now ensconced in her London apartment, Ranee Margaret realised she needed a bigger house in the country where her grown-up sons could come and go. Her eldest son Vyner was being groomed for his role as the next rajah, second son Bertram was serving in the army and youngest Harry was dabbling in theatre. Ranee Margaret's estranged husband agreed to rent (and later buy) for her Grey Friars, a large mansion among the heathland pine plantations of Ascot in Surrey.

After a time, concerned that her boys were having difficulty relating to or even meeting members of the opposite sex, Ranee Margaret settled on a plan to form the Grey Friars Orchestra, an all-female ensemble. Her base in the Surrey hills might have seemed remote from the action and a difficult place to recruit band members, but it was surprisingly well positioned for some eligible heiresses. 'My neighbours' houses harboured many charming damsels,' she recalled, 'lovely young girls who philandered and flirted – and quite right too – with my sons.'[16]

Prime targets were Dorothy Brett and her younger sister Sylvia, the daughters of Reginald and Eleanor Brett, aka Viscount and Viscountess Esher. The viscount was an important figure, responsible for organising major state occasions such as the funeral of Queen Victoria and the coronation of King Edward. The Bretts lived 4 miles away, beyond the orderly ranks of pines, in a house called Orchard Lea, to which address Ranee Margaret wrote to the viscount with details of her scheme.

Resistant to her charms, Viscount Brett was strongly against the idea, but the viscountess persuaded him to allow their girls to join in. The Brett sisters had led sheltered lives till then, so the weekly cycle to Grey Friars opened up new worlds, not least because Ranee Margaret liked to relate exotic tales of Sarawak, of her life among the fabled Dyak head-hunters of Borneo, and of the famous artists and other public figures she knew. Rehearsals were twice a week, and Sylvia Brett recalled that Ranee Margaret 'would often talk about her life in Sarawak' and even shared details about her estranged husband: 'their marriage had ended when he destroyed her beloved pet doves and served them in a pie for her supper.'[17] Ranee Margaret stopped short of disclosing this detail in her memoirs.

In Grey Friars, and its charismatic owner, the Brett sisters (Dorothy in particular) had found a sanctuary from the often perverse and unpleasant atmosphere of Orchard Lea, where their father maintained 'the presence in the household of a series of young men whose role was not well-disguised'. And Dorothy seems also to have suffered the unwanted attentions of at least one notorious friend of her father.[18]

In sharp contrast, Grey Friars and its hostess offered an environment in which they could express themselves. Sylvia recalled the large macaw with scarlet wings that greeted visitors in the entrance hall. In the room beyond, the ranee would sit regally on a high-backed chair or recline in a blue armchair, a green parrot on her wrist, telling her tales and extolling the virtues of her sons to her rapt young audience. Sylvia recounted:

> There was nothing very feminine about her, either in
> her manner or in the way she decorated her ordinary
> and almost ugly home, but she could dominate a

room with her personality and her magnetic eyes, and
enchant everyone in it … Her mind was never still,
and all the time her lovely blue eyes compelled us to
adore her. She was like Mary Queen of Scots, in her
love of being adored.[19]

In the orchestra – 'all of us extremely bad' – Sylvia banged the bass drum
and Dorothy the side and kettle drums. Ranee Margaret was soon forcing
the pace on the match-making front, frustrated at the lack of tangible
progress, and pressed her eldest son Vyner to propose to Dorothy. She
gave him a necklace with which to back up his offer when he popped
the question. One day after rehearsals, as Dorothy lingered in the music
room, Vyner was pushed in through the door to do the deed. He managed
to hand over the necklace, but words failed him, and he stumbled back
out of the room, flustered. It was no bad thing as, in a twist to the plot,
Dorothy – who had little reason not to find men repulsive – was secretly
harbouring an adolescent crush on his mother, not him. 'I can see my
sister, Doll', wrote Sylvia, 'her great dark eyes fixed on the Ranee's face
like a girl in love.'[20]

Vyner now turned his attention to Sylvia, and eventually summoned
the courage to sidle up and offer to tune her drum. Sylvia's good luck
was her bandmates' disappointment. 'Nineteen hostile faces glared at
me,' she recalled, having been the chosen one, 'and wished I would
drop dead.'[21] Vyner was quite a catch, and who – having heard such
enchanting stories about the place – wouldn't want to be the next Ranee
of Sarawak?

Sylvia didn't get much mail (only *Peter Pan* author J.M. Barrie and
Irish playwright George Bernard Shaw were writing to her at this time, she
lamented), so when Vyner's love letters started to appear, Viscount and
Viscountess Brett cottoned on to the hidden agenda of Ranee Margaret's
scheme, and that her son and Sarawak heir had designs on their daughter.
Sylvia was promptly packed off to their shooting estate in Scotland. They
already thoroughly disapproved of Dorothy's escalating fixation with the
so-called ranee. It confirmed all of their prejudices, and they became
increasingly insulting about Sarawak and uncomplimentary about Grey
Friars and its owner.

Some of Dorothy's letters to her Viscount father survive, and in them
she reveals her strong adolescent feelings for the ranee, and her shock at
the 'disgusting insults' issued by her mother, Viscountess Brett, towards
her mentor.[22] 'The Ranee is my friend and I must be loyal to her … I can't

forget all her kindness to me … she is the greatest friend I have … she is everything to me … I will never give up trying,' she pledged.[23]

Dorothy was grounded by her parents and forbidden to see the ranee, even on the rare occasions Margaret visited Orchard Lea. It only served to make Dorothy even more 'obsessed by Margaret Brooke'.[24] Dorothy remained infatuated, and found various ways of defying orders not to go over to Grey Friars. Ranee Margaret knew that the visits by Dorothy had become problematic. Piecing together the different accounts,[25] it seems that the ranee was summoned by the Bretts to a meeting at Orchard Lea and, while Dorothy was locked in her room, agreed that Dorothy's visits must stop. It is clear from Dorothy's notes later in the year (1906) that she had not been able to see the ranee for three months, which the younger woman had found 'rather hard'.[26]

Unable to bear it anymore, Dorothy resumed her visits to Grey Friars. Finding her way barred, she took to clambering in through open windows and hiding in cupboards. Although she had obviously been very sympathetic to Dorothy, the ranee had to write to Viscount Esher about her behaviour. 'Doll turns up everywhere. I never know where I shall find her. The other day when I opened the little trap that leads from the kitchen to the dining room … there, smug and smiling was your daughter.'[27] It is also said that on one of these uninvited visits Dorothy ended up chasing her with a carving knife.[28]

This was the emotionally charged world of the ranee and the English upper classes into which Hudson walked when he first visited his new friend at Ascot. Although no trace survives of his perspective on it all, he seems to have become embroiled in a bizarre triangle, of which he may even have been blissfully unaware.

Ranee Margaret's memoir has kind words for Dorothy and no mention of this real life pantomime. She prefers to recall idyllic summer days spent with Dorothy and Hudson in Wiltshire than the capers at Ascot. 'We spent some happy weeks during the late summer in the lovely little riverside village called Wiley [Wylye],' she wrote. 'Hudson often came over to see us, as he was staying only a mile or two away. We all three used to go for walks.'[29] Sometimes, according to Dorothy Brett biographer Sean Hignett, Hudson and Dorothy would walk or cycle together. For her part, Dorothy was in awe of the man, noting in her diary that she was 'frightened of Hudson – frightened of anyone intellectual',[30] while the venerable naturalist did his best to nurture her interest in and knowledge of the wild nature around them.

Ranee Margaret relates another extraordinary anecdote about one of those country walks with Hudson in Wiltshire. Dorothy wasn't there to witness this incident, having gone to Salisbury. While walking in a meadow, Hudson and Margaret met several horses – hunters – and as they patted one of them it seems that Hudson had a rush of blood to the head, suddenly announcing 'I shall get on him!'[31]

To her astonishment, he mounted using the animal's tail, in true gaucho fashion. The stallion, taken aback by this unexpected intrusion by a stranger, reared and bucked and kicked out. All the while, according to Ranee Margaret, 'Hudson sat like a centaur, lightly handling his mane. Then he was borne away, round and round the field until his hat fell off.'[32]

Having circuited the meadow a few times Hudson brought the horse to a halt, calmly dismounted, collected his hat and re-joined his astonished friend. She was speechless for some time, until finally able to whisper, 'Do you often do that sort of thing?'[33] Perhaps alarmed by his own impulsiveness, his response seems curious:

> 'What a question!' he snapped; 'why bother about it! Please don't allude to the matter again.'
> And there the subject dropped. I remembered that as a boy he had been brought up in the pampas, not far from Buenos Aires, where once upon a time his father had a ranch and where I knew that youths were accustomed to break in wild prairie horses in the saddle. But, after all, it was a wonderful sight to see![34]

Given his apparent reluctance to ride horses in all his years in England, the story seems entirely out of character – yet plausible. Hudson was capable of this, even at this advanced age, but while he knew better than most how to handle a horse, for some reason he seems almost never to have ridden one since he left the Pampas, nearly four decades earlier. We only have Ranee Margaret's version of this story, as for some reason Hudson never mentions it.

Younger Brett sister Sylvia, who would become the next ranee, later wrote a memoir, *Queen of the Head-Hunters*, somewhat less constrained than Ranee Margaret's (the title of the book provides a clue), and with a racier recollection of the events of this period. Dorothy, perhaps as a result of jealousy, was of the opinion that Ranee Margaret was in love with Hudson – 'tall and thin, with a big hooked nose and greying hair'[35] – and thought that he 'withstood the Ranee's advances by ignoring them'.[36] Was Ranee Margaret really suppressing a secret passion for Hudson? We may never know.

There is another plot twist. Had Dorothy now switched her affections to Hudson? Her biographer Sean Hignett wrote that Ranee Margaret began to suspect that Dorothy 'was trying to steal Hudson from her. The matter reached a head one evening when the Ranee, in a fit of temper, hurled a sliver statue of St Christopher at Doll and ordered her out of the house. The friendship and crush that had lasted almost a decade came to an abrupt and messy end.'[37]

It is recorded that Viscount Brett eventually sent Dorothy away to Scotland to keep her out of trouble, like her sister before her, and it is at this time that she wrote to Hudson, from the family's sporting retreat at Callander (close to Gartmore, as it happens) in September 1910, offering to send him a grouse and a speckled trout, conscious all the while that he might 'hate' such gifts. Hudson was polite and pragmatic enough to say he would accept them.[38]

Things may have been a bit strained when, on 21 February 1911, despite the long-running disapproval of her parents, the wedding day arrived for Sylvia Brett and Vyner Brooke, after some years of plotting elopements and then losing their nerve. Although presumably in favour of his son and heir securing a wife, Rajah Brooke was no great fan of such occasions – nor of the Eshers – and returned from Sarawak to make a grudging appearance at the wedding reception at Orchard Lea. Having shown his face, he couldn't get away fast enough, and made a dash for the exit door, which in his haste he couldn't find. 'How the hell can I get out of this damned house?' he yelled at the first person he met,[39] without realising he was talking to the father of the bride, Viscount Esher himself.

Wedding presents flooded in, of varying scale. While Hudson gave the happy couple a signed copy of his *Idle Days in Patagonia*, King George and Queen Alexandra sent a pearl and diamond brooch. The newspapers were full of the story of the new glamorous young rulers for Sarawak. 'Queen of the wild men of Borneo', ran the headlines about Ranee Margaret's successor, heralding a new era for that distant kingdom. While very few photographic traces of Ranee Margaret survive, Pathé News footage survives of the new rajah and ranee performing their royal roles.[40]

There was also a happy outcome for Ranee Margaret's second son Bertram, who struck lucky while watching the Greyfriars Orchestra perform. His eye was caught by Gladys, triangle player and heiress to the Huntley and Palmer biscuit empire. They were married soon after.

Dorothy Brett had also found happiness, and her calling in life, and had enrolled at the Slade School of Art. She stayed in touch with Hudson for some years afterwards, which I think is noteworthy as she was still uneasy with men in general. At Christmas in 1911 she sent him a silk scarf, and wanted to see him, whether in London or at her studio. He responded to 'my dear Dolly', thanking her for the gift and explaining that he couldn't see anyone as he couldn't leave his wife unattended, as by this time Emily's health was beginning to fail.[41]

Corridors of power

As Hudson was establishing himself as part of the loosely constituted Mont Blanc luncheon club, Sir Edward Grey was moving in a separate literary and intellectual circle, with a proper name: it was called the Co-Efficients by its ringleader Sidney Webb. In his 1958 book of essays *Portraits from Memory*, philosopher Bertrand Russell recalled that while there were about a dozen members of this group, including H.G. Wells, 'the most distinguished was Sir Edward Grey'. Russell and Wells were out of step politically with the rest, whom Russell calls imperialists, and who, he said (perhaps a little simplistically for so great a mind), 'looked forward without too much apprehension to a war with Germany ... Matters came to a head when Sir Edward Grey, then in opposition, advocated what became the policy of entente with France and Russia, adopted by the Conservative Government some two years later, and solidified by Grey when he became Foreign Secretary.'[42] Russell and Wells argued fiercely against this with their fellow Co-Efficients, although Russell recalled that by 1914 Wells would be persuaded of the necessity of war against Prussian militarism and the supposed 'war to end all wars'. Despite their shared acquaintance with Grey, and Joseph Conrad (of whom Russell was said to be extremely fond), I can trace no direct links between Hudson and Russell.

In their regular get-togethers, Sir Edward Grey and Hudson may have had a policy of not discussing politics, partly because Grey most enjoyed their chats about nature, but also because he needed time out from 'talking shop'. There were certain issues on which they differed fundamentally. As a 'Liberal Imperialist', Grey for example supported British action in the Boer Wars, which put him for a time at odds with many in his party – and Hudson, had they ever discussed it. Hudson was not in sympathy with this British military action, which did little to enhance the country's reputation abroad, even if it had done the Conservative government no harm in the so-called khaki election of 1900.

Five years later, Grey's Liberals won the 1905 general election by a landslide. Sir Edward 'reluctantly' took office in December, as foreign secretary in the new government led by Herbert Asquith. Of course, Grey regretted that this would reduce even further the time he could spend in nature. His days would now be taken up with fostering friendly relations with the USA, creating alliances with Japan and Russia, and the Anglo-French entente. The latter was soon tested by German interest in Morocco. Not long in office, in January 1906 Grey authorised military conversations with France but didn't tell his Cabinet colleagues. It was a taste of something that would surface later in his ten-year term, with wider ramifications. But fate was about to intervene. Just weeks into Grey's tenure in high office, personal tragedy struck.

In early February 1906, news of an accident reached the Hudson household from the Greys' Fallodon estate. Lady Dorothy had been riding in her dog cart, pulled by a horse, and somehow it had overturned. She had sustained a serious head injury. Grey had been interrupted in a meeting at Westminster to be given the news, and he had rushed north to be by his wife's bedside. She remained unconscious for several days.

Hudson feared the worst, and he confided in Edward Garnett on 4 February:

> The last two days I have been miserably anxious about
> poor Lady Dorothy. Her death would be a frightful
> blow to him [Grey]; every good he possesses and
> whatever else the world may have for him would be as
> much dust and ashes. No two that I have ever known
> were more like one. But it would be just as dreadful if
> she were to recover with the loss of something in her
> character which made her what she was. I could not
> wish for her to come back from that dark place where
> she is in such a case.[43]

Despite all efforts to revive her, Lady Dorothy never regained consciousness. Grey was shattered by the shock of her death. Hudson too was devastated: his respect for her had been deep. 'She was a glorious woman and it was horrible that she should have been crushed out of existence in that way,'[44] he told friends. He wrote a tribute, with Sir Edward's blessing, which was published in The Speaker and reprinted in the Bird Society journal Bird Notes and News.

Hudson recalled time spent with Lady Dorothy, and one incident that exemplified her character: 'she was resting in a canvas chair one evening and we were indoors deep in talk when we were all at once startled'. Dorothy had seen a man snatch a young song thrush from a hedge and stuff it into his pocket. Incensed, she challenged him, and demanded that he let the bird go. He complied, before sloping off. 'In that moment,' Hudson recalled, as he and Edward rushed to see what the commotion was about, 'something new had been added to a face always gracious and noble.'[45]

The Greys' cottage nature diary would see no more new entries. Edward closed the book, but resolved to use it as the basis for a commemorative edition for his beloved wife. He contacted Hudson and other friends, requesting to borrow any letters and mementoes she had sent to them, for possible inclusion. But Hudson was already in the habit of burning correspondence, including most of the large quantity he had received from Dorothy: 'all but twenty or thirty which have survived by chance',[46] he told Garnett. To modern eyes, a pile of this many letters might seem like ample contribution to such a project. We may imagine how many letters had been received and destroyed.

In 1909, Hudson was the recipient of one of the limited edition of 20 copies of the tribute. Sir Edward said he 'hoped the book might help friends of Dorothy to keep free spaces in their lives in which they could find things which cannot be found in work or in society'.[47] Far from derailing him, the trauma of Dorothy's death had the effect of propelling Grey full-tilt into his political career. It seems to have marked the end of Hudson's visits to the cottage, but he remained close friends with Grey and they would continue to have regular and occasional meetings, Grey always relishing their diverting chats.

Not long after his two terms of office as US president had ended, Theodore Roosevelt embarked on a world tour that would culminate in Britain. He wanted to go birdwatching while there, and an arrangement was made that Grey would be his guide. He told Grey that of all his appointments on this world tour this was the thing he looked forward to most, besides his Oxford lecture and a safari in East Africa.[48] In June 1910, they followed the river Itchen route well known to Hudson, whose influence was felt as Grey was able to identify for his guest 40 different species and the songs of 20 of them. Roosevelt pointed out which of the birds he thought had relations in America, and how the songs compared.[49] Of course they had politics and international relations to discuss, otherwise Hudson might have been invited along.

Grey later reflected on the visit. Overall, he said, 'it was a most delightful day: I enjoyed it immensely and so did Roosevelt'.[50] He had thought Roosevelt's speech at the Guildhall 'a very great compliment to our work in Africa … I knew there would be a row about what he proposed to say with regard to Egypt, and I told him so.'[51]

Roosevelt was about to become a major factor in Hudson's conquest of the American book-buying public. Hudson was going west; how far he would get remained to be seen.

Sanctuary in Cornwall

Considering Hudson's love of the rural environment, and his often fragile health made worse by smogs and bad air generally, it is remarkable that he survived for nearly half a century in London. But while he never moved out permanently, he did find a home from home in the far south-west, which may have helped him to live for as long as he did. He had resisted earlier encouragement from friends to go there, and so discovered Cornwall later than he might have done. In finally doing so, he probably added those extra years to his life. It became his sanctuary, especially in winter, allowing him to escape from sooty London, about which he continually grumbled. In 1904 he lamented 'a literally black fog causing suspension of traffic and unutterable misery to every one … Such fogs we have not known for years.'[52]

The following year, he was writing from St Ives on Cornwall's north coast, his tolerance for London winters having finally snapped: 'I found I couldn't work in that nasty cold grey desolate atmosphere. I suddenly rushed away for a month's visit to these parts, and am very glad I did. Out of London fog I feel a great relief,' he wrote, revelling in the gusts of Atlantic air.[53]

From scrutiny of his surviving letters, I sense that it was Ranee Margaret who finally persuaded Hudson of Cornwall's merits. She had a cottage at Lelant, and was often the life and soul of the community on her sojourns there.

Cornwall could take some getting to. Hudson described to Garnett the misery of one train journey west in 1906 when there was a breakdown on the London Underground Metropolitan line, then no cabs to be had, then no seat on the train until it reached Bristol. His habit became to break the long haul with a night or two in Exeter, Devon, before catching what was known as the 'Riviera Express'. Occasionally, Ranee Margaret's chauffeur would drive them all the way there. Hudson could then enjoy

A precious and previously unpublished 'snapshot' of Hudson and Ranee Margaret Brooke in the garden of her Lelant cottage (Badger's Holt) in Cornwall. The small dog under Brooke's arm may be Koko, mentioned by Hudson in a 1916 letter: 'his attachment to you inclines me to think better of his intelligence than I do of most dogs'.[54]
Image courtesy of Jason Brooke and the Brooke Charity.

local excursions from the back seat, although the car struggled on steeper hills. Hudson described on one occasion having to get out and walk (if not push) while taking the opportunity to identify interesting wildflowers along the roadside.

The mild temperatures and fresh air of the south-west came as a revelation to the often sickly man, now approaching his 70th year. He found that even on a December day he could sit at the window of his lodgings on The Terrace in St Ives, with the window wide open and sea breezes stirring the curtains. It was not unknown for butterflies to still be on the wing. And while it was wetter beside the ocean, he could more than cope. 'It has been raining all the time and everyone exclaims against such detestable weather. I tell them to go to London and try that and that it is a million times better to be here in the rain than in that dreadful place.'[55]

He described St Ives as a 'little old huddled town, the small harbour and fleet of over a hundred fishing boats going out at sunset and returning at dawn, and the vast congregation of gulls that form a whirlwind of white wings about and among the boats each morning'.[56] As usual, he lodged with local people. 'I was lucky enough to get into a house of pure Cornish folk,' he reported of his early hosts, 'mother and daughter; the father a mining engineer away in Alaska.'[57] Asked why he always tried to lodge in humble surroundings, he replied: 'the hotel atmosphere would drive me mad in a week'.[58]

One of the fishermen he befriended brought Hudson some freshly caught pilchards for breakfast and supper, as he had never tasted them before.[59] Hudson thought the locals 'utterly unlike'[60] those of the other English counties with which he had become familiar. There were plenty of incomers too. He reckoned there were 40 artists squeezed into the little port, holding their weekly gatherings, including his close friends Alfred and Alice Hartley. While Ranee Margaret relished such events, and continued to gather artistic types around her – Hudson remaining her favourite of course – he remained selective, still shy of social occasions, as a rule. 'I was invited but declined, as I think one can get too much of picture talk. I can at all events.'[61]

Hudson loved the storm-tossed sea-cliffs. He describes in one letter of late November 1905 waiting for the storms to abate a little before exploring the wild, dramatic clifftops towards Land's End. He mentions the fishing fleet venturing out the evening before, during a lull, but being forced back before midnight when the winds picked up again, with a man lost overboard. In later years, if stuck in London in midwinter, he found himself 'longing for the pure fresh air and granite cliffs and blue seas'.[62] The Cornish landscape reminded him of Patagonia. 'It is a rude harsh land but attracts me and I daresay anyone who loves wildness. But the people are not rude … I should like to know it better.'[63] To this end, he stayed for over six months on an early visit, until the end of June 1906, and was soon hankering to return in late summer, in time to see the heather flowering. When he did so, Hudson took a train to Helston, and 'engaged a man and trap to drive me to St Keverne over the Goonhilly Downs just to see the Cornish heath in bloom'.[64] Before long he was having his first thoughts of writing about the county: 'I think by going on writing papers on that part of Cornwall I could make up a book in time.'[65]

Although generally not as severe as they had been in the mid- to late Victorian period, winters could still be much harsher in Edwardian times than they are now. Sometimes even Cornwall suffered. 'The great snow and frost at Christmas drove the bird population of southern England to this end of all the land,'[66] Hudson wrote. He witnessed bird flocks fleeing the brutal conditions to arrive exhausted and starving along the coast, without the energy to attempt the sea crossing. He heard descriptions of the fugitive birds lining windowsills, many of them nearly frozen stiff. Some cottagers brought the moribund birds indoors, but invariably found them dead in the morning. When at last the brutal winter weather relented, Hudson gave thanks for the 'blessed rain that's making the frost-caked earth green again and allowing the remnants of the birds to enter on a new lease of life'.[67] He might have been talking about himself.

Many birds also fell victim to teagling, a local practice in which 'each person buys a handful of small fish-hooks, manufactured for the purpose and sold, a dozen for a penny, by a tradesman in the town … Ten to twenty baited hooks are fastened with short threads to a string,' he explained about these bird traps, which he noted were set by 'the most religious people in Britain!'[68] Some of the locals implored him to help: 'we can do nothing', they assured him. 'They abuse us because we forbid them putting their traps and hooks on our ground – but YOU can perhaps do something.'[69] Hudson was soon on the case.

Hudson knew what he might be letting himself in for if he took a stand. He had been told about Mr Ebblethwaite, who came from the north to live out his late years in Cornwall, and campaigned against cruelty to gulls. 'Finally he succeeded in getting a certain number of boys summoned for cruelty before the magistrates,'[70] Hudson was told. And although no convictions were possible, as there were no regulations in place, and he made himself unpopular, and was jeered at and shunned by the community, some of the local people supported his views, and the culture changed. 'In a little while it came to be understood that, law or no law, the gulls must not be persecuted.'[71]

The Land's End, Hudson's evocative book about the county, was published in 1908. Once again he was courageously frank in his views, and in the book he dared to call for an end to the teagling. 'There are some who are revolted by it', he reported. 'A few of these have begged me to do or say something to put a stop to these disgusting barbarities.'[72] The controversy was soon picked up in the local and national press, and he found himself again in the spotlight. 'The St Ives people are very angry with me for exposing their bird-torturing practices, I see in today's *Telegraph*,' he told friends. 'I'm very glad they are angry: perhaps they will now mend their ways a little.'[73] He was invited to a debate in the Cornish village of Camborne. His would-be hosts, he was assured, 'will be delighted to meet me face to face and hear what I have to say to justify my charges against the people of Cornwall. They want me to know what Cornishmen really are … They have a name for roughness; they very nearly killed Will Thorne [union leader] when he went to meet them face to face, and I shall have to ask you to come with your gun to back me up.'[74] Fortunately – or this story of Hudson might be ending sooner – he was persuaded not to attend.

Notwithstanding its occasionally unflattering content, *The Land's End* was serialised in local paper *The Cornishman*, despite the editor having been one of the most prominent critics of Hudson's book. And legislation was rapidly brought in to ban teagling. It was probably Hudson's quickest political win for nature protection. Meanwhile, other campaigns were being measured in decades rather than weeks.

His book cost him at least one friendship too, Hudson reporting that 'we were great friends until my book on the *Land's End* came out and then he dropped me like a hot potato but let me know by writing to a friend of mine that he was unspeakably disgusted'.[75] The same lapsed friend later wrote him a flattering letter hoping for a testimonial – unluckily for this correspondent, Hudson hadn't forgotten.

It's fascinating how Hudson was prepared to confront the occasional harshness of the locals and the climate in the west. The same cannot be said for the north, which is a whole other story.

CHAPTER 10

Wider Horizons

The North

The seabird cliffs of Yorkshire remain arguably England's most memorable places for the nature lover to visit. It isn't surprising that they occupy a pivotal place in the history of bird conservation in Britain. As Hudson became more upwardly and outwardly mobile, and perhaps as new chairman of the Bird Society thinking that it was a gap in his experience that he had to fill, in late June 1894 he embarked on

Emma Lawes and Hudson's Wiltshire shepherd James Lawes, to whom he gave the alias Caleb Bawcombe, dressed up to be photographed in 1902. 'I never saw any one look so unhappy at an operation,' Hudson wrote. Image courtesy of Angela McAllister.

'birding expeditions' first to the Norfolk Broads, then to Bridlington by boat, then Flamborough Head, then on to Sunderland, Bamburgh and the Farne Islands. Not surprisingly, he enjoyed a 'good week' doing all this, and saw 'tens of thousands of guillemots, puffins, razorbills, cormorants, gulls, terns'.[1] He couldn't fail to be impressed by the astonishing numbers of birds on the islands, packed so close together it was almost impossible to move among them.

He also visited York for a day, principally to see the minster, and Monks House. 'When I spent a little time in York I was mostly interested in the swifts, they were in such numbers.'[2] He mentioned this many years later, when writing in 1921 to Morley Roberts who was in York at the time. It being summer, Hudson envied him the relative cool there, 200 miles north of London. But overall he had curiously little to say about this one earlier visit to the far north of England, and he seems never to have repeated it.

'Somehow I should not be able to settle down at Wadhurst [East Sussex] and live with a contented mind. I like open worlds – downs or great moors like the Cornish or Yorkshire ones,'[3] he wrote in 1906. What's odd is that if he had been to the North York Moors he seems to have left no other trace of the visit. Notwithstanding his claim to like moors, if there was a field guide to Hudson, complete with British Isles distribution map based on available data, it would show him as confined to southern latitudes and lower altitudes, and mainly pastoral habitats (with occasional forest forays), with what would look like just the occasional overshoot taking him out of this comfort zone.

Although he never went back, Hudson retained his impressions of Northumbria until late in life. Writing in August 1919 to Mary Trevelyan, a young admirer with whom he had struck up a regular correspondence, he said: 'I wish I could go once every year to the Northumbria coast by the Farnes and Holy Island.'[4] He makes a further reference to his earlier trip up north in response to a letter of 1920 from Gemma Creighton, whom he had known since she was a child, as she was the daughter of his friends Bishop and Louise Creighton. Her 12-page letter describing a holiday in the hills on the Scotland/England border prompted another bout of nostalgia. 'A mountainous district seen at a distance always produces the effect of a land of mystery and enchantment and fills one with the desire to explore it and the Cheviots – always seen far off – stirred me strongly in that way.'[5] Another Trevelyan letter soon after 'revived memories of the glimpses I have had of all that wild north land of hill and moor and the old longing to spend months in rambling in it. I never got the chance.'[6]

In late 1921, Trevelyan, now 17, wrote to him again describing long bicycle rides and walks. He replied in April 1922, saying that her words had now given him an 'intolerable craving' to revisit the moors and coast of Northumbria.[7] It was a forlorn, belated hope: he had just four months to live.

Hudson may not ever have had the time – or made the time – to revisit Northumbria, but he couldn't claim he didn't have the chance, or the invitations. The mystery is that he would never take up the offers of his friend in the north Sir Edward Grey, whose Fallodon estate is about as far north as England gets.[8] Hudson could have stepped off the train and walked to Grey's front door, as it had been made a condition of the East Coast mainline rail track being laid across the Fallodon estate that any passengers wishing to alight there could do so. Lack of funds was a factor in Hudson's limited range, but I don't think it was the only one.

When in April 1894 Hudson was invited by Don Roberto and Gabriela Cunninghame Graham to visit them at Gartmore in central Scotland, it might have been in the knowledge that he was making this rare expedition to the north of England, and most of the way there. He declined with regret. After the northern expedition of 1894, Hudson would make just two sorties beyond East Anglia in the rest of his life. The first was in 1903, when his doctor ordered him to Harrogate in Yorkshire. Hudson offered an alternative plan and suggested the south coast, arguing that 'the ozone on one hand and a ride on the downs on the other would do me good'.[9] His doctor won the argument.

And so it was that Hudson was writing soon afterwards from Roker's Hotel in the fashionable spa town of Harrogate, where he was taking the waters. Of the waters he seems to have been unconvinced, but he had detected *some* health benefits from the trip: 'I suppose that four hours' shaking in the train from King's Cross did my system some good,' he wrote, sounding unexpectedly upbeat, despite a change in weather to 'rough and cold'.[10] He seems to have gone there with an open mind, not ruling out that there might be 'some miraculous quality in these waters'.[11] He also seemed content with his surroundings, at least to begin with. 'The people are very obliging, and there's a nice quiet room to sit and smoke in.'[12] But his spirits dipped as he tried to summon the enthusiasm to work on a book, a revised edition of *The Purple Land*, requested by his

publisher now that demand for his writing had grown. It stirred bitter
memories of austere days.

> What revolts me is the thought that when I had not a
> penny and almost went down on my knees to Editors,
> publishers and literary agents I couldn't even get a civil
> word … and now that I don't want the beastly money
> and care nothing for fame and am sick and tired of the
> whole thing they actually come to beg a book or article
> from me.
>   When I get away from this dreary country and back
> to the south or west Nature's ministrations may bring
> me back the desire for production.[13]

Perhaps this unhappy thought served to sour his view of Harrogate:

> I should say, judging from its fine appearance and the
> numbers of fine people frequenting it, that Harrogate
> must be highly esteemed by town-loving folk; it is a
> parasitic town nevertheless, and on that account alone
> distasteful to me.
>   To make matters worse, I there found myself in
> a numerous company of the sick … Pilgrims from all
> parts of the land to that pool in which they fondly
> hoped they would be cured of their ills. Perhaps they
> did not all hope for a complete cure, as there was a very
> large proportion of well-nourished, middle-aged and
> elderly gentlemen with hard red or port-wine faces and
> watery eyes, who walked or hobbled painfully, some
> with the aid of two sticks, others with crutches, while
> many were seen in bath-chairs. I took it that these
> well-to-do well-fed gentlemen were victims of gout and
> rheumatism.[14]

It was a small bird that saved the day.

> In this crowd of sufferers mixed with fashionables I was
> alone, out of my element, depressed, and should have
> been miserable but for a small bird, or rather a small

bird voice. Every day when I went to the well in the gardens to drink a tumbler of magnesia water and sit there for an hour or so I heard the same little bird, a willow-wren, which had taken up its summer-end residence at that spot.

I do not mean a song; a little bird when moulting concealed in a thick shrubbery, has not heart to sing: it was only his familiar faint little sorrowful call-note.

The people sitting and moving about me had no real existence; I alone existed there, with a willow-wren for a companion ... and was sitting not on an iron chair painted green but on the root of an old oak or beech tree, or on a bed of pine needles, with the smell of pine and bracken in my nostrils, with only that wandering aerial tender voice, that gossamer thread of sound, floating on the silence.[15]

Blackbird in song by a church window, epitomising two of Hudson's favourite things. CMJ.

Hudson's impression of people from the north of England was based more on those he met in the south than on any sustained interaction on their own ground:

> These rough fellows from the north, especially from Yorkshire and Lancashire, are always surprising us with their enthusiasm, their aesthetic feeling! And you, said I to myself, born in a hideous grimy manufacturing town, breathing iron dust, a worker in an ugly material engaged in making ugly things, have yet more poetry and romance, more joy in all that is beautiful, than one could find in any native of this soft lovely green south country!
>
> Does not this fact strike every observer of his fellows who know both north and south intimately? How strange then to think that well-nigh all that is best in our poetic literature has been produced by southerners – by Englishmen in the southern half of the country! Undoubtedly the poetic feeling is stronger and more general in the north, and we can only conclude that from this seemingly most favourable soil the divine flower of genius springeth not.[16]

Hudson's unexplained aversion to the north also means that in all his half-century in England he never visited the Lake District. Any chance he might have had to do so had passed by the time, at the end of summer in 1918, he wrote to his young friend Mary Trevelyan: 'I am glad to know you had such a fine time in the Lakes and saw so much bird life. I envy you the buzzards most. It is so good to see great soaring birds. Here we have aeroplanes every day and nearly all day long. But it is not the same thing.'[17]

I don't know if Hudson ever got to watch and describe a buzzard, such was his resistance to venturing to the mountains, where a few such birds might have been able in remoter parts to evade the shooting and trapping that had wiped them out across lowland Britain. It's not as if he hadn't read inviting descriptions of the Lakes. He occasionally references the early Romantic poets, Wordsworth, the Shelleys, Coleridge and all. He was also a fan of John Ruskin, who had lived there. Beatrix Potter, who attracted so much further attention to the Lakes in the era, was a contemporary of Hudson. Visiting her cottage, I couldn't shake the question of why he never travelled to that region, given how much it had come into vogue as a destination for lovers of nature and the outdoors. Friends urged him to. Morley Roberts went there in 1905 and encouraged Hudson to follow suit, but perhaps it was too soon after his ill-starred sojourn in Harrogate. Although he would never make it to the Lake District, Hudson's third and final venture to the north of England would take him more than halfway, to the slightly less far-flung and mountainous Peaks.[18]

An expedition to the Peaks

The Edwardian decade had been the golden age for Hudson, who belatedly found fame and some fortune from his writing. That era ended on 6 May 1910 with the death of the king – who was exactly Hudson's age. A week later, and as his 70th year approached, Hudson was suddenly gripped with the desire 'just to see and hear a bird I couldn't get a sight of anywhere nearer home than the Peak District'[19] – the ring ouzel, which is migratory, and found in spring and summer on the higher hills of the British Isles. It was also known to some people as the mountain blackbird. Apart from its white bib, it looks for all the world exactly like its familiar relative of woods, parks and gardens. Hudson had seen this bird before, but had never heard it sing. Nor had he been able to find the song described in any books. It meant a very rare excursion up north and uphill.

'The soaring figure reveals to sight and mind the immensity and glory of the visible world,' Hudson wrote, of the raptors now missing from most of the land. 'Without it the blue sky can never seem sublime.'[20] By restricting his movements to the lowlands of southern Britain he never had the chance to see or describe any surviving birds of this type. A rare visit to the hills might change that.

The Peak District in a sense is to London what Patagonia is to Buenos Aires. After poring over a map in his Westbourne Park belfry, Hudson plotted his route, then took the train to his base in Buxton, a fashionable and thriving spa town nestling in the hills, nearly 200 miles north of London, and 30 miles south of Manchester. It is not far from Kinder Scout, the highest point in this upland spine of northern England – almost 2,000 feet above sea level. It would be just his third trip 'up north' in England and the highest hills he had explored.

Hudson brought his Sunbeam bicycle, and for accommodation he chose a typically basic farmhouse, Goyt's Moss Farm, a mile or two out of town on the moors, at Axe Edge. It was mid-May, and he found that springtime was several weeks behind the one he had left behind in the south. He also found Buxton 'much tortured with motor cars',[21] and blighted by the local quarrying activity which had coated the town in dust. Keen to integrate, in typical Hudson fashion he soon befriended and had lunch with a local vicar, 'a man whose fine features, classic physique and magnificent beard filled me with a great admiration'.[22]

Hudson wrote a lot of letters while on this trip. His base was so isolated he would not have had many distractions in the evenings. Remarks in the letters he wrote from Axe Edge give a flavour of Hudson's chat with the vicar, who would of course have asked him his business there:

> 'I am here to watch a certain species of bird common in some parts of England, and nowhere nearer to London than the Peak District, so I've come 130 miles just to look at one little bird!'[23] The vicar confessed himself to be no expert on birds. He had, however, noticed one recently, and should like very much to know what it was. 'Perhaps it was a carrion crow or a rook', he conjectured, 'but it was exceptionally big – and very black.'[24]

Hudson knew only too well how witness testimony in these matters can be woefully unreliable. The vicar might have been describing a raven, but ravens had been exterminated locally, the thought of which only irked Hudson more. 'One meets with many disappointments when asking for information about the bird life of any locality. One is apt to forget that such knowledge is not common, that it is easier to find a poet or a philosopher in any village than a naturalist.' In any case, 'there was no time for brooding on such problems; my quest was birds, not men'.[25]

Between showers, Hudson explored on foot and bicycle, returning in the evening to write. This is a selection of observations from his letters to Alice and William Rothenstein:

> Goyt's Moss is a hamlet of three or four cottages in a valley or cleugh of the mountains, a very lovely wild desolate spot. There's no cultivation and one can't even get a cabbage to eat. Cycling is a frightful labour on these high places too, and I'm half-killed with fatigue. Then, too, it is foggy and raining every day.

However I'm here and so must put up with it for a space.

Perhaps you know all about Axe Edge and its rockiness and desolation. At all events here is no road, only a rough old stony track, and the postman comes only three times a week.

I think you would consider it a wretched place to be in—treeless, dark, stony, bitterly cold, always foggy or raining, or both; no cultivation, so that you can't have a vegetable to eat, and of course the house is very very small. I am afraid of hurting my head when I stand up in my bedroom. There is no road leading to the place, only an ancient stony track.

Well, much as I like nature and solitude I don't find it very satisfactory and don't think I shall remain very long.

Yours affectionately, W.H. HUDSON.[26]

Hudson was usually an early riser, particularly so on this holiday.

At the low-roofed stone cabin where I lodged a few wind-torn beeches had succeeded in growing, and these were a great attraction to the moorland cuckoos and their morning meeting-place. From half-past three they would call so loudly and persistently and so many together from trees and roof as to banish sleep from that hour.'[27]

He found the terrain challenging for his bicycle, having to drag it over rugged paths and streams. He scanned the moors with his binocular, lamenting the birds that were missing. The fluting, far-carrying sound of a curlew raised his spirits, just a little.

In all this Peak District you will not find a larger bird than the curlew ... The only one left alive by the Philistines and destroyers. Not a buzzard, not a harrier, not a raven, or any other species which when soaring would seem an appropriate object and part of the scenery in these high wild places.[28]

In fact all this mountain exists for sport and
nothing else. The poor small farmers can hardly
make a living and 'if we say a word the Duke's agent
tells us he can do with our land very well – and we
can go'.[29]

He wrote these last words to Wilfrid Blunt, himself an author and the
owner of a great estate, and a shooting enthusiast. On a happier note,
Hudson did find the bird that he came for.

I was here on a special visit to this species; he was
more in my mind than the golden plover or any other.
I came to be more intimate with him – to have my
ring-ouzel day and mood.[30]
    I got a lodging at a hovel in a place where I have
him close by.[31]
    The bird was quite common on the hill where
I stayed – one pair had their nest within a few minutes'
walk of the cottage and I got to know the bird
intimately during my stay.[32]

After listening to his local ouzel's recital, Hudson explored farther afield,
and found 40 or 50 breeding pairs. He was able to report that the ring ouzel
was not an uncommon bird in mountainous districts. It is not so now.
The species has disappeared from many places, and declined by nearly half
overall, in my lifetime.[33]
    What is exemplified by his description of this trip is how patient
Hudson was as an observer, how long he was prepared to give to being
the audience for this one species. Perhaps that is in part reflective of the
era, but I sense that even for his day Hudson was a dedicated watcher and
listener, if not a tolerant one when it came to resorts and spa towns. His
patience did not extend to tourist honeypots. His summary trip report
is withering (he did say that the terrain had left him 'half-killed with
fatigue'): 'I spent five days in the ugly pretentious town of Buxton and
liked it as little as any town in the kingdom. All these parasitic places
are more or less detestable and Buxton is one of the most unpleasant,
I fancy.'[34]
    Undeterred by Hudson's review, I had an opportunity to revisit Buxton
in summer 2021 (I had been there ten years before at the literary festival),
stopping off on my way to visit Emily Williamson's home in Manchester,

birthplace of the Bird Society in the north. I noted the station hotel in Buxton's bustling town centre, and pictured Hudson taking tea there with the vicar. As Hudson did, I cycled up a hill lined with houses on one side and open fields on the other to almost the last house, where I had an appointment with historian Diana Donald and her husband to discuss Hudson and the Bird Society. Diana has become a firm 'pen-pal' and a very helpful consultee through this exploration of Hudson and his times. We sipped tea in the garden and compared notes.

After our chat, as I rolled downhill on my bike in the warm sunshine, something serendipitous happened: I bumped into an old friend, Mark Cocker, nature writer and campaigner. I had happened to catch Mark at a major life milestone moment, as he unpacked boxes and established himself in his new house near his old home. Mark was born just a meadow pipit song-flight distance away – he could point to the house from where we were standing, and all a far cry from the Norfolk Broads he had just left behind. Among many other things we talked Hudson, and Mark gave me a few pointers about exploring Axe Edge and the few places where ring ouzels cling on today.

We could have chatted all day but I had to leave Mark to his home-making, and headed on towards Manchester and my appointment in Fletcher Moss Park, where several sculptures of Emily Williamson were unveiled by the RSPB and Tessa Boase in the dappled shade of the trees beside Williamson's former home. I returned to Buxton in the evening after a hectic and memorable day.

The following morning I looked online for Goyt's Moss Farm, and found an old painting of it; a white-washed building, under a clump of trees that must have been Hudson's cuckoo-filled beeches. Looking at old census records, a John Shufflebottom and his wife Sarah lived there, he a stonemason, with their servant Jane Mycock. By the 1930s the farm was a ruin. I found a photo of it in this state of dereliction, in the snow.

I later visited Goyt's Moss to get a stronger sense of Hudson there, and found on the site a small car park: the farmhouse has been replaced by a visitor centre and toilets for the Peak District National Park Authority. There had been three other houses there, the only trace of them now a cluster of mossy stones, half-buried in vegetation. The appearance of a great spotted woodpecker was a surprise, albeit a small clump of trees has arisen on the spot, none of them beeches.

I left there to go in search of any surviving ouzels, crossing a quiet valley, with only pipits and stonechats for company, to reach and explore a boulder-strewn hillside with a long-abandoned quarry, reputed to be a final stronghold of the ouzel. I found the place as quiet as an empty theatre. The task of saving and restoring migrant songbirds such as the ring ouzel is a shared, international one. The Anglo-Argentine Hudson knew that better – and sooner – than most.

Norfolk with the geese

> When they rose and floated away they were no longer
> shining and white, but like pale shadows of winged
> forms faintly visible in the haze. They were not birds
> but spirits – beings that lived in or were passing
> through the world and now, like the heat, made visible;
> and I, standing far out on the sparkling sands, with the
> sparkling sea on one side and the line of dunes, indis-
> tinctly seen as land, on the other, was one of them;
> and if any person had looked at me from a distance he
> would have seen me as a formless shining white being
> standing by the sea, and then perhaps as a winged
> shadow floating in the haze. It was only necessary to
> put out one's arms to float. That was the effect on my
> mind: this natural world was changed to a supernatural,
> and there was no matter or force in sea or land nor in
> the heavens above, but only spirit.[35]

One of the passions that Hudson developed in later life was for Norfolk, and its migratory geese in particular. But there were early traces of his interest in the county in 1897. Writing to Eliza Phillips, he described a letter he'd received from a Norfolk man: 'He tells me that ten marsh harriers have been shot this summer: and that is one of the rare species we desire to preserve. Again, he implores me not to say that another specimen exists in Norfolk, since if its existence there became known it will be immediately exterminated by collectors! What a state of things!'[36]

The correspondent had also been to Australia, and had seen rare and beautiful birds being killed there too. 'Alas! It is of small use his vexing his soul at the destruction of the lyre bird of the Australian bush when the destruction of rare birds goes merrily on under his very nose on the Broads,' rails Hudson. 'Cowardice – moral cowardice is partly to blame for it. If these Norfolk bird-lovers would openly denounce by name the wealthy private collectors who pay for the rare birds that are killed a change would take place. But they are afraid.' The correspondent had admitted his fear of upsetting the powerful landowners. 'They are great men and we can't afford to make enemies of them.' Hudson promptly made a vow. 'Some day, if I live a little longer – another year or so, I may be able to strike a better delivered blow at these selfish wretches who are robbing the country of one of its best possessions.'

The letter gives a vivid sense not only of Hudson's mission, but also of his fragile heart and prevailing sense of his own mortality. There was some better news. 'I am glad to hear that an effort will be made next year to get an extension of the close time in Norfolk. How the framers of their Bird-protection Order came to omit so necessary a thing I can't understand.'[37]

His interest in Norfolk was piqued by the correspondence he had received, and the reports of bird persecution there. He decided to visit and see for himself, on 'a few days ramble in that district':[38] not by train this time, but to Yarmouth by sea – an 11-hour voyage from the quay at London Bridge.

Hudson spent an afternoon at Breydon Water in the Broads, with a friend – 'a poor naturalist' who had a wife and seven children.[39] Breydon is today a nature reserve, but in those days it was protected only by a 'watcher', employed by the Bird Society to safeguard its rare nesting birds from the collectors. The rarer the bird, the greater its value stuffed and mounted, or displayed in a cabinet. Norfolk was the front line for much of this battle to save and restore rare birds, with species including the crane, bittern, marsh harrier, avocet, ruff, spoonbill and great bustard making attempts to recolonise, but invariably doomed unless carefully supervised.

With his growing success as an author, Hudson became a more regular Norfolk visitor in the Edwardian era, joining the crowds arriving by train at Kings Lynn and Hunstanton, and bringing his bicycle. He made a rare visit at the height of the holiday season in August 1904. It didn't take long to bring out his disdain of resorts – Hudson was never one to gloat on a postcard:

> One marvels at our modern craze for the sea ... the
> result of a life too confined and artificial in close
> dirty overcrowded cities ... towns have sprung
> up everywhere on our coasts ... with their tens of
> thousands of windows from which the city-sickened
> wretches may gaze and gaze and listen and feed their

sick souls with the ocean … It was not so formerly,
before the discovery was made that the sea could cure
us. Probably our great-grandfathers didn't even know
they were sick.[40]

Finding Hunstanton too crowded, he sought solitude at Holme-next-the-Sea: 'An ancient person who turned out to be the vicar came in and began to show me about.' The vicar helped him find a room at Near Farm: 'a farmhouse in the village with green marsh-land at the back, and a line of sand-hills close to the sea half a mile away'.[41]

He walked with Emily from Hunstanton to Thornham and back. Returning to their farm, he realised he had left his 'binocular' on the clifftop at Hunstanton. It could not be re-found, but he consoled himself and Emily that he might not have much need for it anymore (he really must have been feeling poorly, even by his standards). He was well enough to take a walk the following morning to Ringstead Downs, a site managed nowadays as a chalk grassland nature reserve by the Wildlife Trust.

At their Holme farmhouse, Hudson had found the tranquillity he was seeking – but there was a hitch:

> It's nice here too, but there's one drawback – the
> golfers have possession of a long stretch of the sand-hill
> between the village and the sea. You must keep to the
> deep sandy road or rut to go to the beach, and then
> you are warned with yells to stand still until someone
> makes his stroke. When they wave their sticks and
> shout at me I shout back – 'What the hell are these
> swine yelling about' – and go slowly on and chance
> being hit.[42]

Hudson thought golf absurd, and gave this earlier description of the golf enthusiasts he had observed on Hampstead Heath: 'the golfer, arrayed like the poppies of the cornfield, and visible at a vast distance, strolls leisurely about as his manner is, or stands motionless to watch the far flight of his small ball'.[43] He thought birdwatching 'a better outdoor game than golf, as it really does get you a little forrarder, and does not make you swear and tell lies and make you degenerate from a pleasant companionable being to an intolerable bore'.[44]

At Holme, he found that the alternative route, avoiding the golf course, was even worse, owing to the growing volumes of tourist and other traffic on the local roads. 'If one goes by the high road he is covered with clouds of dust from the loathsome motor-cars – they appear to swarm here. How long will the slaves of England endure this brutality I wonder?'[45]

Despite the irritations and the conflicts over rare birds, Hudson was relatively kind about the locals. 'I don't think the Norfolk folk quite as bad as some of our 'realistic' fictionists make them,' he reported, 'but I can never like them as much as I like the people of Wilts, Dorset and Somerset – the west generally. There is less of sweetness in their blood and they are less pleasant in their manners.'[46]

Despite these minor misgivings, Norfolk began to vie for Hudson's time and attention with his more traditional south-westerly trajectory, such that by 1908 he was torn between the two: the oak woods of a southern county and the wild geese of Norfolk in the east. He chose the latter, 'which I don't care for very much except in late autumn when the wild geese come'.[47] By November, he was back again to 'hold communion with nature in the form of wild geese newly arrived on our shores'.[48] He spent long hours out on the salt marshes and sea fronts to greet the incoming flocks in their long diagonal lines, and identified a place where the birds knew they were safe from the guns and could graze in peace.[49]

In October 1910 he wrote: 'It is wonderfully quiet here – no motor hoots and no sound is heard except "the prayerful crowing of the cock," and the clanging of the geese as they come in of an afternoon.'[50] He found particular comfort on this North Sea coast at a time when his physical and mental health were declining. 'Its fascination for me is its solitariness and the wild birds,' he wrote. 'I thought these things would help me in my present state of health and depression.'[51]

He also enjoyed further tours of the interior of East Anglia. In July 1912, taking his bicycle on the train, Hudson returned to explore around Thetford. He knew it was late in the season but he still hoped to find woodlarks. 'There is reason to fear that these few pairs have been or will be exterminated by the collectors, as all that portion of Norfolk is their happiest hunting ground, including the Elveden estate,'[52] he wrote in a letter to Harding, no doubt conscious that Elveden had been the childhood home of Professor Alfred Newton.

He came to Wells-next-the-Sea again in the autumn of 1912, to put the finishing touches to his book *Adventures Among Birds*,[53] mainly a compilation of already-published articles. He wrote some new material while there, and posted it to Linda Gardiner, who was typing it for him.

Unusually for Hudson, he had brought with him on this trip a male companion, the young Scots author Richard Curle. In Hudson's account, they got along reasonably well, arguing only about Scots poetry, which Hudson maintained did not exist.[54] They talked a lot about books and publishing, as Curle received news while on the trip that his publisher had absconded with a lot of other people's money, and a female editor.

Hudson had recently read Curle's latest book, and told Garnett that he had enjoyed it. Curle's father – whom Hudson called 'the fiercest old Tory in the country'[55] – had no interest in his son's writing, so the approval of Hudson was especially important to him. But Hudson was soon weary of book talk, and Curle left early, as his friend was in such a low mood.

Hudson was glad, in Curle's wake, to be free to find a keeper or fisherman to chat with, and a hearth to gaze into, uninterrupted. And of course there were the geese to commune with, and to hope would evade the wildfowlers waiting for them to leave their roosts before dawn.

Forty years later, Curle wrote a touchingly candid account of the trip. 'In one sense the visit proved a failure. It was soon evident that Hudson, to whom solitude was essential in the country, regretted his invitation.' But Curle retained a fond image: 'I shall never forget the figure of Hudson, alone on the barren saltings in the falling dusk, watching and listening.'[56]

If Norfolk was Hudson's autumn place, as Cornwall had become his winter one, Wiltshire for a phase would be all about springtime.

The shepherding life in Wiltshire

Hudson had his first look at the county of Wiltshire within weeks of his arrival in Britain in 1874, visiting family friends at Malmesbury. But it wasn't until 20 years later that he was able to spend significant amounts of time there. In 1896, he described a week lodging in a Savernake Forest cottage. His host told him proudly that he had never been more than 10 miles from home.[57] He also told Hudson about a fearsome-looking bird of prey that had turned up in the forest one day, exciting the locals who for days were in pursuit of it until someone finally shot it. Hudson was sure it must have been a goshawk, although only passage birds from abroad or escapes from captivity were possible at that time.

Hudson's classic account of the end of an era for the shepherding communities of the Wiltshire Downs was eventually published in 1910. It took many years of patient observation and integration for Hudson to get properly under the skin of his subject. The central figure in A Shepherd's Life is an elderly shepherd to whom Hudson gave the alias Caleb Bawcombe.

His real name, it emerged later, was James Lawes. Hudson first acquainted himself with the Lawes family in around 1901, slowly earning the trust of the old man and his wife Emma. Perhaps a condition of their contract was that Hudson gave Lawes the pseudonym, and even renamed the village of Martin, where they lived, to preserve its identity. He called it Winterbourne Bishop. Of this place he said, 'I don't think there can be a lovelier out-of-the-world village in Wiltshire,'[58] and James Lawes he said had 'no thought in his mind, I might almost say, which was not connected with the village'.[59]

Lawes had been 'forty years a shepherd on the Wiltshire Downs,' and at first Hudson found it 'not very easy to understand him'.[60] The old couple were by this time the carriers for the community and did a run to Reading with their horse and cart on most days. The looming figure of Hudson had a job navigating their cottage – his head touched the ceiling and the winding staircase threatened to collapse under him.[61]

In November 1902, Hudson wrote of having been back to the cottage of the 'old people' and, finding that a professional photographer was visiting the village, got him to take their photograph, to give them as a present. Old Mr Lawes needed time to get ready for this. 'He had got himself up in all his best clothes, and stiff hat and polished boots, and was posed in a chair out of doors,' Hudson reported. 'I never saw any one look so unhappy at an operation,' he added. 'I wouldn't have mine taken that way. I made him put on his old cap and my old cloak, and go and stand leaning on the gate with his stick, when he looked like an old shepherd. Of course I shall not have it for some days.'[62]

Surviving Hudson letters reveal other pleasing glimpses of his time there. The Lawes's pet canary took a shine to him, but sadly died soon after. And there is further evidence of the deep humanity of the Bird Society's founder Eliza Phillips. When the Lawes's horse died, and Hudson mentioned this to her, she helped pay for one to replace it.[63]

Besides being embedded in the shepherding community Hudson also spent many hours in the company of gypsies:

> 'Settle down! I'd rather be dead.' There spoke a true
> gypsy; and they are mostly of that mind. How infinitely
> more perfect the correspondence between organism
> and environment in his case than in ours, who have
> made our own conditions, who have not only houses
> to live in, but a vast army of sanitary inspectors,
> physicians and bacteriologists to safeguard us from

that wicked stepmother who is anxious to get rid of us
before our time! In all this miserable year, during which
I have met and conversed with and visited with many
scores of gypsies, I have not found one who was not in
a cheerful frame of mind.[64]

Hudson also made time to revisit the old couple he had stayed
with seven years earlier in Savernake Forest,[65] where he spent a day
patiently exploring for larger birds. His search was fruitless, such was
the intensity of hunting in that era, and for collecting everything for
display in cabinets and glass cases. 'Savernake is extensive enough,
one would imagine, for condors to hide in, but it is not so,' Hudson
lamented. 'The biggest forest in the county now affords no refuge to
any hawk above the size of a kestrel. The larger hawks and the raven,
which bred in all the woods and forests of Wiltshire, have, of course,
been extirpated.'[66]

Freezing spring weather did not deter Hudson from exploring the
exposed downs on his bicycle. He visited 80 Wiltshire villages assessing
the birdlife of the county. He wanted to amass as much local knowledge
as possible before visiting the local authorities. It was typical of Hudson's
tireless efforts for this and other counties' birds. Later in 1903 he was
trying 'to find out and talk to as many councillors as I can to try to induce
them to get a proper bird protection order for the county. It is behind
the others in the southern counties'. He petitioned Salisbury Council in
person to call for stronger bird protection, and visited Charles Hobhouse
who, besides being a Member of Parliament, was also an alderman on the
county council. Hobhouse would duly introduce a bill to ban the import
of wild bird plumage.[67]

Hudson reported that he was getting on quite well, 'one difficulty
being the Hawking Club,'[68] who liked to hunt stone-curlews. Wiltshire
was a remaining stronghold for the stone-curlew, a species of open country,
fond of its flinty terrain. Thanks to the efforts of Hudson and co., the
scarcity and vulnerability of this species had been recognised elsewhere.
It was by this time strictly protected throughout the year in neighbouring
Hampshire, for example. To this day they are subject to intensive conservation action, to keep their small populations going.

The following year, 1904, he was back again in early spring at Martin.
'I think of staying on some days here. The wind blows hard on the bare
pale grey-green downs, but it is wholesome.'[69] His Bird Society colleagues
were following up on his local fact-finding. 'Miss Gardiner has written to

the Clerk of the Wilts CC about their Order.'[70] She was told they were waiting for it to be sanctioned by the Home Secretary.

Hudson gave one of his very rare nature talks to local children at Martin, showing his audience 80 lantern slides over an hour and a half. It seems to have gone well, judging by the response. 'After the lecture this evening a lady, a Mrs Herbert, who lives near, came up to be introduced to me,' he wrote, 'her hat covered with birds and feathers, and asked if she, her husband and a friend could be members of the [Bird] Society!'[71] Mrs Herbert had obviously missed a pamphlet or two.

Hudson became very interested in the harshness of the law as applied to the rural poor in earlier years, and spent time scouring old court records and also spectating at trials, taking a close interest in one court case concerning disputed territory on common land.[72]

In June 1905 he was back. 'It is very peaceful here, and when I go out on to Martin down I find no human creature – only the rabbits and magpies and peewits … and from a distance comes the many-toned bleatings of the sheep.'[73] He found this rural peace in Wiltshire a blessed relief from the 'everlasting brain-worrying noises' of his London base.[74]

'For over two months down in the two or three small villages I stayed at I saw no literary paper and no book. On the other hand I got acquainted with every person in Hindon, where I spent most of the time – a village of four hundred inhabitants.'[75] He sometimes stayed at the Lamb Inn there, and befriended its resident parrot, which, like him, was a settled immigrant, and had once been fluent in Spanish. His efforts to reteach it were only partially successful.[76]

Hudson had grown to love the county, and often hankered after it. 'I dream of south Wiltshire in spring.'[77] He was drawn to open, pastoral landscapes, and was in his element on the open downland, although he had long given up horseback for the bicycle. It recalled for him his carefree days as a youth on the vast Pampas, riding his horse. 'When I got up to the highest part of the road and could put my feet up and let my bike run swiftly on for two or three miles … I had the feeling that I was coming home.'[78]

Hudson fulfilled a lifetime ambition to visit Stonehenge when he cycled there one midsummer night, arriving at one o'clock in the morning and joining a crowd of 500 people waiting for the sunrise. He had known and dreamt of Stonehenge since he was a boy. Sadly, his mystical communion was slightly tarnished by an unruly element in the crowd, and all were frustrated by a bank of cloud that obscured the horizon at just the wrong moment. 'As usual a cloud no bigger than a man's hand appeared at 3 o'clock and blotted all the east out'.[79]

While Stonehenge was much smaller than he'd always imagined, since he was a child and saw a picture of it in a book, Salisbury Cathedral was a revelation to him, 'not only because it is so big compared to Stonehenge, but because of the birds',[80] he enthused. The cathedral had jackdaws – Stonehenge had only sparrows. Of all the cathedrals he visited, Salisbury was his favourite.

By 1908, his full focus was on Wiltshire, to finish his book: 'I must stick to that county now,'[81] he told friends. In May of that year he revisited Malmesbury. 'Climbing to the roof just now I saw 5 miles away the very tall and very thin spire of Tetbury Church. A place of memories for me – it was there I went on first coming to England on a visit to the family of a S. American friend of mine … a very fine fellow who took to drink and went to the dogs. After long years I feel inclined to cycle over and look up these old friends or shed a tear over their ashes if they are all dead.'[82]

Some of his observations nature enthusiasts today could only dream of. 'You can hear the cirl bunting in every one of the villages and hamlets between Warminster and Salisbury,'[83] he wrote in 1909. He later noted goldfinches recovering in numbers, no doubt helped by the new local protections he had helped to bring in.

When *A Shepherd's Life* was published in 1910 it received lavish praise from Ezra Pound, as well as Edward Garnett, and is among the most enduring of Hudson's works, being still in print. A friend sent me a photograph of *A Shepherd's Life* on the shelves of Waterstone's next to a Richard Jefferies title, with a book of my own sandwiched between. I found this a little humbling, but alphabetical order counts for something, I have assured myself.

Hudson visited the statue of Jefferies in Salisbury Cathedral in April 1905 – he had obviously been before, and hadn't thought much of it then. He wrote to Edward Garnett: 'I just now had a look at that Richard Jefferies bust in the cathedral and dislike it worse than ever. The expressionless face of it!'[84] In the same letter to Garnett he wrote: 'There is a deadly want of humour in Jefferies. Yesterday I laboured at *Bevis* until I dropped to sleep over it.' It seems he preferred Mark Twain's American counterpart: 'one chapter of *Huckleberry Finn* was worth more than all that long book'.[85] It is worth noting here that some critics – I think unfairly – have accused Hudson of lacking a sense of humour.

There was a compliment too, though qualified: 'I have been told that I have some affinities with Gilbert White. Jefferies too was a great observer and – unlike White – a poet and an artist, but his work is also a little mannered. He is at times perhaps too consciously an artist, too consciously an observer and student of nature.'[86]

King Edward and Hudson were contemporaries in age, but the fast-living king, embodying the hedonism of La Belle Epoque, predeceased him and died in early May 1910. As one era ended, a new one was approached, with some trepidation.

CHAPTER 11

Shadows of War

End of Edwardia

The end of the Edwardian decade, during which Hudson's fortunes had been transformed, coincided with a dip in his health and spirits. 'Since I finished my book on Wiltshire Downland in November I've done nothing – except try to make myself well with sour milk,'[1] he wrote gloomily. He had persisted with this regime for three weeks over Christmas and New Year before giving it up. 'I've been most careful about diet, and have not touched wine or stout or spirits all the time I've been using it, and eating the most simple and wholesome food – very little meat.'[2]

If Hudson's health was fragile, Emily's was failing. They escaped from London at the end of winter, heading to Brighton and the south coast in search of better air and to escape the gloom. He was desperate to get out on his bicycle.[3]

The malaise continued for some months. 'As time goes by I am less and less able to do what I want or try to do',[4] he wrote to Algernon Gissing in July. At one point he was forced to give up cigars for three months. He did make it to the January and July council gatherings of the Bird Society, meeting Linda Gardiner and the Lemons at the Guildhall. Etta Lemon had joined the British Ornithologists' Union, now that it was admitting women.[5] There had been developments in Parliament with the plumage campaign, although the society was lukewarm about the detail of Percy Alden's latest version of a bill, and withheld support altogether from Sir William Anson's, even if rising Labour politician Ramsay MacDonald was a backer of this.[6]

To follow progress of the bill, Hudson summoned the energy to take his seat in the viewing gallery in Parliament, on 15 July. He also saw a debate on votes for women and a wider range of men. 'There were a good many fine speeches delivered,' he reported to Roberts about the latter debate. 'The weakest were certainly those in favour of the bill … but what brilliant speeches on the other side! … Well, we are on opposite sides on this, so I'll say no more as it might hurt your feelings.'[7] Hudson was also at odds with his close friends Sir Edward Grey, Ranee Margaret and Don Roberto on the issue, although not, it might be added, with Etta Lemon.

Later in 1910, the Liberal government's gradual reforms ran aground, with the Conservative-dominated House of Lords refusing to pass the proposed 'People's Budget', which sought to increase taxes on the wealthy. A general election would be held later in the year. 'I think it would do me a lot of good to see this damnable government turned out', Hudson harrumphed, '– Ananias and the Heavenly Twins and the rest of them.'[8] We might only guess which Liberal politicians Hudson had in mind.

The Liberals and Labour worked together on the election, and although the Liberals lost seats, the emergent Labour Party ranks swelled to 40, enabling them to form a coalition with the Irish Nationalists. Grey would therefore continue in Cabinet, playing a prominent role in ongoing discussions with the Kaiser, although like Hudson, and despite his role, he seemed to have an aversion to foreign travel, and ventured abroad only once. Secretary for War Lord Haldane did the shuttling between London and Germany, and the interpreting. Germany was seeking a commitment from Britain that they would remain neutral if there was war in Europe.

Grey was also endeavouring to rein in the naval arms race. An agreement was made with France to protect the Channel coast and the Mediterranean. 'Grey and the Cabinet insisted on an exchange of notes, stressing that the agreement did not commit either government to action. But whether a moral, if not legal, obligation had been created would be tested later.'[9]

There were further developments on the plumage campaign front at the start of 1911. Percy Alden had introduced a new bill to Parliament to prohibit the sale, hire or exchange of the plumage and skins of certain wild birds, and the importation into Great Britain of the plumage of species that were being destroyed solely for their feathers. Hudson had also been monitoring its progress through the pages of *The Times*, and lunched with Sir Edward Grey in late February, two days after Alden introduced the new draft legislation.

'The Anti-plumage Bill has been read in the Commons for the first time', he reported to Ranee Margaret. 'That means some progress I hope tho' it's doubtful when it will get a second reading.'[10] Hudson's pamphlet, *A Thrush That Never Lived*, was printed to coincide with this latest attempt to establish plumage law. There were demonstrations by placard-carriers against the plumage trade in the West End of London in August and at Christmas. The placards illustrated 'the story of the egret' in photographs, showing the beautiful birds that were being killed for their nuptial plumes, their nestlings left to starve.

John Galsworthy mentioned his support for the campaign in a letter to *The Times* decrying animal cruelty generally. Hudson registered his approval: 'One of the barbarities he cries out against – the plumage business – will be stopped very soon, I think. The second reading of the Bill will come on in a month or so.'[11] Galsworthy even wrote a pamphlet for the Bird Society, called *For Love of Beasts*.[12] He also found time, alongside Don Roberto, to launch a campaign at Kensington Town Hall against cruelty to performing animals.[13]

At lunch with Sir Edward Grey, Hudson got the inside track on the Plumage Bill's chances of navigating parliamentary process, in turbulent political times. It left him in uncharacteristically optimistic mood. 'There is a fair prospect of a law being obtained',[14] he told others soon after. But the industry was still resisting doggedly, and had published some propaganda: a booklet called *The Feather Trade*.

'The "Trade,"' Hudson wrote, 'alarmed at the threatened loss of their profits, are industriously engaged in scattering their letters, circulars and pamphlets broadcast over the country.'[15] With public and political pressure mounting, the industry was keen to deflect, pointing out that wild birds faced greater dangers than the fashion business, namely the spread of civilisation generally, the widespread rearing of pheasants to be shot, the hunting of 'big game' in Africa.

The Bird Society retaliated with *Feathers and Facts: A Reply to the Feather Trade*, acting also on behalf of 'kindred Societies in Europe, America (North and South), India and other parts of the world'.[16] Conservation campaigning was now officially an international business. Grey and Hudson were of course only too conscious of this. Hudson had other business to discuss with Grey: the question of a petition to government seeking a pension for his friend, the famously impoverished Welsh poet and author W.H. Davies. 'I wanted to see E. first,' he explained to Garnett in a letter. 'I lunched with him on Friday and then had a long talk in which I said everything about Davies.'[17] Grey asked him to put the case in writing. No doubt they also touched on the unrest in Ireland, the campaign for wider suffrage, and matters foreign. By 9 March, Hudson had lost confidence in the plumage legislation succeeding.

The regularly far-flung Don Roberto now wrote from Rome, where he was staying with his mother. It was a year of civil unrest and industrial action, which kept him busy when he wasn't overseas.[18] Home Secretary Winston Churchill would order troops in to quell protesting dockers in Liverpool. Don Roberto wrote again in late summer, to relay news of the death of his beloved rescue stallion Pampa. 'He says he never felt anything more',[19] Hudson reported.

Emily Hudson's condition was deteriorating. Her husband spent much of 1912 'attending to her wants, and have only relief at night when the nurse comes on',[20] he told Roberts. He had been 'a close prisoner' in London for the last ten months. 'In all that time I have not been so far as Richmond nor Kew. An eternity practically.'[21] He had to decline with regret Ranee Margaret's offer to take him again to Cornwall, for some respite.

He did make it to the south coast, at least. In summer 1913, the Hudsons based themselves at Seaford in East Sussex for nearly two months. Emily was now losing her sight. 'That's the greatest trouble as she was a constant reader before her illness and now has no way at all to occupy her time.'[22] While there, her husband became nostalgic about those happy days 15 years earlier spent on the South Downs, which looked the same as they did, from a distance at least, as he scanned them with his binocular.

But Hudson was soon restless. 'I see too much of Seaford. It is a naked shadeless or treeless town with too much new building ... the people I knew are mostly vanished.'[23] He was gratified to find some birds at least being resilient in the changing landscape:

> Here I find one curious thing, where buildings,
> golf-links, etc have driven out the birds there used
> to be ... A few wheatears have refused to quit, and
> two pairs have nested in some chalk emptied into a
> depression in one of the building-lots not yet sold.
> I look right down from my window at the nesting spot
> and the birds flitting about ... all day long people are
> going and coming or sitting close to them, and they go
> on with their business just the same. One would hardly
> expect to find wheatears breeding, and getting their
> young off, in a town![24]

Hudson of course kept abreast of developments with the Bird Society while indisposed. There were renewed attempts at plumage legislation. Prime Minister Asquith supported the Hobhouse Bill to prohibit the import of the feathers and skins of wild birds. The first reading had taken place in February 1913. Questions about plumage imports came up 18 times in this session. Meanwhile, the industry considered its next moves, with the International Congress of Plume Dealers meeting in Paris in June. The government also convened an interdepartmental committee of enquiry into bird protection more widely, with representation from the society.[25]

In late June, news reached Hudson that his sometimes uneasy ally Dr Philip Sclater had passed away in Hampshire. Around this time, the Zoological Society and the Bird Society were not exactly seeing eye to eye on the plumage issue. There were further developments on that front in August: 'Just when the session is going to end the Government have finally brought out their anti-plumage Bill,' Hudson groaned. 'What chance of passing will it have, alas!'[26]

The same month, the ailing Hudsons retreated from Seaford inland to Maidenhead in Berkshire, near Cookham Dean, the village that had been the focus of his first book on British birds 20 years earlier. 'I have brought her down here as an experiment, the air at the seaside failing to do any good ... It is a poor spot for a naturalist, but all we wanted was a house with a garden and lawn where my wife could be out of doors all the time in good weather as the doctor ordered, and we found it here.'[27] He must have been feeling better because he was putting his 14-guinea Sunbeam through its paces on this excursion:

> I've got my bicycle here and take runs about the open
> spaces – the Thicket principally, which is perhaps the
> best common in England, as it extends over two miles
> and is mostly a perfect wilderness and tangle of thorn
> and bramble, interspersed with big trees. Our Bird
> Society is now trying to get it made a protected area,
> and I think we'll succeed this time. We tried before.[28]

While roving on his wheel, Hudson returned to Cookham Dean, 'a village of pleasant memories for me', but he was shocked at the rate of urbanisation since his last visit. 'Alas, the whole country round about here is being built over with nasty little red brick villas in which people who go up to the city live and play tennis in their gardens.'[29] He blamed the urbanising influence of Reading, which he had noted before: 'That biscuity place has spread a kind of blight of vulgarity over the country round it.'[30]

Hudson's emotions at the time of Sclater's death are not recorded, but we do know his response to news of the passing of his friend Alfred Russel Wallace, unsung hero of evolutionary theory, who died in November 1913. Hudson lamented the loss of Wallace's 'still vigorous intellect and power to work! He had more work to do and was full of hope for the future of humanity.'[31] Hudson was reading a Thomas Hardy novel when this sad news reached him, and poor Hardy compared unfavourably, at that moment, with his old ally. 'What a pitiful creed is his, compared with that of our grand old Wallace,' he wrote to Ranee Margaret.[32]

Meeting the 'Bard of Bengal'

In 1912, Hudson was invited to Hampstead by William and Alice Rothenstein to meet their latest VIP guest, the Indian poet, musician and artist Rabindranath Tagore. Tagore had been dubbed the 'Bard of Bengal', and would receive the Nobel Prize a year later. 'A public dinner had been given to Tagore, to which many distinguished men and women came,' William Rothenstein later recalled.[33]

According to Rothenstein, this distinguished guest had at least one fellow author above others in his sights: 'Tagore wanted to meet Hudson, for he had read *Green Mansions*; it was his favourite modern book, he said.' Sadly for Tagore, Hudson's days of carefree hobnobbing in north London appeared to be over. 'Hudson had written to give his apologies for declining the invitation,' Rothenstein adds. He had kept the following letter among his many from Hudson, and published it some years later:

> July 13, 1912
>
> Forgive me my dear Rothenstein for not replying to the card about a dinner to Mr Tagore, for days past I have been so much troubled with palpitations I left letters unanswered. But you know I never dine out now—I can't go to a dinner at the conventional hour and eat & come home at some late time without paying for it heavily.
>
> I should have liked to hear Yeats read the Tagore poems; I hope he has got a poet to translate them. Not many of our poets know Hindustani; but these things can be managed another way. For example, Blunt's splendid translations of the Seven Golden Odes of Arabia were not done direct from the originals—he doesn't know Arabic; but he took them from Lady Blunt's literal translations into English and turned them into poetry.

If I could stand being chloroformed I would go to some surgeon & ask him to cut me up in pieces & take out as much as he thought proper, then sew up the remnants, in order to see if that would give me a little more life. But these be idle thoughts.

With love to you all
Yours ever,
W.H. HUDSON.[34]

According to Rothenstein's description of proceedings, Hudson missed a lively evening, and some other very big names:

Yeats and I arranged a small dinner in his [Tagore's] honour. After dinner we asked Tagore to sing *Bande Mataram*, the nationalist song. He hummed the tune but after the first words broke down; he could not remember the rest. Then Yeats began the Irish anthem—and his memory, again, was at fault; and Ernest Rhys could not for the life of him recollect the words of the Welsh national anthem. 'What a crew!' I said, when I too stumbled over *God Save the King*.[35]

It is recorded that Tagore composed the national anthems of India and Bangladesh, and influenced that of Sri Lanka – perhaps we can excuse him an off-night. Don Roberto's biographer Cedric Watts lists the names of those who came to pay homage: 'The young poets came to sit at Tagore's feet; Ezra Pound the most assiduously. Among others whom Tagore met were Shaw, Wells, Galsworthy, Andrew Bradley, Sturge Moore, and Robert Bridges.'[36]

Rothenstein later took Tagore to his place in the Gloucestershire countryside, but it was a rainy summer and the weather threatened to put a dampener on their trip. His guest was philosophical: '"A traveller always meets with exceptional conditions," said Tagore, when I apologised for the cold and rain, and the absence of sun. When kept indoors, he busied himself with translations of more poems and plays.'[37]

Curiously, Hudson may have missed these evening soirees but he later claimed that he did meet Tagore at Hampstead, in his own time, and way. He wrote: 'On two mornings I met there [at the Rothensteins' house]

Rabindranath Tagore, and renewed the acquaintance of years ago when he first came to England. He is the finest specimen of an Indian gentleman I have ever met, and has a wonderful charm in his manner.'[38]

Hudson's dalliance with the freedom of artistic speech campaign had not quite prepared him for the works of D.H. Lawrence, the latest of Garnett's protégés. In autumn 1913, he read *Sons and Lovers*, by what he called Garnett's 'favourite author'.[39] Garnett had met Lawrence two years earlier in 1911, and had been mentoring the ambitious young author. Garnett had persuaded Lawrence to cut around a hundred pages from the book. Hudson had mixed feelings about this work describing working-class life in provincial England, calling it 'a very good book indeed except in that portion when he relapses into the old sty – the neck-sucking and wallowing in sweating flesh. It is like an obsession, a madness'. On the other hand, he found some of the characters 'extraordinarily vivid … Only they seem more real than most of the human beings one meets.'[40]

On another occasion, Hudson wrote: 'Lawrence makes one angry as a rule with most of his things, but he has great talent.'[41] Nevertheless he thought Garnett overstated Lawrence's genius.[42] Garnett meanwhile encouraged Lawrence to read Hudson, giving him a copy of *Nature in Downland*.[43] Whether this subsequent comment by Hudson is linked to Lawrence's verdict on his book we can only guess: 'Certainly this and other thrusts will not hurt my friend, D.H.L., who is himself addicted to passing the most scathing verdicts.'[44] Garnett's alliance with Lawrence was relatively short-lived, and they were barely writing to each other by 1914.[45]

There had been no moves to legislate against the plumage trade until 1908, nearly two decades into the Bird Society's life. Several bills then came and went, shot down by opponents and stymied by politicians who knew how to hobble the progress of a paper. Objections to the idea of banning plumage imports to Britain took many forms. Members in the Lords pointed out that the trade would just be diverted to rival European countries, and thereafter the merchandise would find its way to Britain anyway. And how would the authorities know if a feathered hat was new, or old, pre-dating any new legislation?

Despite the turbulent progress of the plumage campaign, a mood of celebration surrounded the annual meeting of the Bird Society in 1913. After two decades or more of lobbying by the Bird Society, the Home Office had finally granted a departmental committee of enquiry into the feasibility of creating a so-called White List, a system whereby all bird species would be considered protected unless a need to remove this protection had been proven; that the species was demonstrably problematic in some way. Winifred, Duchess of Portland took great delight in announcing the news to the assembled crowd at the annual meeting in Westminster Hall.

Archduke Franz Ferdinand

It was an auspicious year for the Portlands at home too, as they had the honour of welcoming to Welbeck for the shooting season Archduke Franz Ferdinand, heir to the Austro-Hungarian throne, and his wife Sophie, Duchess of Hohenberg. The couple had earlier been received in London by Prime Minister Herbert Asquith, and treated to a royal shooting party at Windsor with the king and the Prince of Wales.

While Winifred was also vice-president and president of the ladies' committee of the Royal Society for the Prevention of Cruelty to Animals, her husband had other priorities. It was a bumper year for keen shooters like him, with Lord Burnham's Beaconsfield estate in Buckinghamshire recording what was thought to be the highest number of birds ever shot on one day. A week before Christmas 1913, King George and his son, later to be King Edward, shot over a thousand pheasants in six hours – 'about one bird every 20 seconds'.[46] The prince recalled that his father was curiously silent on the train journey home, alone with his thoughts, until he finally, glumly, articulated them, conceding that 'perhaps we overdid it today'.[47]

After Royal Windsor, on 22 November the archduke and Sophie caught a special train at King's Cross station for the journey north to Nottinghamshire, alighting at Worksop where 'the platform was covered with crimson carpet and the station decorated with shrubs and flowering plants. Crowds had gathered to greet the royal couple.'[48]

They stayed for a whole week at Welbeck, and must have become well acquainted with the Portlands in that time. Perhaps Franz Ferdinand and Sophie told them about how his uncle, the Emperor of Austria-Hungary, had disapproved of their marriage. 'We constantly marry our relatives and the result is that of the children of these unions half are cretins or epileptics,' the archduke once said in defence of his life choices.[49]

During their stay at Welbeck, an incident took place that might have altered the course of modern history, when one of the staff employed to load the guns and pass them to the dignitaries to fire off at flushed pheasants fell over. Both barrels of the shotgun were discharged, 'the shot passing within a few feet of the archduke and myself', the duke recorded in his memoir, *Men, Women and Things*. 'I have often wondered whether the Great War might not have been averted, or at least postponed, had the archduke met his death there, and not at Sarajevo the following year.'[50]

I don't think Sir Edward Grey met the archduke during this visit but he was busy on the diplomatic front in December 1913, and his reputation was enhanced during the so-called Balkan Crisis. He had called the ambassadors of Europe together for a conference in London – the last act of the old 'Concert of Europe' – and was working on Germany to maintain a fragile peace.[51]

Relations were also increasingly strained on the plumage campaign front. Zoological Society chief Chalmers Mitchell had become president of the Committee for the Economic Preservation of Birds, a coalition convened by the industry to protect the bird plumage trade. They held a meeting on 16 February 1914 to set out their position, and issued a short press statement two days later. In another curious twist, this meeting was presided over by the Countess of Warwick, Daisy Greville, who is best remembered as a mistress of the Prince of Wales, later King Edward VIII. She was such a fan of Hudson that she later created a shrine to him in the grounds of her Essex home.[52]

Feather wars

In 1914, with the latest draft Plumage Bill on the table, the RSPB and its allies gathered at Burlington House in Piccadilly with three representatives from the Natural History Museum onside. Opposite, Chalmers Mitchell 'chose to appear on the platform in support of the plumage traders', implying the support of his Zoological Society. C.E. Fagan of the museum asked him to state publicly if that was the case, and pressed him for a definitive response, at which point a humiliated Mitchell 'was forced to admit that he had no mandate from the ZS'.[53]

After the meeting, Mitchell, 'livid with rage, stepped down from the platform and, coming up to where we were sitting, stopped in front of Fagan and positively hissed out "I demand of you a public apology!" he got an appropriate reply but no apology, public or private'. Thereafter Mitchell refused to shake hands or even to speak to Fagan,'[54] one witness at the meeting recorded.

Fellows of the Zoological Society of London, of whom Hudson was one, were outraged that Mitchell had implied his organisation's support for the plumage trade. One – Sir Harry Johnston – wrote to *The Times* on several occasions early in 1914 to stress the point that Chalmers Mitchell was not representing the views of many Fellows, and that the society should be siding with the campaign to *end* the plumage trade. One letter in *The Times* included the following:

> Unless he [Chalmers Mitchell] disavows the statement made in the pages of the *Selborne Magazine* that his committee deprecates legislation for the protection of birds or would postpone it to some vague future period, he is not acting fairly towards the numerous Fellowship of the ZS; for there are, I am convinced, in that society a great many persons besides myself who consider that, if the ZS is to intervene in the question at all, it should be to range itself on the side of the Government and the advisers of the Government in the NHM, and assist in bringing forward as soon as possible a measure which shall on behalf of GB close the now open market for the traffic in the skins of rare, beautiful or useful wild birds.[55]

Hudson mentions in letters that lawsuits were threatened against him in connection with this campaign/counter campaign on plumage,[56] and it is most likely because of what he may have suggested about establishment figures being bribed by the plumage merchants. I don't think it ever went to court, and I don't suppose Hudson's allegations were specific enough for it to get that far.

It was most likely Sir Edward Grey who had 'the blue book with full debate on the Plumage Bill' sent to Hudson in mid-March 1914.[57] 'We hope it will be law by April!' Hudson wrote, the untypical exclamation mark signalling his excitement.[58]

A report in the *York Press* of the debate around the Hobhouse Bill gives a flavour of proceedings in the Commons. Member for Orkney and Shetland Mr Cathcart Wason – a 6 foot 6 inch farmer who liked to knit while listening to debates – withdrew his amendment asking for ostrich feathers to be banned as well, after being assured that there was no more cruelty on a South African ostrich farm than on a sheep farm. Sir Edwin Cornwall then objected that licences would be granted to allow the import of plumage for museums or for scientific research. He reasoned that this would leave the door open for collectors, who were more of a threat to wild birds than the plumage trade.[59]

In response, it was cited that 5,000 birds of paradise had been exhibited for sale by three traders in June 1913. The postmaster-general said he did not think that collectors shot anything approaching that sort of number, although whether those birds of paradise were destined for hats or display cabinets is unclear. He promised that if this law passed he would set up an advisory committee to decide who might be permitted a licence.

But the truth is that the protection of wildlife was slipping down the political agenda. Political trouble was coming to the boil, at home and abroad, and Hudson's earlier optimism was soon tempered by concern that unrest in Ireland would scupper things, as a groundswell of nationalism there was met with the rise of loyalist volunteers. 'The Ulster fight may begin any time within a month from now,' Hudson wrote to Garnett in early March, 'and once a shot is fired, farewell to any legislation this session.'[60] He was right to be cautious, although it was the escalating tension in Europe that would shortly overtake Parliament.

Life in a Surrey pinewood

Even the arrival of spring in 1914 could not prevent Hudson from slumping into depression. Flowers sent to him by well-wishers, including Ranee Margaret, could not lift his gloom. For a change of scenery he took Emily out in one of the rapidly growing fleet of taxi-cabs that had recently become a feature of London life. They were driven around Hyde Park, pausing to watch the riders and horses at Rotten Row. There were rays of sunshine between towering cumulus clouds, but to Hudson all seemed dark. 'It was a pity I had not slipped out of life during the last few dreadful days,' he later reflected. He had been experiencing '"visions," flashes, colours and strange figures appearing all about me, day and night'. His doctor had tried to assure him it was just a phase and would pass.[61] Hudson had ground almost to a halt, unable to write. There were two letters from Sir Edward Grey to which he felt unable to reply. He had his doctor kindly write back to Grey on his behalf.[62]

In late summer, Hudson took his seat in Ranee Margaret's car for the long journey back to Cornwall. They went first to Grey Friars, her country house near Ascot, to collect some things, where they both realised that he wasn't up to the long journey. She summoned a new doctor. After examining the patient, the new man disapproved of his current treatment and prescribed a new one.

All the shuttling between London and Worthing, where Emily was now in permanent care, had taken its toll, and Hudson was forced to accept that he was seriously ill. He stayed at Grey Friars for long-term convalescence, where he could enjoy exploring its own patch of the area's extensive conifer plantation. He also had access to Lady Stepney's larger estate next door. Lady Ponsonby lived nearby too, with many a tale to tell from her time in the court of Queen Victoria. Sir Walter and Lady Palmer were also neighbours, and a little further afield were Lord and Lady Esher, aka the Bretts, whose daughters Dorothy and Sylvia had featured so prominently in Hudson's life before they left the family home.

Besides half a dozen servants, Hudson was also sharing the house with a menagerie of pets led by Tootles the parrot, and so many dogs he couldn't attempt to remember their names. Ranee Margaret was busy attending to the needs of her three now grown-up sons, and her ever-hectic schedule of social engagements, including visits to Edinburgh and Liverpool to address meetings, and attending the opening of Parliament by the king. She returned to Grey Friars from time to time to check on the convalescent, and recounted in her memoir an incident that gives a strong flavour of her rapport with Hudson at this time. He had news of some unexpected visitors.

> He was much excited at the appearance of a pair of
> large hawks, somewhat rare birds for that part of the
> country, which had built their nest some forty feet from
> the ground in a branch of a fir tree nearby. One pouring
> wet evening, on coming home to Grey Friars for tea,
> I met Hudson at the hall door in the greatest state
> of excitement, pointing to a ladder fixed against the
> hawks' tree.[63]

Here we might imagine the glint of mischief in Hudson's eye. "'Look here,' he said. "I sent to Sunningdale for this ladder. Do just run up it and see what the nest is like inside. I can't trust anyone else. They would see nothing, and I can't go myself because I've hurt my knee.'"[64]

I can picture him pointing rather feebly at his trouser leg, with a pout and pleading look. It's not clear whether this was the old war wound, from the self-inflicted bullet in Patagonia, or a new injury. Not that his hostess could care: 'He evidently thought the request quite an ordinary one, and was somewhat surprised and annoyed when I declined to comply with it. Next day, when the ladder was called for and carried away unused, he was very disappointed. I was sorry, but no! at my age I dared not venture such a gymnastic feat!'[65]

So, on this occasion at least she refused to fall in with his scheme. He might have been in his 70s, but she was hardly in her first flush of youth either, being nearly 60 years old by this point.

I wondered if it were possible that Hudson had found another fugitive pair of goshawks, there at Grey Friars? The ranee does describe them as large, after all. Her anecdote doesn't tell us much more, but a surviving letter from Hudson to another friend gives his version of events, and reveals that he found another way of identifying the contents of the nest.

Unsurprisingly, they turned out to be sparrowhawks. He would have been pleased nonetheless. 'Landowners and their keepers have for many years been engaged in trying to exterminate first the goshawk and after it the sparrowhawk,' Hudson railed in *Lost British Birds*. 'In the case of the latter they have not quite succeeded yet.'[66] He described the nest site and the activities he'd observed:

> It is a tall tree but with little foliage and standing on a hillock close by one can look at the nest with the binocular and see all four young birds quite plainly and watch their doings. Yesterday two of the young birds looking like owls sat side by side in the middle of the nest apparently surveying the scene and elbowing one another; a third stood up on the rim of the nest shaking his wings as if wishing to fly, and the fourth had dragged a small bird to the opposite side and was tearing at it.
>
> Sometimes they drop a bird, and it looks as if it had been plucked and prepared for cooking at a poulterer's shop, so well do the old birds clean it of feathers. They also take off the head, which I fancy they swallow themselves … By-the-by, the old hawks here are excessively shy and slip away and vanish when we go near the tree.[67]

Hudson had a lot of time to properly consider his new surroundings, and decide that he didn't like them so much – the wall of conifers in their plantation rows hemming him in and obstructing his views. Exploring, he found the ground below the trees frustratingly bare and lifeless, bar the wood ants and their impressive mounds of pine needles.[68]

He described his existence in a satirical essay 'Life in a Pinewood', which became chapter 1 of *The Book of a Naturalist*, in which he lampooned the fashion for owning such properties among the Surrey pines, which were purported by some to have health-giving properties. His piece was admired so much for its evocation of pleasures for all the human senses that the Royal National Institute for the Blind asked permission to reproduce it, which he happily granted. He discussed this with Ranee Margaret, who was magnanimous enough to let pass its rudeness about one of her homes.

He was also uncomplimentary about parrot keepers, and wrote of 'hating the bird before me because of the imbecility of their owners'.[69] Luckily for Hudson, she was by now well used to and tolerant of his brute candour. It illuminates their unusual camaraderie, her forbearance, and that maybe she didn't take his complaints too literally or seriously. I think that the starkness of Hudson's printed words on a page sometimes masks the irony or jest that infuses them.

Whether the pines were a help or not, Hudson's health slowly improved, but the papers were full of talk of war. His anxiety about this may have been a factor when he had a relapse while on a visit to the south coast. 'I was gradually getting well under the Ascot doctor,' he wrote to Edward Garnett, 'but unfortunately on a visit to my wife here caught cold. I am laid up with bronchitis and other troubles. I have been confined to my room a month and no prospect of getting over it yet.'[70]

As June neared its end, news rang round the world that Franz Ferdinand and Sophie had been shot dead in Sarajevo. Four days later, Sir Edward Grey read a telegram from the British ambassador to Serbia, Dayrell Crackanthorpe, who wrote describing the mood in Belgrade as one of 'stupefaction rather than of regret'.[71]

On 22 July, Hudson lunched again with Grey, and was picked up and returned to Tower House in Grey's car. Although preoccupied with bigger issues, Grey told him that he still expected the Plumage Bill to be passed. Hudson was not so confident, although not because of the threat of impending war in Europe, but rather thinking that the ongoing unrest in Ireland would eat up the parliamentary time required. 'I fancy the cabinet people are all convinced that there will be no war in Ireland and that things will go on much as usual,' he reported to Ranee Margaret.[72] Hudson was right – the Plumage Bill was dropped, but for a different reason.

John Galsworthy made the following diary entry on 14 July,[73] demonstrating his continuing support for the plumage campaign: 'Sorrowful petition from Birds Protection Socy. to do what I could for the Plumage Bill. Fate to be decided on Thursday. Wrote letter to the Times'.[74] Galsworthy then dined alone at the Savoy, hoping that *Peter Pan* author J.M. Barrie might join him. The next day his latest letter was published, with others on the same subject. It is curious to find on the same page his comments on the dread approach of war.

> July 29. These war clouds are monstrous.
> August 1. The suddenness of this horror is appalling.
> August 2. Too ghastly for words.[75]

That 2 August was a Sunday, and Don Roberto was back in Trafalgar Square making a fiery speech at a peace demonstration, speaking from the plinth of Nelson's column alongside Keir Hardie and others. Meanwhile, Galsworthy continued to plot the inexorable slide to war:

> August 3. A miserably anxious day. I hate and
>   abhor war.
> August 4. Belgium's neutrality violated by Germany.
>   We are in. All happiness has gone out of life.[76]

Anxious to help with the war effort where he could, Galsworthy enquired about ambulance training.

Hudson's 73rd birthday was on 4 August 1914. This is also St Dominic's Day, and his mother's Catholic neighbours had held strongly to the view that this child should have been named after the saint (Hudson sometimes signed himself Dominic in his letters to Ranee Margaret). But he had never been keen on the occasion, or the fact of ageing. Sir Edward Grey, looking out from his Westminster office above (perhaps appropriately) Birdcage Walk, made his now immortalised comment that evening: 'The lamps are going out all over Europe. We shall not see them lit again in our lifetime.'[77]

In the context of today's political soundbites it seems a surprisingly pessimistic note for Grey to strike, but it was typically candid. He was no cheerleader. The prevailing mood in the country was that it would all be over by Christmas, but of course no one really knew what the world was in for, and history would prove him right to fear the worst. La Belle Epoque (1871–1914), as it would later be dubbed, had come to an end. Europe was at war.

CHAPTER 12

Lamps Going Out

The First World War

Two weeks into the conflict, Hudson described his situation. 'I have been staying at the Grey Friars, Ascot, the Ranee's country house. It is a beautiful place in a pine-wood with another larger pine-wood (Lady Stepney's) in which I walk every day and watch the birds'.[1] 'I am absolutely alone, except for the six servants in the house; a silent house surrounded by silent woods ... the only visitor I have is Miss Ponsonby who trots in from old Lady P.'s house close by to see me and discuss the news.'[2]

Keeping pace with the times, Ranee Margaret had a telephone installed at Grey Friars, but he had 'heard no tinkle of the tele-bell', and all was 'marvellously peaceful'.[3] But Hudson was glad of the occasional company of neighbours. On one of her regular visits, Maggie Ponsonby told him about a recent adventure, when she had travelled in the middle of the night to Aldershot to wave goodbye to her brother as he boarded a train with the other recruits. The rumour was that 150,000 of them would provide a second line of defence in northern France, in case the enemy broke through from Paris. But of course the War Office were keeping details out of the papers. She was well connected, with one brother a radical aristocrat member of Parliament, another the King's equerry, and a third Major Ponsonby, 'now perhaps getting killed in Belgium' as Hudson put it.[4] Hudson joked that he would not be able to protect Grey Friars, as he lacked a revolver – perhaps no bad thing, considering his track record with pistols.[5]

While Hudson tried to appear sanguine, Ranee Margaret was becoming a little frantic. She called the war a 'dire catastrophe',[6] and retreated to her estranged husband the Rajah's house, near Cirencester in Gloucestershire, partly in case of an enemy invasion and to put more miles between herself and the south-east coast where they might be expected to land. But the manoeuvre had evidently not eased her anxieties, and she found the Rajah's Chesterton House lacking in a few home comforts, such as curtains. I was surprised to discover in one of Hudson's letters that Bird

Society founder Eliza Phillips, now frail and in her nineties, was staying with her there. The ranee's ageing husband was also by now in need of support, as his beloved Sarawak's future was in great peril, and his son and heir Vyner must be made ready to take over the reins, assuming he could safely travel halfway round the world to get there. Ranee Margaret's two younger sons would also enlist, one being posted to Egypt. 'The old Rajah is half crazy about the war',[7] Hudson reported to Roberts.

Meanwhile, Hudson sat alone in the porch at Grey Friars, surrounded by gardens and beyond them the regiment of pines, as he looked over page-proofs and assessed schoolchildren's essays for the Bird and Tree Competition. Tootles the parrot was a constant companion, placed on his favourite tree beside the porch, along with several dogs – Licky being Hudson's favourite – a couple of macaws and a cat called Caesar. The magpies, jays and even a green woodpecker that visited the garden when they thought no one was looking must have seemed exotic, like fellow refugees. War planes occasionally droned overhead.

Hudson wrote from here in mid-September 1914: 'My wife is going on all right and I hear every day – she urges me to stay on here where I'm getting every comfort and help and half a dozen servants to look after me.'[8]

He was able to make visits to London and the south coast, and while visiting Emily he witnessed Worthing emptying, as holidaymakers rushed back to London, fleeing from the coast to the city as rumours spread and fears grew of imminent enemy landings on the coast. The now quiet resort was instead filling with army volunteers: 'a herd of 28 thousand recruits, labourers, clerks, hooligans from London slums, and all sorts', as Hudson described them, 'half of them wretched-looking beings, thin and pale'.[9]

But despite his relatively comfortable situation at Ascot, before long Hudson had had enough.

> The talk is war – war, war, and I'm weary of it … We
> have six or seven papers here a day and I search in
> vain for something not about the war … my hope and
> prayer is that we may crush the mighty war lord, God's
> friend and favourite, utterly before long and so have a
> normal life for the world once more.[10]

It had also dawned on him what war would mean for him and his fellow writers. 'What a deadly thing it must be for literary folk! What chance will any book stand now? Perhaps there is to be no season at all for publishers,' he railed. 'Editors tell me they can't promise to use anything they accept for months to come – perhaps not till after the war.'[11]

Like many people, Hudson had affected some bravado about the war at its outset, but quickly reined this in as the grim reality hit home. He told Garnett that he prayed for it to be over and for normality to resume. Thirty months in and Garnett provoked him into this reaction: 'You hope I'm satisfied with the war! Well, it's quite useless our discussing that subject – we can't understand each other.'[12]

Nearby, Ascot's famous racecourse grandstand had been commissioned for a war hospital, with Hudson's doctor in charge. It soon began to fill. By mid-November 1915, 70 or 80 war wounded were being treated there.

Edward Garnett served for a time as a medic, in Italy and France. 'He has written to me,' Hudson reported, 'and enclosed a letter from a local who says he loves my writing, and that of Richard Jefferies.'[13] When he showed the letter to Ranee Margaret, she arranged to send more books by Jefferies, including The Story of My Heart – 'a revelation to those who have never read Jefferies',[14] Hudson called it. Hudson also received a letter about birds from a ten-year-old child, whose father was serving alongside Garnett.

Later in August, Hudson made a brief return to London by train to collect some belongings and see his doctor. As he waited on the platform at Ascot station, a train-full of soldiers sped by – prompting spontaneous hurrahs and hat-waving from the platforms. He found London looking unfamiliar, and eerily quiet. He now saw for himself the streetlamps painted out, with night-time air raids expected. He hoped that 'the blood that is being shed will purge us of many hateful qualities – of our caste feeling, of our detestable partisanship, our gross selfishness, and a hundred more … It may be that before another fifty years the human race will discover some means of saving itself from rotting without this awful remedy of war.'[15]

Just before Christmas, he was with Emily again at Worthing. 'The war has been playing havoc with several of our young poets,' he wrote from there. 'Rupert Brooke is fighting somewhere in the trenches, I hear.'[16] His young friend Edward Thomas had not yet been sent out.

As a momentous year ended, Hudson was grounded again at Worthing on the south coast, lifting the newspapers each morning, braced for bad news. He saw news of his writer friend Maurice Hewlett's son being lost in the Channel when his sea-plane was shot down on Christmas Day. A few days later he read an update, that Hewlett junior was still alive and had been rescued. Hudson almost jumped out of bed for joy. Never at ease in Worthing, he dreamed of finally escaping to Cornwall come April, when the gorse-covered cliff-tops and hillsides would be in bloom.

With the civilian population continuing to evacuate, thousands of raw recruits were filling the seaside resorts of the south coast. 'Shanty towns' of wooden huts were appearing on the slopes of nearby hills. Warfare raged just across the channel, the gunfire of northern France audible in some places. 'Two months shut up in one room',[17] Hudson groaned. Emily was able to take short walks outside, with the aid of her companion-carer, while Hudson tried to keep himself occupied with his writing. 'I'm just in the middle of a book I must finish,'[18] he told Roberts as spring arrived. By the end of the month, his report gives some idea of the recruits being prepared for the fighting:

> Still full of troops here ... but what a change in
> their appearance since they came last September ...
> Now they are marvellously 'plumped out' brown and
> jolly. There are about 12,000 of them billeted in the
> Worthing houses ... on Wednesday next they are going
> to march off to Maidstone ... the aeroplanes are about
> here every day – generally flying very high now.[19]

Given their high spirits, the new recruits may not (unlike Hudson) have heard, as he confided to Roberts,

> the truth of things from the front, which is kept out of
> the papers. About Neuve Chapelle, a most lamentable
> business. The French military acted against the advice
> of Joffre and Kitchener, but would have been all right
> if his officers had not blundered terribly. Our men
> were mowed down wholesale by our own guns. Smith-
> Dorrien and two other unhappy generals have been
> recalled. Not a word of all this is allowed to appear in
> the papers.[20]

Hudson had other sources. Edward Garnett returned from France where he had been serving with an ambulance unit, and brought with him a bleak view of the situation on the front, and prospects for the winter ahead – 'awful days of carnage and disaster',[21] as Hudson then described them. He was starting to find names he recognised among newspaper lists of the fallen. But he did have some better news to report, as Emily was now 'going on fairly well – well enough to be working at woollen knitted garments for the soldiers, like everybody else … To be able to do that much has given her a kind of happiness … There's nothing like it [work] to keep one going – unless it tempts one to smoke too much.'[22]

Spring brought little respite. To add to the misery of war for combatants and civilians, April 1915 was said to be the coldest since records began. Hudson was for a time back in Tower House as the mercury finally began to rise in thermometers. 'The leaves of my elms have come out with a rush the last few days. It is like the sudden change from bleak winter to spring in the high latitudes of the Siberian tundra.'[23] He had been reading more Russian novels, but would also suffer the effects of reading too much news.[24] This was now driving him to distraction. He would look briefly at a newspaper before 8 am, then try to put thoughts of it aside, then look at the final edition of *The Star* in the evening. He was trying to give up his habit of reading too many papers, then indulging in talk with others about what they had read. He likened himself to a man in a restaurant who has had everything on the menu and still wants more. He had what he called a 'mental dyspepsia', worried he might become one of the people he knew who had been driven mad just thinking about the awfulness of it all.[25]

Hudson returned to Ascot, but it didn't help much. 'I want to be in Cornwall – at the Land's End!' he wailed, the appeal of the Surrey pinewood now waning as he dreamed of his wild Cornish coast. The enclosing wall of trees sometimes unnerved him, this man of the Pampas: he spoke of it triggering a 'forest fear'.[26] For the sake of his sanity, he finally set off with the ranee for their long-awaited trip west, electing to drive rather than take the *Riviera Express*. 'I got about a good deal in the car during the five weeks', he reported later, 'visiting all my old haunts as far as I could and seeing old friends.'[27] The ranee would pick him up in the car and take him out to find familiar places and faces, and maybe even some new friends. 'At one or two or three other houses I have been to tea and met a crowd of people,' he joked, 'mostly uninteresting.'[28]

Bird Notes and News was publishing increasing numbers of letters from soldiers in the trenches, describing their rare moments of solace when birdsong distracted them from the horrors unfolding around them. In late April 1915, news came through of the death of poet Rupert Brooke. Hudson and Ranee Margaret received the news while on their extended break in Cornwall. Hudson 'bitterly lamented' the loss. Brooke died from illness on a ship in the Mediterranean, where the horrors of Gallipoli had yet to unfold.[29]

It was with mixed feelings that they set off for the drive back to Ascot in late May. There would be delays. 'On our way back we tried the Tavistock route but found the hills too steep for the rather weak car and at some hills had to get out and walk.'[30] Hudson made sure the time wasn't wasted:

> During one of these ascents I discovered a most lovely
> wild flower I had never seen before – the Bastard Balm,
> the Melisse des Bois of the French, a flower of the same
> tribe as the dead nettle but much larger – cream white,
> with a purple lip. Finally we had to abandon that
> route and go twenty miles or so round to the other one
> which goes by Okehampton.[31]

Despite the disruptions of war, he found there was still some demand for his writing. 'I get a few article in magazines and weeklies from time to time and have half a dozen out now,' he told Roberts. 'As for books I couldn't get one placed while the war lasts. I'm not surprised you are trying to get some job with the forces: that seems to be what everyone is trying for.'[32]

He found some solace in the storms and heavy rain of late summer that had given London's parks an unseasonably fresh, green look. The acacia tree outside his Tower House window was revived. His forlorn hopes for peace were poignantly expressed in his account of a walk through the city on 13 August 1915:

> Yesterday was our best day this long time past … a
> magnificent rainbow appeared and lived in the black
> eastern sky from 5 to 6.30 – I never saw one last so long
> or show such brilliant colour. I went down to Hyde
> Park … & in Edgware Rd. this crowd of people were all
> gazing at it as if it was an unknown thing, something
> unimagined, a message of peace perhaps. But there will
> be no peace.[33]

In the shorter days of autumn he groped about in the darkness of the streets, if he dared to venture out to post a letter. The last streetlights were now being painted out, and he went to bed each night expecting Zeppelin air raids. The first came in October, and he maintained that if raids happened he'd rather be in London than away, so he could witness them, and have some idea of their real impact.

A year into the war, John Galsworthy sent Hudson his new book, *The Freelands*. Writing to thank him, Hudson explained that he had hoped to receive it from Ranee Margaret, being too poor himself (a situation exacerbated by the war) to buy his own. He had just finished reading Tolstoy's epic *War and Peace*, 'particularly interesting just now when the Russians, as almost always, are being so terribly beaten'.[34] Galsworthy wrote to his publisher to say that his 'friends of letters seem to like the book more than usually'.[35] He mentioned Hudson's name, along with Garnett and Conrad – a further measure of the store he set by Hudson's opinion, and the weight he thought it would carry.

It has been enjoyable piecing together traces of Hudson's alliance with such a literary giant as Galsworthy. The biggest revelation is something oddly coincidental.

Among the angels

In January 1915, Hudson suddenly mentions in a letter to Ranee Margaret Brooke that Galsworthy has known Bird Society founder Eliza Phillips since he was a boy, way back in the 1870s, when the young arriviste Hudson was still finding his feet in London. Phillips had been a close friend of Galsworthy's parents, to the extent that they asked her to be godmother to his older brother Hubert, 'but her favourite was John',[36] Hudson had established. Their love of animals – especially dogs – helped them to bond. Hudson mentions it as he intends to give Phillips a copy of Galsworthy's latest book, a tribute to his beloved cocker spaniel Chris.[37]

I ordered one of the many biographies of Galsworthy, hoping against hope that it might contain a photograph of Phillips. No image of her is known to survive, and finding one would be precious. Drawing a blank with the biography, I made contact with his grand-daughter, the artist Jocelyn Galsworthy, who was very helpful but alas had no knowledge of Phillips. I then tried Birmingham University's archive, which houses some of Galsworthy's collection. The bulk of this was sold to Forbes in New York some years ago. The Birmingham archivists were extremely helpful, and there are quite a few photographs from the Galsworthy collection viewable online, but sadly I could find no trace of Eliza Phillips.[38]

For part of the war years, Galsworthy worked in a hospital in France. Besides animal rights he was a renowned social reformer, an advocate for women's voting rights and for prison reform, and president of a society that sought to foster international relations through literature. I've seen no record of Hudson attending this, although his friend Joseph Conrad did.

By late 1915, Galsworthy had piqued the interest of US publishers (unperturbed by war at this stage, before America joined in), and he was invited to draft the foreword for an American edition of *Green Mansions*. He wrote the long and effusive testimonial while at his Devon farmhouse in September:

> Hudson, whether he knows it or not, is now the chief
> standard-bearer of another faith … All Hudson's
> books breathe this spirit of revolt against our new
> enslavement by towns and machinery … are true
> oases in an age so dreadfully resigned to the 'pale
> mechanician' … [he is a] very great writer; and – to
> my thinking – the most valuable our age possesses. Of
> all living authors – now that Tolstoi has gone – I could
> least dispense with W.H. Hudson … Hudson is not,
> as Tolstoi was, a conscious prophet … his spirit is
> freer, more wilful, whimsical – almost perverse – and
> far more steeped in love of beauty. I would that you
> in America would take him to heart. He is a tonic, a
> deep refreshing drink, with a strange and wonderful
> flavour.[39]

Galsworthy had first written in 1912 on Hudson's behalf to his contacts in New York publishing, sending them a copy of *Green Mansions*, and extolling the virtues of its author. It was the start of something big in America for the humble Huddie, as Galsworthy and his wife Ada fondly knew him.

Late in 1915 Hudson was established in Cornwall again with Ranee Margaret and her staff, but the prognosis wasn't good. 'We thought the change in milder climate and change would have done me good, but I got worse in health almost day by day.'[40] In early December, he heeded the advice of Dr McClure and was taken to St Michael's Hospital in Hayle, linked to St Theresa's Convent. Ranee Margaret, a devout Catholic, was an almost daily visitor, the Hartleys and other local friends slightly less regular.[41] They would find Hudson propped up on pillows, fringed with his notes and books, or feeding the birds at the nearby window. From there he enjoyed a splendid view, and a balcony overlooking the pastures and Hayle estuary beyond – today an RSPB nature reserve.

The self-styled 'religious atheist' noted the mix of faiths around him.[42] 'The next room to mine is now occupied by a priest, and the Bishop of Southwark is a guest this week,' he wrote, ten days in to his stay. 'But religion doesn't come into the nursing part of the business and the doctor, Hamilton, is a Free Kirk person.'[43]

He warmed to his theme, as he was fond of needling Ranee Margaret about her Catholic faith. 'The doctor is a Scotsman,' he joked to Garnett, 'and staunch protestant so no harm will come to me spiritually. Be assured of that.'[44] Luckily for Hudson, the nuns – 'the kind nurses' in his words – 'don't mind a protestant'.[45]

The care of the nuns soon had a positive effect. On Christmas Day, Hudson announced cheerfully that he hadn't felt so well for a year. He was also able to get a few more articles published in weekly and monthly magazines. As word spread of his situation, the celebrity patient evidently became an attraction locally, and he even hinted at some enjoyment of the attention. 'I'm having quite a run of visitors to see me these days – rectors, priests, library ladies, and so on.'[46]

Early in 1916, he exchanged letters with John Galsworthy.[47] The Galsworthys were at Hudson's hospital bedside soon after, and John made the following brief entry in his diary:

> February 17. By train to Lelant, Cornwall. Saw W.H. Hudson in his nursing home at Hayle. He has aged, got softer and whiter. Very fine still. Thence to the Ranee of Sarawak's rented house, at Lelant. Much talk.
>
> February 18. At the Ranee of Sarawak's. Woodside, Lelant, Cornwall. Sou-West gale. Motored with her and saw St. Ives. Motored to Hudson's; found him better. Much talk.[48]

The following day, Galsworthy returned to Wingstone, their farmhouse near Manaton in Devon. 'Found Ada well. Walk. Rode. Very sleepy.'[49]

Not long before Hudson was released from the convent hospital, he reported to Roberts that his friends were going to be lodging nearby. 'The Galsworthies – he and she – are coming to the Tregenna Castle Hotel for 3 weeks, to be near me, they say, and I can see them much oftener if I am at

Lelant.'[50] The plan was somewhat disrupted when Galsworthy was struck down with influenza. As soon as his health improved, he moved with Ada to stay at St Ives, and they became regular visitors at Ranee Margaret's cottage.[51]

'A great pleasure, Huddie,' Galsworthy wrote in his diary after one visit. 'He is not really so ill, I believe, as he thinks himself.' He then scribbled a typically perceptive summing up of the old man by whom he was so obviously captivated. 'Quite the strangest personality in this age of machines and cheap effects. Like an old sick eagle.'[52]

When spring came round, Hudson was finally deemed fit enough to be set free from the convent hospital, to join the rooks and jackdaws he had been feeding from his window sill. 'I shall grieve to leave this home, tho' a religious house and stronghold of the Pope (and Satan),' he admitted, tongue in cheek. 'I'll miss it mainly for the splendid view from my window and the balcony of the country and the green fields.'[53]

The nuns, for their part, were (he said) sorry to see the 'only heretic among them' go. 'They were anxious to keep me and when I said I was hoping to 'economize' the Head Sister said they would gladly reduce their terms to meet my views, so I had to say I wanted a change!'[54]

Of the staff who had been caring for him, he was going to miss one in particular – Sister Mary Cornelia: 'my dear and intimate friend: who has been from 17 years of age in this order … and is now about to take the last and perpetual vow or vows'.[55] Detecting her uncertainty in the matter, and perhaps to provide some balance in the guidance she was being given, Hudson had been encouraging Sister Mary to think twice about committing to the life of a nun. 'I've done my best to try and shake her resolution, but she seems bent on carrying the thing through.'[56]

Ranee Margaret disapproved of him 'meddling' in the life choices of Sister Mary, and it would become a source of friction between them as her patience with his secularism – and needling – wore thin.

When the former Sister Mary later wrote to him from France, he relished relaying to Ranee Margaret how grateful Mary was that his 'providential words' had 'set her free' from the Church.[57] He must have known he was testing Ranee Margaret's faith, and eventually patience. 'The God you feelingly describe is not the God of nature,' he wrote to her in May 1917, 'the Ranee of Sarawak's giant shadow.'[58]

There had been wider tension in the convent because of wartime paranoia about anyone with German heritage, a description fitting a few of the nuns. Ranee Margaret picked up the rumours that were spreading.

Things came to a head when the Lady Superior, overcome with grief about the anxiety it was causing her staff, asked Dr Hamilton to call in the head of the police who, when he came, tried to assure them all that it was nothing but idle chatter. Happily, all the nuns were able to continue with their work.[59]

When Hudson was discharged in early March 1916, Ranee Margaret sent her car to collect him, and he was driven the short distance to Badger's Holt cottage, at nearby Lelant. She, meanwhile, was 'deep in snow' 250 miles to the east, at Grey Friars. Mostly alone again, he sometimes missed the company he'd enjoyed in Ascot, of staff and pets and neighbours. 'How is Maggie [Ponsonby]?' he asked Ranee Margaret in a letter, before hinting at his difficulty in keeping track of her entourages. 'And how is Amy? And how is everybody else including Lassie and Nancy, Caesar and Tootles, and Bootles and Ootles and Mootles and Gootles.'[60] Perhaps inspired by Hudson, Maggie Ponsonby had written a play. 'I'm afraid to ask her to show it to me', Hudson confessed, 'as I'm so brutal in saying just what I think.'[61]

In Ranee Margaret's prolonged absence, and perhaps missing her entourage of pets, he made the acquaintance of what he described as a 'very important rat' at the Lelant cottage.[62] He scribbled an essay about it and posted it to The Times, not expecting it to be taken seriously, but to his amusement 'they took it at once'.[63] Even for Hudson 'it was odd to have a long article in The Times about a rat in these days, but...'.[64] Soon after, it was also published in a Cornish paper – a matter of great pride locally.

Following his extended stay in hospital, Hudson would finally heed medical advice and try to be careful from now on not to overburden his fragile heart, and pick his battles carefully. When he encountered another owl in chains on one of his outings, he refused to get involved, to avoid further heart strain.[65] His latest doctor put him on digitalis, to regulate his heartrate. Unexpected developments meant that he was going to need yet another doctor. 'I'm always hearing from everyone I know that they hear I am in good spirits and better health,' he told Garnett. 'So were my doctor's spirits good the other day when he called to see me and chuckled and coughed over his own funny stories, and next morning getting into his motor-car and lighting a cigarette, he dropped dead.'[66]

Hudson kept himself busy with his writing, needing the income apart from anything else to pay for Emily's care, while others he knew were giving up. 'One of my friends says he has written five books since the war began – one to order – all the others not accepted, and at last in despair he has enlisted.'[67] War would at least pay a salary.

William Rothenstein provided this fascinating insight into Hudson's continuing poverty, and the generosity and esteem of his friends, that might otherwise have been lost:

> Hudson's books brought him but little. Even as late as 1916, the Ranee of Sarawak told us that, Hudson's wife being seriously ill, he was hard put to it to send her away to the sea-side.
>
> Sir Edward Grey was then Foreign Minister; when I wrote to him of Hudson's difficulties, he replied in his own handwriting; a delicate precaution, I thought, on the part of a man so beset as he was.
>
> A sum of money—£200—was collected among his friends, and, through Edward Garnett, was discreetly placed to Hudson's account without his ever discovering the secret.[68]

Hudson's fortunes in this regard were about to change. He had been eagerly awaiting publication of the American edition of *Green Mansions*, complete with Galsworthy's preamble. Wartime surveillance measures often meant delays or interceptions of post – the publisher in New York told Hudson his letter had been opened by the British censors. 'I hope they didn't think *Green Mansions* meant something mysterious and dangerous', he joked to Ranee Margaret.[69] When two copies of the book finally arrived, he thought them 'quite nicely got up'.[70]

Hudson and Ranee Margaret were deeply saddened when Bird Society founder Eliza Phillips passed away on 18 August 1916. 'Vicar Widow's Death' ran the headline in the Croydon local paper *Norwood News*. Hudson had remained a faithful correspondent and occasional visitor at her Croydon home, where she had lived contentedly until the end, reading her daily paper and the Greek poetry she loved. She had lived to be 93, similar in age to Emily Hudson. She left in her will the enormous sum of £100,000 – half of it to support the welfare of non-human animals. She also left £500 to Hudson, £200 to John Galsworthy, and £1,000 to Catherine Hall. The affectionate obituary in *Bird Notes and News* spoke of her 'unwearied devotion and tactful management', and of her 'nobility of character, a bent of mind perhaps somewhat masculine, and a deep sincerity of purpose'.[71]

Next to the obituary for Phillips is a shorter notice of the death of J.A. Harvie-Brown of Dunipace House in Stirlingshire – with whom Don Roberto had corresponded about rooks. Although not a champion of rooks, Harvie-Brown had given the Bird Society a donation of £10 in 1892, with which it set up its first banking account. Another death notice is for Captain J.M. Charlton, killed in France on day one of the Battle of the Somme. It was his 25th birthday. He had been a young medal winner in the society's essay competition only six years earlier. The battle was still raging at the time these notices went to press, and would do so until mid-November.

Hudson's health had improved by the autumn of 1916 to the extent that he suggested taking a detour with the ranee in her car to visit the Galsworthys in Devon, while on their journey west returning to Cornwall. But Galsworthy advised that the road would be too steep for their vehicle, perhaps having heard about its earlier travails in Dartmoor.[72] It seems unlikely that Hudson ever got there.

With *Green Mansions* now published in the States, John Galsworthy provided updates on its progress, passing on press notices and notes from the publisher Knopf, who was satisfied with early sales. Exposed now to a marketplace many times larger than Britain's, Hudson's mailbag started to bulge. Galsworthy also felt the impact, and passed on one letter he received from a professor in Ohio thanking him for introducing Hudson to readers in the USA.[73] Moves were also afoot to have Hudson's other books on South American subjects translated into Spanish, besides being repackaged for the North American market.

Hudson wrote to Galsworthy soon afterwards, impatient to know if Knopf wanted to republish *El Ombú*. This would give him an opportunity to include a story about the feathered women of the mythical Patagonian lost island city of Trapalanda, omitted from the original edition on Garnett's advice. It was a campfire tale he'd heard told by an old gaucho many years before. Garnett's objection had been that he'd seen a similar story in another book. *El Ombú* would eventually be repurposed as *South American Sketches*, complete with Hudson's version of the Trapalanda legend.[74]

Galsworthy's efforts on Hudson's behalf didn't stop there. He now encouraged his friend to work on a memoir. With *Green Mansions* increasingly capturing the imagination of the American book-buying public, Knopf was only too happy to be at the head of the queue for this one too.

Far away and long ago

Etta Lemon later recalled Hudson's frustration at not being able to enlist in the army, but if he indicated that to her, I have not seen it in writing anywhere. In any case, he was by now 30 years too old: active service was restricted to men under 41 (raised to 51 in the final year of the war). And Hudson hadn't carried a gun since he left the Pampas.

Don Roberto, meanwhile, only 20 years past his enlist-by date, didn't let his age get in the way of signing up for the war effort. After pulling some strings, he was soon given a mission by the War Office – to lead a team to South America to procure horses. Not long after, he was writing to Hudson from Buenos Aires.[75] 'A young gaucho yesterday rode forty-eight horses for me, all bare-backed. I remember the time when I could have done it myself. The gauchos are still there though rather tamer than they used to be, but still most self-respecting men, carrying a large knife. (I am eminently self-respecting).'

Don Roberto described for his friend a world moving on:

> Here things are changed, and yet they are not. That is, there is plenty of wild life and lots of gauchos. (I have been with them this week, eating meat roasted on a spear and drinking mate, and nothing else, just as of yore.) The difference is that most of the ranches have telephones ... There are no roads, but, instead, fenced tracks on which motors CAN run ...
>
> I am camped out under some Paraiso trees and roasted some meat. All around me are hundreds of horses, in front pampa, and to the left the great woods stretching down to the River Uruguay and in the distance Entre Rios. Like in the old days, we ate the meat with our knives ... and then we lay down on our saddles and had a siesta. There was a little warm wind blowing, making pleasant music among the green tufts of Pampas grass ...
>
> Last night I and the gauchos drove some five hundred horses through the plains, in high grass. It was a wonderful sight, but sad to think it was their last happy day on earth.[76]

With a heavy heart, he supervised the loading of these 500 horses onto ships at Montevideo, destined for the battlefields of northern Europe. They left Uruguay in early May 1915, for the long Atlantic crossing. Crew and passengers were only too aware of the risk now posed to shipping by German submarines. Just a few weeks earlier the *Lusitania* was sunk on this route, 10 miles from the south coast of Ireland, with the loss of nearly 1,200 lives.

Any fears were well founded. Just as Don Roberto's ship reached the English Channel, it rocked with the impact of a torpedo. They had been struck. The crew managed to run the stricken vessel to a sandy beach, where all passengers and the cargo of horses were safely landed.[77]

The despatches from Don Roberto had set Hudson's mind wandering, as he contemplated his own background and the landscapes conjured again so vividly by his friend. Don Roberto was soon given another overseas assignment. 'Now you talk of Venezuela!' Hudson wrote in November, from Cornwall. 'How I envy you all these wanderings and the adventures you must have had! And I at Lelant talk so of going to Land's End or Godolphin!'[78]

In the quiet of evening, he read again Don Roberto's evocative descriptions of boyhood haunts revisited.

> In those latitudes there is no twilight, night succeeding day, just as films follow one another in a cinematograph … the horses came out of the corral like a string of wild geese …
>
> The falling sun lit up the undulating plain, gilding the cottony tufts of the long grasses, falling upon the dark green leaves of the low trees, glinting across the belt of wood that fringed the River Uruguay, and striking full upon a white estancia house on the opposite bank…
>
> Not far off lay the bones of a dead horse, with bits of hide adhering to them, shrivelled into mere parchment by the sun. All this I saw in camera-lucida, seated a little sideways on my horse, and thinking sadly that I too, had looked my last on 'Bopicua'.
>
> It is not given to all men after a break of years to come back to the scenes of youth, and still find in them the same zest as of old.[79]

Hudson hoped his friend would write a book about his travels, when the war was over. 'And when will that be?' he mused. 'Shall I live to see peace in earth again? To me nothing is left but memories, and I'm here putting some of my boyhood's days in a book which will have a certain interest because it gives a sort of picture of the country and people before it began to be civilised.'[80]

A combination of Don Roberto's missives, fevered dreams and possibly medication had sparked Hudson's visions of his own Pampas childhood. He asked his nurses for paper and pencils and began to scribble what in time would become his acclaimed memoir, *Far Away and Long Ago*. He may also have been inspired by his fondness for one book over a lot of others he had by his bedside. 'I have neglected them all to read a three-penny book I picked up on a cheap stall a few days ago – Leigh Hunt's Autobiography.' It reminded him of his adolescence, when he first took up reading, and the time when, as a 14 or 15 year old, 'owing to being struck down with a fever which made me a prisoner for a couple of months, I first began to look at books'.[81]

Leigh Hunt's book helped him relive his youth. 'The book charms and disarms me, especially the early chapters and most of all those about his mother. What a marvellously beautiful picture he gives of her! Well, she was an American and must have been strangely like my mother, who was also American,' although 'Hunt's mother's people were loyalists while my mother's forbears were furiously anti-English'.[82]

I think it likely that this book influenced Hudson's decision to write his own memoir of childhood, and portrait of his own mother, among the formative experiences and influences he would describe. Guillermo Enrique – as he was known to the neighbours back then[83] – learned to ride from around the age of six, and spent much of his childhood and young adult life in the saddle, if not bareback, living the Pampas gaucho life; wild, close to nature, and frequently hazardous. His parents had migrated south from New England in the 1830s, lured by the prospect of owning land and ranching on a new frontier.

Hudson described early moments that inspired his life-long love of nature, such as when as a six year old, trailing his older brother on an outing, he saw his first flamingos.

> An astonishing number of birds were visible—chiefly
> wild duck, a few swans, and many waders—ibises,
> herons, spoonbills and others, but the most wonderful
> of all were three immensely tall white-and-rose-
> coloured birds, wading solemnly in a row a yard or so
> apart from one another some twenty yards out from
> the bank.

I was amazed and enchanted at the sight, and my
delight was intensified when the leading bird stood
still and, raising his head and long neck aloft, opened
and shook his wings. For the wings when open were of
a glorious crimson colour, and the bird was to me the
most angel-like creature on earth.[84]

It was in many ways an idyllic upbringing for the nature-loving boy, which
he evokes so colourfully in this memoir. For his parents, meanwhile,
this life was turbulent and unpredictable. Hudson recalled the day that
the defeated rump of General Rosas's army passed the family estancia,
his father hiding his horses, and then walking calmly towards them,
unarmed, to discuss the situation with the defeated soldiers. Meanwhile,
other grown-ups cowered inside the house. Overall, despite some family
hardships and traumas in a challenging environment, and the disruptions
of civil strife, the unschooled but self-taught young Hudson had fond
memories of that far away former life.

Don Roberto received a letter from Theodore Roosevelt in June 1916,
the former president writing from Long Island, New York: 'You know,
I think, how much Mrs Roosevelt and I like your writings,' Roosevelt
began. 'We feel that from you, and from some of the sketches of Hudson,
we get the South American as he is given nowhere else. What you and
Hudson have done for South America, many have done for our frontiersmen in Texas, Arizona, and New Mexico. Others have written of the
Mexican frontiersmen.' He urged Don Roberto to now write about such
men in Brazil.[85] Don Roberto wrote back to Roosevelt in March 1917,
mentioning the force that the former president had offered to raise to join
the war effort, and which the current president Wilson had vetoed, much
to Hudson's annoyance.[86]

Green Mansions was having to be regularly reprinted in America to keep
up with demand. It was even reissued with a prologue from Roosevelt. The
new US edition of A Crystal Age, meanwhile, had a new preface from the
literary editor of the New York Times, but sales didn't catch fire in anything
like the same way. 'El Ombú [aka South American Sketches] doesn't go so
well in America',[87] Hudson was told, having 'moderate' sales, along with
the new-look Purple Land.

Don Roberto was handed another transatlantic mission. In January
1917 he was sent to Columbia, his mission this time to procure not horses,
but cattle. 'Bogota, set in its plateau in Columbian wilds, is a kind of
Athens where all men write,' he reported.[88] 'I should have liked a year

or two in such a land,' Hudson replied. 'It would have given me a keener appreciation of the "Vicissitudes of the season" in our "brumous island".'[89] Hudson was also harbouring idle thoughts of less far-flung foreign lands. 'I only wish I could go to Portugal – a nice warm wintering country … in the valley of the Oporto, which they say has a delightful winter climate.'[90]

At the climax of another successful mission, Don Roberto returned to Britain from central America, via Jamaica, but after the excitement of his latest adventures he was struggling to muster much enthusiasm for domestic life: 'To return again to all the cares of life called civilised, with all its listlessness, its newspapers all full of nothing, its sordid aims disguised under high-sounding nicknames, its hideous riches and its sordid poverty, its want of human sympathy, and, above all, its barbarous war brought on it by the folly of its rulers.'[91]

When war came, poet Robert Frost returned to the USA and Edward Thomas nearly uprooted his wife and three children to go with him. As a mature man with a young family, Thomas would not at this stage have been conscripted, but he began to agonise about signing up for the army. Soon after the fighting started, he wrote his celebrated essay *This England* in which he articulated this anguish.

Corporal Edward Thomas

In early 1915, Thomas told Hudson that he was engrossed in writing a biography, and had thought better of enlisting. This would change. Interestingly, he didn't consult Hudson on his final decision. In August of that year, Hudson was asked by Ranee Margaret Brooke if he had heard from Thomas. Hudson replied that he hadn't, although there were rumours. Hudson had felt his young friend's dilemma. 'What a tremendous struggle must have been going on in him, that he would think of exchanging his pen for a rifle.'[92]

In the end, Thomas had enlisted with the Artists' Rifles and would be given a long period of training, and another spell teaching map-reading to young officers. In September 1916, a now uniform-clad Thomas visited Hudson, taking leave from his barracks north of London. 'You would hardly know him in khaki now, in his bold soldier stride,' Hudson reported to Ranee Margaret. 'He used to rather slouch along.'[93]

As the months of war passed, Hudson became increasingly anxious for news of his friends, particularly Thomas when he was finally sent to France. They exchanged letters at the start of 1917, Thomas asking about

Helen Thomas aged 36, with her dog, c.1913.
Image courtesy of the Estate of Helen Thomas.

progress with the memoir, which Hudson reported was on hold, his ardour for writing cooled by the bitter wintry weather at Falmouth in Cornwall. Thomas invited Hudson to visit Codford in Wiltshire, where he was billeted, but this was out of the question. Hudson hoped that he would still be able to hear from Thomas if he was sent to France. Thomas's call to arms came in February.[94]

Despite his army commitments – and perhaps even helped by them, and his soldier's salary – Thomas was having a volume of poems published and promised to send one on. 'I am always wondering what an effect this strange military experience will have on your literary mind,' Hudson told him. 'I only hope that the reply will come in some book while I am still alive.'[95]

Thomas saw action for just two months. In early spring Hudson received the news he had been dreading. The letter came from Roger Ingpen, another of Thomas's close friends. Thomas had asked, in his last letter to Ingpen, that a copy of his book of poems should be sent to Hudson when it was published.

Hudson wrote to Edward Garnett the next day: 'It is a great grief to me as he was a most lovable friend – I feel his death more than that of anyone I know who has fallen so far.'[96] And to Morley Roberts on 17 April: 'Alas!

A careworn Edward Thomas in 1914, aged 36, with Europe on the cusp of war.
Image courtesy of the Estate of Helen Thomas.

My poor young friend, Edward Thomas … has just been killed.'[97] He also shared his grief with Ranee Margaret: 'Fighting was so hateful to his soul, and yet he went because he could do no other, because, as he said, he would never dare to say that he loved England again if could not fight and die for her.'[98] 'Poor Thomas hated war.'[99] Hudson's sorrow was deepened when he thought of Thomas's 'young wife'[100] Helen and their two young children, and an older child now aged 17. Hudson would remark that in Thomas 'he had seen the son he wanted'.[101]

When the war ended, Hudson was sent the manuscript of the poems Edward Thomas had left behind when he was sent out to the Western Front, 'to be killed at Arras'.[102] He also gave the letters he had received from Thomas to James Guthrie, who was compiling the tribute A Literary Pilgrimage in England.[103]

'I have a letter today from Walter de la Mare,' Hudson later wrote, 'who says an effort is to be made to get a Civil List Pension for Mrs Edward Thomas.'[104] He provided a letter in support of this.[105]

Concerned for his friend's future well-being, John Galsworthy encouraged Hudson to start wintering abroad. Failing that, he also recommended

Kingsbridge in Devon. But the needs of Emily, if nothing else, would keep Hudson's horizons narrow. Ranee Margaret wanted to know if Galsworthy could find a suitable winter refuge for him. Hudson was shocked at the suggestion, pointing out that Galsworthy was much too important a figure to ask such a thing: 'people hold their breath when they hear his name', he reminded her.[106] Hudson had heard rumours that Galsworthy could now command advances of £4–5,000 for his books. He was nearly right, but it seems he was unaware of the extent of Galsworthy's philanthropy: he was giving all his spare earnings to charity while the war lasted. 'Literature is its own reward,'[107] Galsworthy wrote in his diary. He remained a man of acute sensitivity and compassion, and a loyal ally to the occasionally but increasingly irascible Hudson.[108]

With south coast resorts rapidly filling with war-wounded and their families, Hudson would struggle to find somewhere safe – and relatively warm – for the winter.

Winter refuges

With few other options and in the belief that it was the 'warmest village in England',[109] Hudson spent the winter of 1916/17 in Falmouth. Penzance and the other main coastal towns had filled with the casualties of war and their carers. He promptly found himself in the grip of sub-zero temperatures, and even the toe-end of England – normally insulated by the ocean currents on three sides – was affected for much of January. 'No such weather has ever been known in this part, is what they all say … Even by the fire one feels cold and all I can do is to sit as close as I can and read and write.' Poor Hudson was 'nearly shrivelled up by the icy blasts'.[110]

The weather must have been partly to blame, but he didn't take to Falmouth, describing it as 'an ugly uninteresting town … Dull old Penzance is quite a fairy city, an Arabian garden of delight, in comparison,' he grizzled, only half-joking. He certainly seemed to struggle for inspiration in his new surroundings:[111]

> This is the first village I've ever stayed at where I've found it impossible to add anything fresh to my notes. I feel like going out and seizing hold of anyone I meet and saying to him – 'Did you ever see a mouse, or even a black-beetle? Well, for God's sake tell me something about it.' They can tell you nothing.[112]

But at least he wasn't stuck in the city. 'London with its weather and explosions must be like hell.'[113] If it was hell in London, the Western Front was beyond most people's powers of description.

With his advanced age and the stress of the war, after two years of it the Rajah Charles Brooke was on his death bed at Chesterton House, where he had always spent time on long leave from Sarawak riding with hounds. Like most of the trials and tribulations of the Brooke dynasty, the declining health of the rajah was national news. Besides his regular updates from Ranee Margaret, Hudson had been following news of the rajah's decline in the papers. Brooke passed away on 17 May 1917, aged 85.

His now widow was sanguine: 'Although I respected him and admired his achievements, I was never in love with him,' she later confirmed in her memoir.[114] Their son Vyner was the new rajah, and Sylvia Brett now the ranee. Margaret Brooke was always publicly diplomatic if not completely forgiving of her late husband's peculiarities. Sylvia, on the other hand, lacking Margaret's powers of diplomacy and revelling in her new persona of 'Queen of the Head-hunters', didn't pull any punches about her father-in-law: 'an unscrupulous and inhuman man',[115] she called him. She was repelled by his Gloucestershire pile, calling it 'that appalling sporting house. Foxes' heads, brushes, whips, spurs and hunting horns were everywhere. The whole place seemed to smell of bran and oats. In the passage there was every sign of slaughter.'[116]

In September 1917, Hudson sent Ranee Margaret a copy of John Galsworthy's novel *Beyond*. 'Some of the pages are uncut because I had read most of the story in the magazine [*Cosmopolitan*] before the book came out.'[117] In fact, Hudson didn't care much for the book, and let his friend know. Galsworthy was used to such unvarnished honesty. 'Galsworthy says he is pleased because I can tolerate *Beyond*,' Hudson later reported to the ranee. 'Perhaps a bit of irony on his part.'[118] Galsworthy didn't care much for it himself, and nor did some reviewers, who would also treat this one 'very roughly',[119] Hudson later noted, adding 'God forgive him for writing plays in this year, 1917!'[120] This is more irony on Hudson's part, even if he didn't know Galsworthy's work in the war years was all in a good cause.[121]

The USA finally entered the war, and American troops arrived in Britain in late summer 1917. 'I'm sorry I couldn't go to see our Yanks come through London yesterday,' Hudson wrote on 16 August. 'I wanted to see them very much. I went instead in the late afternoon to the Fountains and fed a woodpigeon that knows me and always flies into my hand to be fed to the great admiration of lookers-on. Also two broods of young moor-hens all furiously hungry.'[122]

Hudson's fondness for feeding the birds in the parks brought him into conflict with authority once more, in an incident he dared to describe in writing at a time when government censors were opening mail:

> The keeper came up and told me that the Govt. order was a fine of £100 or a 6 months' imprisonment for anyone caught in the act of feeding the water-fowl in the park. I expressed my astonishment and laughed and went on feeding the birds. He walked off angrily off to the other keeper, an old soldier, who knows me and winks at my law-breaking.[123]

The war continued to rage, and in London the enemy bombing raids escalated. In the autumn of 1917, Hudson witnessed his fifth, and there were more to come. One took place while he sat judging children's essays for the Bird and Tree competition prizes. Another arrived when he was dining in Whiteley's restaurant, his meal interrupted while he was ushered to the basement with fellow customers. In November, he was frustrated at missing an air raid that came the evening after he left for Ascot once more. He described the anxiety of waiting the next day for news of the damage done.

After that, there was a lull in the attacks on London. 'I don't believe there will be more raids', Hudson told Roberts on 5 December. 'They have discovered now I fancy that they do practically no harm (at a great cost) and only keep alive the intense irritation and hatred of them as baby-killers: and as they want peace it is a bad policy.'[124]

In a reversal of the human traffic of the war's early months, Londoners were now flocking to the south coast for safety. Hudson headed west again for winter. On his way to Cornwall he found Exeter packed with wounded soldiers, with 3,000 having been taken there. In Cornwall, he wandered among the crowds filling the streets of Fowey, finding that the main hotel there and others had been requisitioned as hospitals for the war-wounded, and for relations and friends of invalided soldiers. Unable to find a room he moved further west, to Penzance.

He also found sanctuary in Fowey and Looe that winter, but when the chance came he was soon ensconced again at Penzance, on North Parade, in what he thought was the warmest spot in town, his room sheltered by buildings and hills to the north and east. 'It looks south and has a front where you can walk by the sea and it is a well-sheltered situation for winter, but I never could endure it in summer. It then gives you a peculiar malady called the "Penzance headache".'[125]

Hudson had a 'quaint old landlady' called Ilfa Roxborough who provided him with butter, cream and jam, and other 'rare luxuries'.[126] His rent of £2 10s per week included 'newspapers, subscription to Library, sweets, perfumes, laundry, etc.'.[127] 'I'm within very easy distance of all the spots in West Cornwall where I have friends … I go out for very short walks, as a rule to the Library when I get two or three or four books to glance over of an evening and return the next day – unless I find one I can read through.'[128]

Even here, near Land's End, he couldn't shake an underlying dread of attack, in this case of the German navy targeting the Cornish ports. 'May the cursed raiders keep away,' he almost prayed.[129]

Meanwhile at Westminster, two years into the war, Sir Edward Grey's sight and health were failing, owing to overwork and stress, as the death toll mounted on the Western Front – and beyond. 'Poor Grey,' Hudson wrote, 'how I pity him in a darkened world when he wants to see so much. He feels it most in London.'[130] It couldn't last.

Grey's war

Exhausted, Sir Edward Grey finally stepped down from his position as foreign secretary and in July 1916 he accepted a peerage and became Viscount Grey of Fallodon. He had become the longest continuously serving foreign secretary. Asquith's ministry collapsed by the end of the year and Lloyd George became prime minister. Grey would now be in opposition, from his seat in the House of Lords.

In London, life carried on, between Zeppelin raids and black-edged telegrams. The plumage war simmered, and the factions continued to make claim and counter-claim, behind the scenes if not publicly. Then push came to shove. On 16 July 1916, Hudson and the Bird Society found themselves in a spot of bother: 'I had been anxious to attend our Bird Watchers' Meeting this afternoon,' Hudson wrote, 'as we are now threatened with a libel suit, but I don't feel up to it in this wet.'[131]

I am not sure which of the opponents had issued the threat, or how far they took it, but if it got to court I am sure we would know about it. Hudson had been typically vocal in accusing the opposition of being in the pay of the plumage industry, but he wasn't daft enough to name names in public.

The anti-plumage campaigners changed tack. Perhaps a temporary ban could be sought in such trying times. In 1917, a petition was raised by the Bird Society and others that succeeded in persuading the government to ban plumage imports for the duration of the war, which we know now would end the following year. But the resultant economic disadvantage felt by Britain and France at this delicate time would see the ban relaxed before the end of the year.[132]

Grey's war years had their share of personal tragedy too. He would receive news from the Western Front that his nephew had been killed, and of course other close friends. Then in 1917 Grey's house on the Fallodon estate caught fire, and part of it was badly damaged. 'It is difficult in these dark days not to become disheartened and discouraged,' Grey wrote to his sister. 'I find that what helps me most is watching the stability of nature and the orderly procession of the seasons.'[133]

In May 1917, Grey made an emotional return to Hampshire and the cottage by the river. 'The beauty has been overwhelming,' he wrote to Louise Creighton. 'The beauty of the season makes the contrast of man-made war more horrible and poignant; on the other hand it gives comfort and support … that evil cannot prevail.' He understood, he said, how Noah felt when he saw the rainbow after the floods.[134]

Hudson was finalising his memoir as he left for the south-west again in December 1917. He was also working on his forthcoming *Book of a Naturalist*. From one of his letters to Roberts, it is revealed that Hudson's typist was also a useful critic, and his unofficial censor. When he was having 'a little fun at the idea of design in nature' in his essay about the evolution of the bat, it 'gave offence to the lady because, altho' in some degree emancipated, she is still under the domination of the old creation idea'.[135]

His typist ('she is a dear person,' he wrote) had responded with the following note: 'Nature is fearful and wonderful after all and I don't like her reduced to a sort of clever pantomime creature bandying words with a sniggling imp from another region.'[136]

'And so on, very severe!' Hudson told Roberts, adding, 'but when she goes on to say – "I can't bear the slightest touch of anything within measurable distance of vulgarity in your writing…" I think it is time to take a sound opinion. She is a dear person and has written books herself, but in this instance I would think more of your opinion than of hers.'[137]

He wanted to know if Garnett also detected vulgarity. If so, the essay had to be altered. Reading this letter closely, I had a sudden moment of clarity. The typist in question was Linda Gardiner, Bird Society communications specialist. It is a rare and perhaps the only surviving example of something she wrote to Hudson, because none of her letters to him were kept. It's also an illuminating glimpse of how he would describe her values, and of her piety. I have read elsewhere that she typed *The Book of a Naturalist* for Hudson, in which this essay about bats later appeared.[138]

In early 1918, Hudson stayed mostly indoors at North Parade, Penzance, to work on his memoir, with the landlady's cat for company. Garnett had evidently loved the part of the manuscript that he had been sent so far. Hudson parried the praise, in typical fashion: 'As to its being a 'masterpiece', that's all your fun … Once it is finished I don't care a straw whether it goes for publication or not: it will come out some day.'[139] Hudson thought it might be his last work: 'If I could finish that I would say "Damn Death, let it come since it must",' as he wrote to the ranee.[140]

The war seemed to be reaching a deadly climax. A series of German attacks along the Western Front began on 21 March 1918 – the so-called Spring Offensive. The Fourth Battle of Ypres raged for three weeks in April. 'I have been in a constant state of intense anxiety', Hudson wrote on 8 April, 'but I am beginning to hope they will never break through: but good God, what an awful waste of life!'[141] In the wake of this latest onslaught hundreds more wounded soldiers arrived at Exeter. A poet he knew drowned himself in the bay there, unable to bear any more of the awfulness of it all. Hudson took consolation in the blossoming apple orchards and the city's magnificent cathedral.[142]

Returning to London, he was greeted as usual by his long-serving Scottish housekeeper, Mrs Jessie MacDougall, a character in Hudson's domestic life who has captured my imagination, from her cameo-style appearances in his letters. She was increasingly making Hudson possible, and keeping him on his toes, despite the challenges and stresses in her own life.

Jessie MacDougall and the pandemic

From her occasional 'walk-on parts' it is just about possible to build a picture of Hudson's 'Mrs MacDougall' and her life. She comes to the fore in the last year of the war, along with a fresh hell – the scourge of a global pandemic. In an undated letter, Hudson mentions Mrs MacDougall having been ill for a fortnight but not having said anything about it.[143]

When he found out, he immediately summoned a doctor for her. 'These are perilous times in London,' Hudson wrote on 23 June. 'A furious epidemic, Influenza, has invaded us here. My housekeeper is down with it, and I had to go for a doctor and a woman to attend her. It took her very suddenly, the temperature rushed up to 100 and the pulse to 130.'[144]

Hudson had gone out on his errand of mercy in the evening, and found the doctor recovering after a hard day of tending to army recruits. The medic took some persuading to come to Tower House. Hudson of course knew the serious risks he himself faced, knowing that if he contracted the virus it would most likely kill him.

Four days later he considered Mrs MacDougall recovered, though she was still too weak to leave her room. According to Hudson, once diagnosed she was resentful that she had been treated by the other residents of the house as though she had the plague. Hudson tried to make light of it all:

> I tell her (to comfort her) that she is a bad-tempered
> unreasonable woman and, worst of all, is lacking in a
> sense of humour, like all her countrymen – the wild
> half-naked barbarians who come to us from the far
> north – from Caledonia stern and wild. That only
> serves to put her in a fresh rage and she asks what
> would have become of England if it wasn't for the Scots
> in this war – especially the Black Watch in which
> her young husband had been fighting since the war
> began![145]

Snows came in late June 1918 as temperatures dropped to near-freezing that midsummer, heaping misery on the troops in the trenches. The unseasonably cold weather no doubt helped the flu epidemic to spread and endure. The spectre of death focused Hudson's mind on sorting out his legal affairs. 'Will you let me put your name down in my will – last will and testament if you please – as Executor – one of two?' he asked Roberts. 'I want to make the will some time before the winter – the season I fear – comes back on us.'[146]

Hudson had long assumed that Emily would outlive him, despite her greater age and her long-term dwindling health. 'Of course I mean to leave the few hundreds I shall die possessed of to my wife, with the exception of two or three trivial legacies: but when she goes I want all that remains to go to our Bird Protection Society; and to see to that part of the business I must have another man.'[147]

He hoped to appoint as his other executor the publisher Ernest Bell, 'a man of the most beautiful character, who spends his life and money in working for causes with which I am in sympathy'.[148] Bell was on the

Council of the Bird Society, and Hudson trusted him implicitly to execute his primary wish to support the cause.

Hudson was back at Grey Friars in Ascot in July 1918. 'I fled from influenza in London as it broke out in the house – two of the inmates had it here, one being my housekeeper.' To his horror, the virus caught up with him. 'Now it has burst on us here!' he wrote to Harding. Ranee Margaret had brought it, and was soon confined to bed. 'She considers herself the most perfectly healthy woman in England,' Hudson cried, yet 'at noon today she was suddenly bowled over and has a temperature of 103!'[149]

Hudson was now caught on the horns of a life or death dilemma: London of course was impossible to get to and move around without mixing in crowds, and Worthing – according to his wife Emily – was still full. 'Must consider the position tonight,' he wrote anxiously to Harding, making a rare admission of fear, as the barely suppressed stresses of the war bubbled to the surface: 'I have known fantastic fears and tremors, and have been frightened at a mole.'[150]

Further misfortune was about to befall Jessie MacDougall. With just a few months of war left to run, she received news that her husband had been wounded in action. She was called to Aldershot to visit him in hospital, having had no details of the extent of his injuries. Hudson hoped against

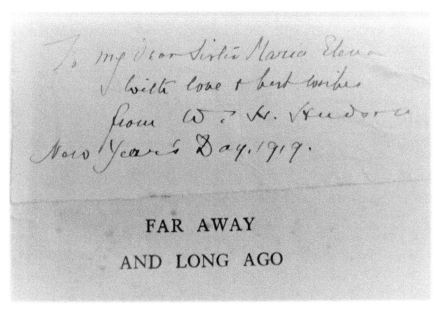

Hudson's dedication on a copy of his memoir of childhood, to his beloved sister Maria Elena. Tragically, she died the following August, the last surviving of his five sisters and brothers.

hope that she wouldn't find him badly mutilated.[151] Mr MacDougall was discharged a couple of months later, on crutches and unable to place his foot on the ground yet, but the prognosis was encouraging, and doctors expected him to recover the use of his leg.[152] Unlike so many others it looked as though he had escaped the war with his life, as the end of the four-year nightmare now seemed at last to be in view.

Peace at last

Hudson left London again to escape the flu pandemic at the end of October, and he remained on edge even when the war finally ended a month later. A letter survives that he wrote to another of his titled friends, Lady Grogan, on Armistice Day, 11 November 1918. He described Worthing as being 'in an uproar since twelve o'clock this morning'.[153] He knew he couldn't join the crowds on the streets, even if he'd wanted to. And with 12 million dead across Europe, 750,000 of them Britons, he was in no mood to celebrate.

While the shooting may have stopped, the flu pandemic meant that the mass mortality had not. A week later, Hudson was still with Emily on the south coast, lying low. 'Influenza is still raging undiminished here and schools have been closed since I came a fortnight ago. A good many die of pneumonia. Still, that has not kept the people from wild demonstrations on the supposed peace.'[154] Still anxious to avoid non-essential human contact, and crowds, Hudson retreated once more to London.[155]

In the last year of the war, Galsworthy rejected his knighthood. Hudson was given the news by Morley Roberts, who was, in Hudson's words, 'jolly glad to see that our friend Galsworthy rejected a knighthood with scorn'.[156]

'I have duly congratulated him on not receiving a knighthood,' Hudson wrote to Garnett. 'Someone says to me: "He might have declined it more graciously, since, after all, it is from the King His Majesty." To which I reply, Fiddle-de-dee!'[157]

I looked into this, and in fact Galsworthy *did try* to decline the knighthood politely. Notice of the award was sent to his London home late in December 1917, while he and Ada were out of town. He wired a week later on New Year's Eve to decline – but too late. Notice of the award had already been released and was published in the press on New Year's Day. He was anxious about what his peers would think: accepting such honours seems not to have been the done thing in his circles. 'We spent an unpleasant day, thinking of all our friends weeping and gnashing their teeth,'[158] he wrote. Smelling controversy, the *Daily Mail* and other papers were soon on the phone.

Within a day, Galsworthy had 175 letters and 20 wires to answer. Ada had to write 'innumerable letters to our friends telling of the mistake',[159] including one to Hudson, 'about the honor and the battle',[160] as Hudson described it.

Galsworthy again wired Prime Minister Lloyd George. 'I must persist in my refusal to accept,' he repeated. Telegrams, like tweets today, might inevitably seem terse and churlish, but his later note to the prime minister was much more expansive and very contrite: 'Men who strive to be artists in Letters ... should not accept titles ... please forgive me.'[161]

As the war neared its end, Hudson had surprise visitors at Tower House. 'Yesterday evening Galsworthy and Ada made a sudden appearance here much to my amazement,' he reported, evidently pleased, 'as I never have any one visit me now.'[162] The Galsworthys had found somewhere to buy and were moving to fashionable Hampstead. Hudson would continue to enjoy their company in times of peace.

Meanwhile, 300 miles to the north, Sir Edward Grey was sitting in his study at Fallodon, with red squirrels appearing at the window, venturing onto his desk to steal a titbit from his outstretched hand before darting back to the ledge. He could see them, but not well. He dictated a note to Hudson about his situation, which Hudson summarised. 'Have just had a long letter from Lord Grey about his blindness. He can't read at all; he says he's having my book read to him. He sees just well enough to walk and cycle but everything appears like a blurred photograph. He can also fish.'[163] Grey's friend Lady Pamela Glenconner was the reader, and the book was *Far Away and Long Ago*.

This memoir of childhood, what Hudson had called his 'small book', was published as the war ended. It soon met with critical acclaim. Besides Garnett's high praise for it, and Don Roberto's,[164] it brought glowing reviews from other respected quarters. 'Passages in *Far Away and Long Ago* will undoubtedly go to posterity entire,' wrote Virginia Woolf, adding that she had greeted the book 'like an old friend'.[165] Perhaps aware that his mental and physical health were on something of a knife-edge, the *Times Literary Supplement* exhorted him to keep writing: 'We must beg Mr Hudson not to stop here, but to carry the story on to the farthest possible limits.'[166]

Most lyrical of all was *The Star*:

> Mr Hudson is more than a naturalist. He is a man of
> genius who transmutes lead into gold—the lead of

knowledge into the gold of feeling … As you hear
the music of his prose … you recapture the delicious
tenderness of childhood with its wistful wonder and
vision … Mr. Hudson is a nightingale naturalist with a
voice that throbs in waves of magical melody.[167]

It was also revealed at the end of the war that Galsworthy had worked in government intelligence. His columns in the US press must have influenced American public and political opinion of the war, as well as Hudson's prowess as a writer. He was a much sought-after opinion leader – an influencer, we'd call him nowadays. His views were published weekly in *The Observer* newspaper, which Hudson always looked for. Britain was becoming a more democratic place. The Representation of the People Act 1918 widened voting rights by abolishing practically all property qualifications for men over 21 and giving voting rights to most propertied women over 30. With war over at last, the world could now begin to pick up the pieces.

CHAPTER 13

Picking up the Pieces

Regrouping

The plumage factions regrouped in 1919. Another petition was raised by the campaigners, with many high-profile backers from the world of literature – where Hudson wielded his influence – including Sherlock Holmes author Sir Arthur Conan Doyle, G.K. Chesterton and Bertrand Russell among a hundred signatories. The government's response was to do nothing.[1]

Colonel Charles Yate brought another plumage bill in February 1920. The trade riddled it with amendments and failed to attend committees, thereby disabling them. The Yate Bill stalled in May.[2] Hudson vented his frustration in a letter to Linda Gardiner: 'I have seen Hammond's very strong letter about the wreck of the Plumage Bill. I know he can write well on any subject as he formerly edited *The Nation* until he found a government appointment, but this did not seem a subject he could deal with so vigorously and to the point.'[3]

Morley Roberts recalled Hudson's feelings at this time: 'He spoke very bitterly about the fate of the Plumage Bill, and presently spoke of Lord Grey, for whom he has a great admiration … It seems that Hudson urged Grey to speak on the subject of the Bill and he promised that he would.'[4] Roberts, aware that Hudson had not long to live, interviewed him on 21 November 1920, and transcribed his responses, to try to preserve something more of the man.

Hudson had not seen Sir Edward Grey for some time, and received a telegram late in 1918 inviting him to lunch.[5] In November 1920, Hudson took up a similar invitation and was collected in a chauffeur-driven car and taken to Lady Pamela Glenconner's home at Queen Anne's Gate, where Edward was waiting, having arrived from Fallodon on 3 November.[6]

Lady Pamela had become as fond of Hudson as Lady Dorothy before her. Her husband had died, and her relationship with Edward could now blossom. It seems to have been an open secret that the two had

been having a romantic relationship, and Ranee Margaret remarked to
Hudson that the couple would now be able to marry. In a rare show of
disapproval of his friend, he let her know he had taken exception to
the tone of her comment.[7] I think it speaks for the depth of his respect
for Grey.

Hudson and Grey had a lot to catch up on, with talk of war perhaps
kept to a minimum, but Hudson anxious to know about the work
being done to rebuild the house at Fallodon. There was also the small
matter of a general election, called as soon as the war ended. Against
his better judgement, Don Roberto had been persuaded by friends to
contest the seat of West Stirlingshire. He stood as an 'independent
liberal' this time, giving him even more licence than usual to say what
he liked. From his platforms he urged audiences to vote socialist and
on no account to vote for Lloyd George, whom he equated with the
Conservatives. But in truth he was now tired of the 'infernal folly' of
elections.[8]

The parliamentary committee set up six years earlier to review the
Wild Birds Protection Act, of which Grey was chair, had been disrupted
by war like everything else. It finally reported, with Hudson of course eager
to know more. It favoured the Bird Society's long-advocated system of all
birds being protected, or what we might think of as 'innocent until proven
guilty' of some antisocial habits justifying their rights being removed. But
the committee recommended fiddling with existing legislation for the
time-being, while setting up another committee to look at introducing
a bill, and gathering more evidence in the meantime on the habits of
individual bird species, about which, it said, not enough was known.
Scientists and policymakers would be needed to provide more research
and evidence. A Wild Birds Advisory Committee was set up for England
and Wales, and a separate one for Scotland. The RSPB's chairman would
sit on these.

With the war finally ended, American soldiers were present in London
in numbers, and US president Woodrow Wilson arrived in December.
Morley Roberts reported to Hudson on the occasion:

> Naomi and I stood in Pall Mall to see Wilson and
> the King go past. For my part I thought I had rarely
> seen a stronger face than Wilson's. He was a 9 point
> 5 howitzer compared with our King who looked very
> tragic. But I wonder whether he's got the patience and
> perseverance and time to deal with the appalling tangle

in which Europe now finds itself. It's very difficult for
the American to understand what national terror is. So
far they have lived in a more or less stable house and
we over here are housed over crumbling and trembling
volcanic ruins.[9]

In the thick of disentangling Europe and rebuilding the ruins was
Viscount Grey. As if he didn't have enough on his plate, Grey would also
be appointed ambassador to the United States, working on the issues of
Irish independence and the Treaty of Versailles, which he was anxious
to persuade the USA to ratify. He would seek to secure a meeting with
President Wilson.[10]

Sad news reached Hudson from America in early January 1919. 'I'm
very much cut up at the death of my old enemy and friend Roosevelt who
has both abused and praised me,' he wrote to Roberts (I'm not sure what
the abuse was – perhaps there was some criticism in a letter long lost):

> He, I believe, did more to rouse the war-temper in
> America than anyone: I doubt if it had not been for
> him that Wilson would ever have come in. After
> Roosevelt's scathing reply to the 'We-are-too-proud-to-
> fight' declaration he had to pocket that kind of pride.
> Still, I love him [Wilson] for what he did do in the
> end – and for his visit to Carlisle and his speech in a
> little chapel. Nothing of the Snob in Wilson![11]

Hudson was now fielding increasing volumes of fan mail and press
clippings from the States. 'I've had a flood of rather gushy review stuff
about my book, but it is no pleasure to me. I'm more interested at present
in the Natural History books I'm writing.'[12]

Letters were also starting to come from readers in Argentina, including
one from Buenos Aires:

> about *Far Away*, etc. from a fellow I knew as a boy. He
> was one of six, they were rich people, and they used to
> visit us – three or four of them at a time, riding from
> their place about 8 miles from our house. We were
> friendly enough but looked on them as Philistine –
> people of means who took no interest in things of the
> mind, but only in dollars.[13]

Hudson soon returned to the subject of this family, it seems in response to an obituary notice he'd seen for one of them, a medic.

> I ... was in correspondence with him a few weeks
> ago about my *Far Away* which he had been reading.
> I knew him as a child and as his family were unadul-
> terated Philistines and illiterate we rather didn't
> like them, though they were getting very rich and
> adding miles of land to their estate, which ran side
> by side with my father's land. Of course when our
> impoverished family broke up and I came away this
> man's father took the opportunity of buying our place,
> and this son inherited the whole estancia when the
> old fellow died, but though a rich man out there he
> appears to have found his whole life in his medical
> researches here.[14]

Alas, they had been reunited too late.

Hudson's latest compilation of essays *The Book of a Naturalist*, was published, and he told Don Roberto he thought it 'quite as good as anything in that line I've done before',[15] as he supplied one for Missy Bontine. It was a rare example of expressing satisfaction in anything he'd written. There was glowing praise for his friend: 'But enough of my own affairs! I do wish that you would take up your pen again! ... as your writing is more refreshing and stimulating to my soul than that of any other living author.'[16]

Unlike Hudson, Don Roberto also wrote books in Spanish, and was becoming known to a wider readership, including in Latin America. After reading some of his work in that language, Hudson told his friend that although he 'can't hit as hard in that softer tongue as you can in English (nobody could) you are as vivid, picturesque and racy as ever ... there is no one in the world who could do sketches like this'.[17]

With the future of Europe being thrashed out by the League of Nations, Viscount Grey's thoughts were ever with the opportunities it might also present to protect nature, and he wrote to Prime Minister Balfour in June 1919 to suggest that the North Sea archipelago of Heligoland be made a bird sanctuary, to stop the millions of migrating birds there being slaughtered.[18]

In August 1919, Grey set off for America on an ambassadorial mission to bring the USA into the League of Nations, 'on which he believed the future happiness of the world to depend'.[19] 'I love the Americans,' he reported. 'They are easier to get on with than people here. I spoke at Harvard to a thousand people I'm told, I could hardly see them. About birds and so on.'[20]

Hudson delayed his winter migration for more than a month as he wanted to see Viscount Grey in London when he returned from his diplomatic tour of the USA. The opportunity to see Grey came in November, and he reported some of what they discussed: 'I lunched with Lady Glenconner and Lord Grey and Prof. Mackail on Friday and feared on meeting the last-named that we would have nothing to say to each other: then I luckily remembered that last May, in Penzance, I had read the whole of his long life of [William] Morris, so I mentioned that and we then talked about Morris.'[21]

When US president Woodrow Wilson was in Britain at the end of the war he made what he called 'a pilgrimage of the heart' to the home of his ancestors – a listed Georgian building in Carlisle called Cavendish House.[22] It had been built in 1832 by his grandfather, and by a strange quirk of fate I found myself with reason to go there too. Today it is the home of Peter Dance, who turned 90 in early 2022, a man with as much to say about Hudson as anyone in the UK I had met at that point – one of very few people able to answer some of my questions, or at least hazard an educated guess. Peter showed me his comprehensive collection of Hudson's works, including some first editions and signed copies, with inscriptions. He has other treasures, such as Hudson's personal map of Hampshire with handwritten notes, and some unpublished letters.

Peter was looking to donate this collection to a suitable institution, to enable public access to it, and on his behalf I made enquiries to potential hosts. The upshot is that at the time of writing we are working with Kings Lynn Library to house the collection. This impressive building was built with money from Scots-American philanthropist Andrew Carnegie and opened in 1905, shortly before Hudson visited the town, as he explored Norfolk and made his regular visits to commune in particular with the geese arriving in autumn.

It is fitting then that a collection of Hudson's many works of fiction as well as nature writing should be lodged there, to mark the centenary of his passing. As an unschooled and self-taught man, he loved libraries and knew their value. Hudson was fond of Kings Lynn, calling it 'a very charming old town',[23] which confers on it a rare and honoured status. Goodness knows he was a hard man to please when it came to urban areas.

Hudson's collected works would extend to 24 titles by the end, and some of these are reworkings of earlier titles, in light of a changing world and a more experienced Hudson, and as one or two old scores were settled.

The second life of Hudson's early books

It seems that the longer he lived, the more feverishly Hudson worked, terrified by the prospect of losing his grip on life. By the end he was engaged in a scramble to buy back the rights to his earlier books, to free them up for reissue, and as much money from American publishers as he could squeeze out of them, to leave to his beloved Bird Society.

His *Birds in a Village*, first written 30 years earlier, evolved and expanded – mirroring the growth of villages all over the home counties – into *Birds in Town and Village*. 'It pleased me when its turn came to be reissued,' he said of this new improved version, as he had 'continued to cherish a certain affection for it.' Some of the original work was out of date, he noted, 'especially in what was said with bitterness ... anent ... trying to save the beautiful wild bird life of this country and of the world generally from extermination ... [owing to] the feather-wearing fashion and of the London trade in dead birds'.[24]

In the preface, Hudson manages to sound cheerful: 'Happily, the last twenty years of the life and work of the Royal Society for the Protection of Birds have changed all that, and it would not now be too much to say that all right-thinking persons in this country, men and women, are anxious to see the end of this iniquitous traffic.'[25] It was an upbeat note for him to have struck, and said more in hope than expectation – with the hand of his publisher upon him, I suspect.

'This book is full of his unsurpassed perception and unique charm', the *Times Literary Supplement* gushed. 'Some of his best passages about birds are equally delightful and vivid sketches of human life.'[26] The *Manchester Guardian* was equally effusive: 'Mr. Hudson loves all birds; they are his work, his recreation, his life; he writes about them as no one else can: he sees what others miss.'[27]

After giving a talk to the RSPB local group in Guildford, Surrey, I took the opportunity the following day to visit Brian Clews, author of a book comparing the birds found in Cookham, Berkshire, today with those at the time Hudson wrote his *Birds in a Village*. Although some species are now lost, there have been some gains too. Brian even has red kites visit his garden: such a sight was beyond Hudson's wildest dreams. Brian had been keen to determine where exactly Hudson had stayed while exploring Cookham. I later did a bit more delving, and found two addresses where he lodged, one being Midway Cottage, on Popes Lane.

Hudson also had the satisfaction of being able to revisit another early project and produce his 1920 *Birds of La Plata*, a reworking of *Argentine Ornithology*, his long-term and often painful collaboration with Dr Philip Sclater, secretary of the Zoological Society of London. Free to revise it how he liked, he now stripped out Sclater's part altogether. 'The original work was out of date as soon as published, and the only interest it still retains for readers is in the accounts of the birds' habits contributed by me,' he wrote, with more than a trace of bitterness.[28]

He may have found this belated parting of ways with Sclater cathartic, but it seems the effect was fleeting. 'It was my poverty that made me allow it to be re-published,' he confided in Roberts in 1920, regarding *Birds of La Plata*. 'It wasn't worth it.'[29] Re-immersing himself in these memories of exploration as a young man, making fresh discoveries in nature, forced Hudson to retrospectively re-evaluate his path in life: 'Now after so long a time the pang returns, and when I think of that land so rich in bird life … the reflection is forced on me that, after all, I probably made choice of the wrong road of the two then open to me.'[30]

But of course Hudson's low moods would often occasion statements like this that sit uncomfortably with the evidence. The other road open to him was a boat back to Buenos Aires. It is fair to conclude that it was an idle thought, a late-life revision of things, as he had never previously indicated any strong desire to give up on his British adventure.

David 'Bunny' Garnett tells an entertaining story in his memoir *Great Friends* about his last meeting with Hudson. The war had interrupted their friendship and they were renewing acquaintance. Garnett was by now a young man, engaged in the book trade, and Hudson was selling off his collection of first editions, to raise funds that he intended to leave to Edward Thomas's widow Helen, and her children, and the Bird Society. Garnett found Hudson at home, and the books laid out on his sofa. It was at this point that Garnett told Hudson that he knew the secret of his long-forgotten novel *Fan – The Story of a Young Girl's Life*, published under the assumed name Henry Harford nearly 30 years earlier. Bunny had managed to find two copies of it by advertising in the trade press.

Hudson seemed shocked. 'He looked for a moment as though he thought I was going to blackmail him',[31] Garnett recalled. But there is a surviving letter from Hudson to Edward Garnett indicating that he had not kept the book secret from him, and as far as he was now concerned if it could be revived to raise funds then so be it.[32]

It would be in large part thanks to the American book-buying public that Hudson had found fame and some fortune in late life. The fortune was never sought for himself.

Big in America

It was the US sales of *Green Mansions* that would, more than anything else, help his beloved society to carry on and step up the fight save the birds. 'It began to sell there about four years ago,' Hudson reported of *Green Mansions*, astonished at the response his book was getting in the USA, 'and has gone into about ten editions in that time'.[33] Meanwhile, 'in England … the sales, the publishers tell me, are improving'.[34] It was this work of fiction in a foreign land, and the movie rights to the story, that would allow Hudson to accumulate substantial funds that he could leave to the Bird Society.

The American success of *Green Mansions* also rekindled interest in his earlier, barely noticed writings. He was soon negotiating with another publisher, Chapman and Hall, over the rights to previously published work. He was determined to secure the best possible terms now, in the interests of the Bird Society. 'I shall take advice about what to do in the matter: but I must wait until I see a member of the Society.' The RSPB would provide legal expertise.[35]

The fan mail continued to flood in from the other side of the world. 'In America they [his books] are now having a great vogue and I receive a perpetual stream of letters from readers over there from all over the country.'[36] 'Among my letters yesterday there were two from California – one from an old lady about *Green Mansions*, the other from a child about *A Little Boy Lost*,'[37] he wrote in March 1919. 'I imagine I have two readers there for every one in England,' he estimated in 1920.[38]

'Those of my books most read in America are *Far Away and Long Ago*, *Green Mansions* and *The Book of a Naturalist*,' he wrote in another letter. Of the latter title he also reported 'that last book of mine … has brought me a continuous stream of letters … from all over America – from Florida to Seattle. Yesterday I had one of about forty pages from Vancouver, B.C.'[39]

He also gives an indication of just how much more lucrative things had become. 'Three or four publishers give me eight or nine times more than I should have ventured to ask for a book a few years ago.'[40] Again, there is a hint of regret at this belated recognition of his value. 'A few years ago what they paid was only enough for bread and cheese. Now that I've finished my life and don't want anything they throw their damn cheques at me. An old story told by many I dare say.'[41]

Hudson's mailbag remained full into 1921, when he reported to Morley Roberts that he was still under 'a pile of books from America, with letters, etc.'.[42] He had invitations to visit Florida,[43] and admirers of his Norfolk geese descriptions in Connecticut.[44] He sparked a debate with his theory that pumas/cougars didn't attack people, and that any records of them doing so were cases of mistaken identity or collateral damage – the cougar intent on something else and the human merely getting in the way. But perhaps the biggest reaction he got was to his article 'Do cats think?', no doubt inspired by the amount of time he had been spending alone with his landlady's cat in Penzance. It resulted in a deluge of letters from pet owners replying in the affirmative, with plenty of anecdotal evidence. He remained fascinated by the minds of animals right to the end.

American conservationists also reached out to him for help. When transcribing, editing and revising *Lost British Birds* after Hudson's death, Linda Gardiner found something he'd written in one of his notebooks, extracted from a letter sent by James Henry Rice, champion of bird protection in America:

> It seems to me England ought to furnish a small army
> of trained men and women, able to handle the bird
> subject anywhere: but it may not be so. You may
> lack them. They are not plentiful in this country by
> any means – not exceeding half a dozen in all North
> America. We have thousands of ornithologists who
> spend their time collecting bird skins and revising
> species.[45]

Hudson had jotted five words in the margin. 'We can say the same.'[46]

Back in the artist's studio

Despite his apparent displeasure at the earlier attempts to capture him in oils, Hudson was back again in the artist's studio chair, William Rothenstein now occupying a 'fine big house on Campden Hill',[47] in

Kensington. It seems that Rothenstein had moved to an even more prestigious location, but Hudson seems not to have believed his wealth could have stemmed from art sales. 'I suppose W.R. came into money when his father died a year or so ago, which may explain why he has so expensive an establishment.'[48]

Hudson reported to Don Roberto about this latest sitting, and meeting T.E. Lawrence there – the fabled Lawrence of Arabia – in June 1919.

> Yesterday I met … a young English adventurer, who during the war joined the Arabs in Syria and although a civilian was made a commander of a force by them and fought against the Turks. Then, the War Office getting wind of his daring sent him a commission, etc., and he was Colonel Lawrence for the rest of the war, then went and joined Allenby in Palestine.
>
> I found him arrayed in the most beautiful male dress of the East I have ever seen – a reddish camel-hair mantle or cloak with gold collar over a white gown reaching to the ground, and a white headpiece with 3 silver cords or ropes wound round it. As he is clean-shaven and has a finely sculptured face the dress was most effective. He said it is only worn in Mecca by persons of importance, and nowhere but at Mecca. He is a worshipper of Doughty, and also told me he had read *The Purple Land* 12 times.
>
> While W.R. worked on my portrait I had a grand talk with Col. L. in his remarkable dress and we argued furiously about Science versus Art. But though he was all for art he had a keenly observant mind and could put one into the East and its atmosphere better than any book I know – except Doughty perhaps. It struck me that *The Purple Land* was just the sort of book that would appeal to a young adventurer like Lawrence – a sort of Richard Lamb himself.[49]

It is a pity Rothenstein couldn't have painted this whole tableau. Hudson and Lawrence had made plans to speak again. 'I am going this morning to have a talk with Col. Lawrence in hopes of getting some information I want about the Arabs which it is idle to look for in the books,'[50] he told Roberts. No doubt conversation soon turned to birds, and Hudson was able to tell Lawrence that in the spring he had received a letter from a reader 'who tells me that the name of the wagtail in Arabic means the "Father of Salutation".'[51]

Hudson was again a repeat visitor to the artist's studio. 'I've been three times of a morning to sit to him and each time he has done a portrait.'[52] And what did he think of the results obtained by arguably the greatest portrait painter of his generation? 'All unsatisfactory.'[53] It seems that Hudson would remain dissatisfied with the outputs. 'Another book of portraits has been published by Rothenstein and one is of me,' he groaned, 'just as bad as all the others he has done so many times.'[54]

I went to the National Portrait Gallery to find Hudson, but not surprisingly he is not on display, like a vast number of the portraits kept there. There simply isn't room. I did find the first English Rajah of Sarawak, James Brooke, however. To be fair to Hudson, Rothenstein's portraits of him are curiously bad, and unflattering. Hudson's experiences as a subject of Rothenstein can have done little to overcome his ambivalence about art. He was also annoyed that there were so many books on the subject now flooding the market. 'The booksellers' counters are piled up with them', he moaned. 'The publishers say they are a drug.'[55]

Had the artists not moved out, John and Ada Galsworthy might have become neighbours of the Rothensteins in Hampstead. After they returned from another long tour, Hudson met the Galsworthys in Kensington Gardens and Hyde Park in 1919, 'both looking very fit and beautiful after their travels in America'.[56] He took them to his favourite spot where he liked to go to 'get cool' on warm summer afternoons, sitting under an elm tree near the Serpentine lake. He had decided this was the 'prettiest spot in London'.[57]

He spent two hours there with his friends, for part of which they enjoyed the company of a small boy Hudson had befriended on earlier visits. Perhaps he relished telling them about his meeting earlier that day at Kew with the former Sister Mary Cornelia, the nursing nun he had encouraged to fly the convent and seek her freedom, rather than sign her life away to the Church.[58] She had visited him for tea before their walk to Kew Gardens with the news that she was getting married, and to an Ulster Protestant at that. They were to honeymoon in the mountains in Ireland, and settle down in a small town near Belfast.

'A blessed thing for poor dear Elsie [Mary] that I didn't obey you!'[59] he teased Ranee Margaret. He wrote to her again the day after the wedding, still congratulating himself on the part he had played in the liberation of this young woman.

It was with Hudson's 80th year approaching that his long-time friend Morley Roberts interviewed him, to capture something of the man while there was still time, with more than half an eye on a biography. Roberts extracted perhaps the most revealing articulation of Hudson's self-identity as a writer:

> I myself would be anything you like except the profes-
> sional showman of the beauties of what people are
> pleased to call the countryside ... My style of writing
> was bare and free from ornament. I was profoundly
> interested in the little things, in trifles. No, it was not
> strange none cared much for my books. They have
> made handsome amends since then.[60]

Hudson returned to the Mont Blanc in July 1920, finding no one there whom he knew.[61] At the end of 1920, John Galsworthy sent him the last but one of his series of *Forsyte* books. Hudson had been reading novels from all over the world, but none, he soon declared, had given him so much pleasure as this one. 'However,' he added, 'the only criticism I am entitled to make is on a natural history subject.'[62] He had picked up on Galsworthy's misidentification of an injurious insect as a gadfly, when it was obviously a hoverfly. 'Not one of them ever bit or stung anyone,'[63] Hudson advised. The error was unlikely to harm sales, Galsworthy may have been thinking with a knowing smile, as he read his friend's feedback.[64]

It seems unfeasible, given Hudson's parlous health by this stage, but he had thought of calling on the Galsworthys while he was passing through Devon heading for his final summer in London, but couldn't be sure they would be there. 'Well, one never knows where you are,' he wrote to his friend. 'It is like trying to keep your eye on one bird in a company of birds all wheeling about in the air together.'[65]

The completed five-books-in-one and best-selling *Forsyte Saga* was one of the last Hudson ever looked at, even if he didn't read it. He received his copy from the author on his return to London in summer 1922, after his final sojourn in Cornwall.

CHAPTER 14

Swansongs

Hollywood calls

Hudson lived long enough to be touched by the advent of moving pictures. Although he doesn't mention it in any surviving letters, I am sure he must have attended a show or two at the Picture Theatre in Penzance, Cornwall, which was opened in late 1912 and proclaimed 'the handsomest picture palace in the country – outside the West End of London', according to the press, with seating for nearly 500 movie-goers.[1]

We do know he was a fan of Charlie Chaplin – he said as much in his interview with Morley Roberts, late in life – and that he was taken to the cinema in London by Violet Hunt, a society hostess and admirer who adopted him in his later years. They saw one film he described as being about nightlife in New York – perhaps a Fatty Arbuckle picture.

It's clear that Hudson knew the value of his story *Green Mansions* when the scramble to secure the rights to his work began in America. He eventually sold the film rights to the book for the then-staggering sum of £2,500, to be added to the fund that he would leave to the Bird Society in due course.

He had written to Ranee Margaret from Cornwall in December 1920, as her youngest son Harry was following his love of theatre into this new dimension of moving pictures. Hudson seems to have been trying to interest Harry in his children's story *A Little Boy Lost*, now also being republished in full colour for the American market. 'I wonder too what sort of a mess of *Green Mansions* they will make in America where a company wants to film it,' Hudson added. 'Poor Rima – what will she look like?'[2]

His words were to prove prophetic, although he can't have imagined it would take another 40 years for the tale to reach the silver screen, by which time it would be in glorious Metrocolor, and Panavision.[3]

Taking care of business

Hudson returned to Cornwall late in 1920, for a long stay through winter and spring. Something dislodged a memory of the first time he had been there. He asked Ranee Margaret if she could help him find the family he had lodged with then, including a 'healthy, strong adventure-loving and book-unloving little fellow I liked very much'. They duly found the place, the boy now grown up, 'with nothing but his trousers and big boots and a thin shirt on out in the rain loading a cart with straw. He knew me.'[4]

The boy had left farm work, gone to the war for two years in the trenches, and against the odds survived. He ended up marrying the lady who employed him, 'to everyone's astonishment and indignation at such a mixing of classes',[5] Hudson later reported.

> When I told him I wanted to be introduced to his wife
> he took me in and I found her a most interesting lady.
> A woman of culture who even as a young girl was in
> revolt against social conventions and more attached
> to animals than to her own species. She has travelled
> too and has been to India. Her father was an artist –
> Scotch.[6]

Hudson was also flattered that of the two books on her piano, one was his memoir, *Far Away and Long Ago*.

Ranee Margaret left him on his own for the winter, her absence more protracted than she'd expected as she became embroiled in the dramas of her family. Hudson felt her absence more strongly than ever now, his independence and mobility reduced. 'Oh, I wish I could go and see you as the dearest friend in the world!' he wrote. 'I miss the nice China tea at Ascot – and of course you. Especially our quarrels, which are so comforting to the spirit.'[7]

It meant she wasn't there to comfort him when his usual sure-footedness failed him, and he found himself back in the care of the nuns; not with illness this time, but after falling off an elevated pavement while trying to find his way in the dark in Penzance. The accident left him with a badly damaged hand, and depressed that it made it difficult to write, but otherwise he was lucky to survive without more serious injury. This, and his worsening health in general, left him unable to travel any distance. 'I think if I must go out again by night I will have a light – one need not carry an old time horn lantern now as it is easy to get an electric lamp,' he wrote to Roberts.[8]

His spirits were lifted a little when there was sudden momentum on the plumage campaign front early in 1921. In February not one but two bills were brought before Parliament, by Mr Galbraith and Captain Brown. In early March came a third, led by Mr Trevelyan Thomson and supported by Viscountess Astor, the second woman elected to Parliament in the wake of the 1918 Suffrage Act, and ten others. The latter bill omitted the clause to prohibit the sale or possession of illegally imported plumage – overcoming the problem of determining what was new and what was old, but also allowing people to continue selling it if they happened to have it. There was another compromise: an appended list of birds that the industry wished to see exempted from protection. But it was progress.

But in truth this was all overshadowed by Hudson's knowledge that he wouldn't see Emily again. Soon after, the news came that he had feared for so long: she had passed away at Worthing.[9] He was unable even to be present at the funeral. 'I was too ill to attend,' he wrote forlornly to friends. 'The Dr said it would be my death. And no doubt he was right, as I think it a wonder I have not gone off one of these last days. My heart has never been worse.'[10]

It would be June before he was well enough to travel back to London and on to Worthing to visit Emily's grave. When he finally got there, he was satisfied with the plot she – and he – had been given. 'I had hoped to acquire a plot for us beside Richard Jefferies, in the old and most rustic part of the cemetery, with its noble old trees, but the heads of this place insisted that this part was full. I – or should I say we – must be content to rest, as they call it, in another part.'

There were consolations that he noted while he sat with her. 'There is a good pine tree by the grave,' he reported to Roberts, 'and the turtle doves were crooning all the time I stayed there. The whole place seemed swarming with birds.'[11]

The passing of the Importation of Plumage (Prohibition) Bill in spring 1921 was for Hudson overshadowed by Emily's death, but his thoughts were recorded later:

> Any gains for animal welfare made so far had been imposed from above, by legislation through the devotion of … the 'cranks', the 'faddists', the 'sentimentalists', of their day, who were jeered and laughed at by their fellows, and who only succeeded by sheer tenacity and force of character after long fighting against public opinion and a reluctant Parliament, in finally getting their law.[12]

The wording of the legislation that finally passed had been adjusted to accommodate various concerns. It allowed plumage and skins to enter the country for scientific purposes, under licence. It also permitted wearers of feathered bonnets to arrive at British docks on board ocean liners without the indignity of having their headgear confiscated by customs officials, as had been happening in US ports, amid much furore in the press. And it exempted from the import ban ostriches, eider ducks and birds ordinarily used as articles of diet. Critically, it didn't prohibit possession for sale. Celebrated ornithologist Max Nicholson likened this to banning the importation of morphine, but not its use or possession.

On the wider bird protection front, Viscount Edward Grey would chair the Home Office-appointed Wild Birds Advisory Committee for England and Wales, with Bird Society chairman Montagu Sharpe also involved. This would be another long work in progress.[13]

No longer paying for Emily's care, Hudson resigned his pension, feeling that now he should do without it. He now began to correspond with Etta Lemon and the Bird Society about his revised will, Emily having predeceased him. He apologised to Lemon for having taken a month to respond to her letter of condolence, a delay due to his 'weakness and depression of spirits'. Lemon later wrote that she had only known Emily Hudson in late life, when she was 'old and sick', suggesting that she and Hudson had maintained a largely professional relationship.[14]

In deploying his estate for the cause of bird protection, Hudson had in his sights the men he detested most of all, and he proposed that his legacy be spent on campaigning for a law to stop private collectors destroying rare birds, once and for all. He reminded Lemon that when he had proposed this in the past she had refused to consider it on the grounds of cost. This would no longer be an obstacle, he assumed. In the first instance he would pay for a circular to be sent to all landlords, shooting tenants and natural history societies to gather opinion. He asked her to propose it at the next council meeting.[15]

A negotiation on the use of his estate was thus set in motion that would continue if and when he made it to London. It is noteworthy that the author of the RSPB's centenary history was unaware of this, presumably not having sight of the Hudson letter that sets it all out.

Hudson's attentions also turned to what would become lengthy ne-gotiations with his publishers over rights to the content of his books, which he wanted to free up for the Bird Society to use as leaflets. One publisher asked him if he received a fee from the society. No doubt taking a deep breath, Hudson patiently explained that, on the contrary, he would be paying for the illustrations as well as the print, that the leaflets were for schoolchildren (and some sold to adults to raise funds for the cause) and that he had given much of the last 25 years of his life to the organisation (it was actually 31 years by this point). He sent a couple of samples of his leaflets, and described his efforts to produce a book version of his *Lost British Birds* pamphlet, paying £60 for the artwork and probably hundreds to print it. 'But I should not mind spending a thousand on it as Bird Protection is my – hobby or craze if you like,' he explained.[16]

His publishers were not making it easy. In August he wrote: 'I am most anxious to get away from London but see no prospect of it … as I can't finish my affairs for some time', so invested was he in his 'long dreary business and fight with publishers'. He was still in London in late November and still hadn't done all he needed to do there. Nor had there been time to get away and see Ranee Margaret. 'Are you still living other people's lives and not your own?' he asked her in a letter.[17] 'Shall I ever see you again?'[18] She was preoccupied with family matters, and there is a sense of some strain between her and her old friend, not helped by their increasingly lengthy periods of separation.

His business in London behind him if not quite finished, Hudson finally escaped once more to Cornwall late in 1921, leaving behind the fogs; the 'deadly London winter atmosphere' that by this stage had 'almost suffocated' him.[19] He recalled a touching 'snapshot' moment of what would be his last journey to the south-west, and wrote to both Morley Roberts and Linda Gardiner about the incident. 'On Saturday morning I set out and caught the fast train at 10.30. It was thick fog for about 60 miles, then I dropped to sleep, and by and bye when I woke seeing hills and woods through the fog I asked a lady sitting opposite me if we had passed Westbury? "Yes, we passed it," she replied with a smile.' He spent a couple of nights at the Queen's Hotel in Exeter, and for recreation attended not one but three services in the cathedral.[20]

He made it to Cornwall, by this stage clinging to life. 'Often at night, lying awake from the time I stop reading till daylight appears, I wonder if I shall live to see daylight. And each morning brings me letters asking for a book, an article, a preface.'[21] He summoned a new local doctor, 'a big dark rough-looking fellow he is, in rough grey tweeds and a cap! The roughness of his bedside manner almost amused me … he wouldn't give me any medicine. "What, not a tonic?" I demanded sarcastically.' 'I try to poison people as little as possible,' the doctor had replied.[22]

Hudson returned to the convent hospital with a gift for the nuns who had cared for him with so much love and patience, through his injury and illness. It was an early sixteenth-century artwork, *Virgin and Child*, attributed to Andrea del Sarto, a family heirloom that had belonged to Emily's father's family in Italy. Hudson had paid for it to be restored and reframed for the chapel wall.[23]

At the turn of the year 1922, Hudson's thoughts were as usual with the Bird Society, and in a letter of 1 January he declared his intention to be back in London in time to attend the May council meeting. His friend John Harding was now on board. 'I am glad you consented to go on the RSPB Council and hope I shall be able to go to the Council Meetings and that we may meet there,'[24] Hudson told him.

Hudson was increasingly desperate to complete his expanded book version of the pamphlet he had produced for the SPB three decades earlier. 'I struggle on with my work,' he added in another letter to Harding, 'and was sorry to see a paragraph in yesterday's Times about my *Lost British Birds* as I don't know when I shall be able to do it. I have other things to do that come before it and little strength to work.'[25]

Mrs Frank. E. Lemon, M.B.O.U, F.Z.S.

Etta Lemon led the discussion with Hudson over the terms of the game-changing legacy gift that helped secure the future of his beloved Bird Society.

'I must see books to do the rest of my *Lost British Birds*,'[26] he declared. Luckily, Ranee Margaret introduced him to a local solicitor friend, whom Hudson was pleased to discover had 'the best collection of bird books in Cornwall'.[27] As if his *Lost Birds* project wasn't enough, Hudson was also working feverishly to finalise a book about animal senses (he thought there were many more than five) and how they perceive the world, three chapters of which were being serialised in *Century Magazine* up to September, scheduled just in advance of publication of the book in October. In the end it would be published posthumously, and titled *A Hind in Richmond Park*.

Hudson was more in demand than ever. He had 'any number' of requests from editors for pieces, even offering him 'extravagant prices'. 'I want only to do the work I began so long ago,' he groaned.[28]

Farewell to the rocky land of strangers

Publishers' representatives were now journeying from New York to London, and then from London to Land's End to sit with Hudson in his winter quarters, more than 300 miles from the big city, and negotiate deals. He played hardball with them – it was all for the birds.

Early in the year, Hudson had lunch there with an agent from the Century Publishing firm. He was pressed for the American rights to the book on animal senses he was endeavouring to finish, and permission to publish it in *Century Magazine*.

The same agent returned in March, and the manager of the Dutton company, who made a special trip from London to see him, to try to head off the rights being snapped up by rival US publishers. There were the film rights to consider too, in this new age of cinema. 'I didn't want him to come as I have no sense of brotherhood with pushy, shrewd businessmen out of big cities,' Hudson wrote, 'but I was agreeably surprised in him.'[29]

The agent had obviously been at pains to establish some common ground with his potential client, with his reputation for guardedness. 'He was born and bred in the blue grass mountain region of Virginia and spent his boyhood and youth breaking colts and shooting wild turkeys and all sorts of adventures,' Hudson reported. 'He knows more about nature than books … Of course he wanted to get the book I am doing.'[30]

Hudson had a bit of fun with this fellow-cowboy-turned-literary-agent, who had, he said, a direct line to the man at the top. 'I suppose you know that Macrae is the one and sole ruler of E.P. Dutton?'[31] Hudson wrote, mimicking the agent. 'He, Dutton, is very old and does nothing in the business. Macrae does what he likes and what he does is great and noble and generous.'[32]

Hudson now had a representative in New York, to whom he had to post a copy of *Afoot in England*, evidently now impossible to get hold of. Luckily Ranee Margaret was willing to loan him the copy from her bookcase at Badger's Holt cottage.

In the context of all this money talk, a return visit to the convent might have soothed his soul. In late February, he wrote:

> Yesterday my second excursion was to the Convent
> with a basket of grapes for the patients which
> I promised long ago and couldn't buy because of their
> dearness. Now we have an unlimited supply of Cape
> grapes at a low price, so I took a lot and had a nice long
> chat with the Sister Superior of the Convent. She said
> she had been thinking of me – wondering why I hadn't
> given her the book I promised her – *Dead Man's Plack*.
>
> I said it was because I had thought better of it: that
> it dealt with history – a dreadful time when Dunstan,
> Archbishop of Canterbury and ruler of England under
> the king, proclaimed the celibacy of the clergy and
> drove the married priests out of their livings and
> homes, and also was guilty of many other monstrous
> crimes ... And as she would have to submit my book
> to Father Ryan, the head priest of the Convent ... he
> would forbid her reading it. She laughed and said the
> priests had nothing to do with her reading ... So I must
> get it for her.[33]

As their dialogue about uses of his legacy continued, Etta Lemon asked him about seeing if they could persuade the London park authorities to create bird sanctuaries, but he thought it was too late. He recalled the report they had asked him to write with advice on how they could improve the city's green spaces for birds and wildlife generally. The irony wasn't lost on him that the same people had threatened him with legal action once upon a time.[34]

The newspapers were carrying reports that Hudson was at death's door. He believed it was because he was turning down the work that was still being offered to him by editors. Reports of his demise were a little premature. Having made it to April 1922, he had lived long enough to see the Plumage Act become operative; and yet Hudson's mind was still fixed on other bird business, although May came too soon for him to make it back to London. He wrote to Etta Lemon: 'I wish I could be in London for the annual meeting to hear Lord Grey'.[35] Grey was due to update the society about his work chairing the Wild Birds Advisory Committee. Hudson also wanted to see Grey to find out the truth about his eyesight, and whether it really was recovering, as people said.[36]

Hudson's negotiation with the Bird Society on how to deploy his legacy resumed by letter. Subjects other than a war on collectors were now on the table (Lemon may have steered him away from it), namely education. 'For the first years the interest of the money must suffice; afterwards the capital may be used as required,' he instructed Lemon. 'But I hope that the capital may be ADDED TO BY OTHERS, so that the publications which are suitable to be distributed on Bird and Tree Day may be kept up indefinitely.'[37]

Much to his relief, for he was lonely and had greatly missed her, Ranee Margaret returned to Cornwall in the spring. 'In the afternoon the Ranee and Lady Helmsley turned up for tea,'[38] he wrote on 18 May. 'I can't get enough exercise to keep me in a decent state. I have to get a car and take a drive instead of a walk.'[39] But all too soon she had to depart again, to attend to further family business. 'The Ranee of Sarawak is leaving next Wednesday. I'm sorry as her company is or has been of some value,'[40] he admitted. He didn't feel ready to go with her, and was suppressing his fears that this would be the last time he would see her.

He remained in Cornwall until mid-June, and described in a letter to Roberts his last visit to the clifftops high above the Atlantic waves that he loved so much. 'Last Sunday was the loveliest day I have had in 6 months and I spent it lying on the rocks of Land's End with a companion. The sea was divine – sea, sky, air, earth.'[41]

His coyness about naming this companion makes me wonder if it was Linda Gardiner who had come to see him. The other candidate is his young pen-friend Mary Trevelyan – but there is no reason why he wouldn't have named her. Whether here or at another location Hudson and Gardiner discussed unfinished business and, acknowledging that he had not long to live, they talked about *Lost British Birds*. She agreed to complete it for him. It is easy to imagine him speaking to her the words she quotes in the book:

> Just now we have a committee appointed by the
> Government to consider the whole question of bird

protection with a view to further legislation. Will
this Committee recommend the one and only way
to put a stop to the continuous destruction of our
rare birds? I don't think so. For such a law would be
aimed at those of their own class, at their friends, at
themselves.[42]

Gardiner wrote on his behalf, in an impassioned way, of 'science heedlessly
looking on' and even complicit in the destruction – 'interested only in
dealing the final death blow'.[43]

Hudson's driving passion, until his dying day, was to end the relentless
slaughter of wild birds. His early pamphlet for the Bird Society, *Lost British
Birds*, had been much longer than most. In it, he described the species
already gone – or on the brink of disappearing – from these islands. He
had been planning an enlarged and updated edition of the *Lost British Birds*
pamphlet for nearly 20 years, since 1904.[44] He commissioned and paid
for the colour plates himself. But he had run out of time: Gardiner would
complete the book from his copious notes. From the turn of the century,
she had been his closest surviving friend and soulmate where the love
of nature was concerned. It would be a labour of love for her. Hudson's
handwriting was never easy to decode, in part because of the sheer volume
of his writing, and the pace at which he poured his words onto the page,
but she didn't need much persuading.

There is a romantic – if tragic – subplot to all this. Hudson had secretly
been in love with Linda Gardiner. We can gauge this from his surviving
letters to her, which she kept, with his blessing. What her feelings for him
were we can only guess, as her letters to him don't survive. But of course
Hudson was married, and even if Gardiner was not, both knew they could
never be together. At stake were her career, and quite possibly the very
foundations of the Bird Society and its campaigns.

The extent of their love has been the subject of conjecture by various
commentators, and we can never know for sure. Two things seem clear,
however: they remained devoted friends until the end, and Gardiner –
like all Hudson's friends and notwithstanding his dark moods, bouts of
remoteness and 'native lack of delicacy'[45] – retained a deep affection for
him, as evidenced in this book. They had it in their gift to destroy all
trace of the 'affair' but chose not to do so, Hudson's love letters to her
re-emerging years later.

In *Lost Birds*, Hudson/Gardiner reiterate the three main threats to
the survival of wild birds – the recreational 'sportsman', the gamekeeper

and most of all another type, whom 'neither law nor education has so far touched … whose number has been steadily growing since 1894' namely, the collector.[46]

Despite more than three decades of campaigning, 'the hunt for rare birds and their eggs has become a craze of an infinitely more dangerous character than ever before'.[47] The Bird Society's watchers had worked courageously and diligently in often remote localities to keep guard over our most threatened surviving species, and thwart the pernicious attentions of these collectors, and their hirelings. But the pressure on them was relentless.

As far as wider legal protection for birds was concerned, the cogs of change were grinding slowly. The government was continuing to ponder the need for stronger bird protection laws. A committee was in place. An effective Wild Birds Protection Act was a protracted work in progress.

I was delighted to find a copy of the later, expanded and colour-illustrated Rare, Vanishing and Lost British Birds in an antique shop in Dunkeld, Perthshire, 15 years ago. I took a hit paying for it, but it was worth every pound. It's a beautiful book, conveying Hudson and Gardiner's undying passion for the birds, their anger at the shocking waste and wilful destruction of the beauty of the natural world, and above all the clarion call that we must do something about it, in the face of overwhelming odds.

Gardiner's at least Platonic love for Hudson shines through. 'I have transcribed every word with the greatest care, and every reference has been looked up and completed,' Gardiner writes in the foreword. 'No one could be inspired by a more intense desire to honour his memory. I probably knew more of his mind in regard to this particular subject than anyone else knew.'[48] The enlarged Rare, Vanishing and Lost British Birds was finally published in 1923, a year after Hudson's death.

If it was Gardiner who accompanied him to Land's End on that June day in 1922, it was their final scene. While they must surely have met again in London in the two months he had left, there is no specific record of it; but he wrote to her two days before he died. It seems the right place to leave them, and their forbidden love, on a Cornish headland in the sunshine, birds wheeling in the ocean updrafts above and below.

He packed up most of his clothes in a big canvas bag – 'or bale', as he called it when writing to Ranee Margaret – and donated it to the relief committee of the poor in St Ives. The vicar of Breage wanted to see him, but Hudson had run out of the time and energy needed for that. 'So goodbye to everyone,' he wrote to his old friend, 'in this rocky land of strangers.'[49]

Hudson's meeting with Helen Thomas in August 1922 to discuss how he could help her and her three children, and contribute to a book of Edward's war poems, was almost his last act. 'I can't get over his loss,' Hudson told her. 'He was as dear to me as if he had been my son.' Image courtesy of the Estate of Helen Thomas.

The last summer

'I really didn't expect to come back ever to London when I went away in November,'[50] Hudson wrote when he arrived back in the city once more at the start of summer 1922. He hadn't made it back in time for the reception hosted by the Bird Society on 9 June at the Zoological Gardens in Regent's Park. The special guest was Dr T. Gilbert Pearson, president of the US National Association of Audubon Societies. Pearson had given an address in the lecture hall after tea in the newly opened pavilion. Hudson had been busy in Cornwall that day, donating his winter clothing to charity, in a big bundle.

Of course Hudson would have been an honoured presence at this event, although I can't find any mention of it in his letters from Cornwall. It may be that he had no wish to participate, even if he could have got there, owing to a combination of fragile health, his usual reticence and a sense from him that his work in this line was done. It was now for others to take forward the fight for nature; people like young H.J. Massingham, whom he particularly respected, and of course Etta Lemon and co.

Pearson had crossed the Atlantic with a packed schedule ahead of him. Besides the Regent's Park event he visited several European countries and then chaired the historic gathering of international bird conservation organisations on 20 June. It was hosted by Reginald McKenna, a former minister of finance in the British government, to discuss the formation of an International Committee for the Protection of Wild Birds. It launched what we know today as BirdLife International.

While Hudson took a back seat, his close friends were well represented, as the now Viscount Edward Grey of Fallodon was there, as well as Massingham, who visited Hudson a week later and no doubt told him all about it. William L. Sclater, son of Hudson's old collabora- tor Dr Philip Sclater, was also present, as well as Earl Sydney Buxton, Dr Percy Lowe, Frank E. Lemon, Honorary Secretary of the Royal Society for the Protection of Birds (RSPB, today the UK partner in BirdLife), Jean Delacour, president of the Ligue pour la Protection des Oiseaux (LPO, now BirdLife in France), and P.G. Van Tienhoven and Dr A. Burdet of the Netherlands.

Their declaration of principles stated: 'by united action, we should be able to accomplish more than organisations working individually in combating dangers to bird-life'.[51]

Hudson does mention that Dr Pearson wanted to meet him, and although I haven't found any specific evidence that it happened, it is likely they met at Whiteley's department store restaurant on 27 June. 'I shall have to lunch there [Whiteley's] every day',[52] he had told Garnett on 15 June. Hudson was trying to find time for everyone, Dr Pearson included.

He was so much in demand that he had to send an apology to Roberts on 25 June. 'I fear I can't make any lunching engagement as I have accepted several invitations for the week. Monday I have to meet Garnett and perhaps someone else in the evening.'[53] The cryptic 'someone else' may again have been Linda Gardiner. Garnett had sent Hudson his recently published book profiling literary giants – Friday Nights – in which Hudson featured, alongside such luminaries as Ibsen, Nietzsche and Tolstoy. He would manage to find time for one last meeting with Garnett, at the start of August.[54]

With the Mont Blanc restaurant in Soho no longer the centre of the action, William Whiteley's department store restaurant in Bayswater had become the new go-to venue for his meetings. Here, he found some luxuries that had been in short supply during wartime. In a July letter to Roberts he made a plea: 'Don't send cream: I get all I want at my W.W. [William Whiteley's] luncheons.'[55]

There were final farewell meetings with Galsworthy and T.E. Lawrence, of Arabia. Besides his restaurant appointments, Hudson had home visits from his Hampstead friends Violet Hunt and the Rothensteins. When he lunched with Viscount Grey, Hudson would no doubt have heard about his recent marriage to Lady Pamela Wyndham, a discreet ceremony that had taken place on 11 July.

One person he wouldn't see again was Don Roberto, who was in Scotland. Hudson opened a letter written on 30 July, from Cardross, and the following day he wrote his last letter to his old compadre, sending 'carinosos recuerdos' to Mrs Bontine. He signed off with some optimistic views about the new Argentine president Alvear, whom he thought had some 'ideas about "progress"'.[56]

There were many letters to answer, too, including from Henry Salt, Garnett and even the Labour leader Keir Hardie. And of course there was another important meeting with Etta Lemon, to discuss more bird business. She now told him that the society wanted to create an annual Hudson Lecture, to honour his name and memory. He gave the idea short shrift. Preaching to adults, he told her, was a waste of time. It was young people they must concentrate on. He also made clear that he wanted no publicity for his gift. 'It made me sick to see all the paragraphs about my giving up my pension,' he told her.[57]

By the time August came around, Hudson was on the point of exhaustion. 'I do want to see you very badly,' he wrote to Ranee Margaret, 'and you never come to town and many things here do keep me from going to Ascot.'[58] Happily, she came knocking at the door of Tower House, and they made an arrangement for him to visit Ascot, and if possible to travel together on his intended excursion to the south coast, via Grey Friars. He told her he'd bumped into Maggie Ponsonby in Notting Hill.[59]

An unexpected but timely drop in temperatures offered some relief in the airless city. 'The weather is still very strange for summer – to-day it is rather wintry,'[60] he observed. On 2 August, he apologised to Roberts for not having time to reply to his card, having been 'half-mad with my anxiety to finish what I was doing before getting out of town'.[61]

I have pieced together his diary of appointments and am as sure as I can be that he never made it out of town, to the south coast, via Grey Friars in Ascot one last time – and to revisit Emily's grave that would soon be his.

Mercifully, there was still time for an emotional reunion with Helen Thomas, widow of Edward. 'Mrs. Thomas comes up from Otford [Kent] to see me next Wednesday,'[62] Hudson wrote. He had delayed yet again his escape to the south coast in order to fit this in. They met on 16 August, and Helen later recalled the poignant occasion:

> I found him there at Whiteley's – a tall gaunt
> eagle-eyed man who talked only about Edward. He was
> in tears all the time and said, 'I can't get over his loss.
> He was as dear to me as if he had been my son. I never
> had a son, but he, I felt, was the son I would have liked
> to have, and I loved him. I have no one else, all my
> love was for him.' He could not speak for weeping.
> I was too deeply moved to speak. He then asked about
> my circumstances and said that he had heard they were
> difficult for me. He said, 'I have first editions of all my
> books and these are of some value. I will send them to
> you.' And so we talked for about an hour and he said,
> 'I should have met you before. It would have been good
> to talk to you. I am a very ill man and I have not long
> to live. We shall not meet again.'[63]

She asked Hudson if he would write the preface for a collection of Edward's tales and sketches, and of course Hudson agreed. 'I must try to do it as he was a friend and most lovable fellow. And was killed,' he wrote shortly afterwards, although afraid he might not be well enough to honour the commitment.[64]

We know what else Hudson discussed with Helen Thomas because later that day he wrote about it to Linda Gardiner, his last letter. Of course he had also spoken to Helen about the plight of wild birds. She had recently been to the Rothiemurchus estate in the Highlands of Scotland, encountering there a collector in search of nests of the rare snow bunting. The man told her he had seen seven crested tit nests sawn out of tree trunks in the forest. Hudson wished that Ogilvie Grant, Curator of Birds at the British Museum, could visit those ancient pinewoods to see this destruction for himself. To the very last, Hudson's fury at the greed and selfishness of collectors was undimmed. The collector also told Thomas that he knew someone in Devon who 'kept' Dartford warblers, and from these captives he would repeatedly take and sell the nests and eggs.[65]

Hudson found a way to summon the energy to write the preface for the Edward Thomas collection, called *Cloud Castle*. It was most likely his last act of writing for publication.

The day after his meeting with Helen Thomas, Hudson passed away in the night. In the morning, when Jessie MacDougall's little girl came to wake him, he was finally at rest.

Another rare photograph of Hudson. He mentions this image to Ranee Margaret in late 1919. It was 'was taken by my doctor here [Cornwall] with his raven on my knee'. 'I know only one snapshot of [Hudson]', Morley Roberts wrote, 'with a raven on his hand, which is truly characteristic.'

A last word

The following lines are from an undated Hudson letter, with no addressee as the first part of the letter is lost. It is almost certainly written from Cornwall in his final days there, to Ranee Margaret Brooke. I love the simplicity and sincerity of it, and offer these extracts for a final glimpse of his life and character.

> Several letters from strangers I've had in the last few days, but the loveliest is from a lady named Imogen Hawthorne Dewing of Connecticut who has just been reading my 'Geese' chapter in *Birds and Man* and said she never hitherto met anyone with her feeling about geese, and then tells me of her childhood at a farm and how she used as a little girl to sit by the side of a goose on her eggs, and how she loved and trusted the goose. A most charming letter. I felt that I could embrace and kiss her for it. But she wasn't here so I could not.

Despite failing health, Hudson's energy and passion in late life for the cause of conservation was extraordinary, until the very end.

Rooks by moonrise. Hudson had a singular affection for the humble rook. 'It will dine on dung in cold weather', he once noted, 'though not cheerfully'. CMJ

I am sorry you are not here in this weather: as the day advances towards its evening it grows more splendid. The sky is a divine crystal blue with a few whitest angel clouds sleeping, scarcely moving on it, and even the cawing of the rooks has taken a new sound – a note of contentment and happiness.[1]

Postscript

I n May 1925, Prime Minister Stanley Baldwin unveiled a monument in London's Hyde Park that forms the centrepiece of a bird sanctuary dedicated to the memory of Hudson. Don Roberto chaired the committee that commissioned Jacob Epstein to sculpt the monument, which depicts Rima, the 'bird-girl' from Hudson's novel *Green Mansions*. The appointment of Epstein and then the partial nudity of Rima caused a furore for many months afterwards. But that is another story.

The Hudson memorial and bird sanctuary in Hyde Park: a meditative place of pilgrimage today, honouring the spotlight-shunning writer who came to Britain to save the birds.
CMJ

Notes

Preface

1. I discovered some kind of family link when my uncle Ronnie Buchanan messaged me: 'I am delighted you are working on W.H. Hudson. He was a hero of Estyn Evans, the professor who taught me and your dad.'

2. Roberts, 1925, p. 254.

3. https://www.proquest.com/openview/a049d8222dfcd52d7dbf0bd690efd0c1/1?pq-origsite=gscholar&cbl=1817849 Wells, R. Neglected Self: W.H. Hudson.

1. Smelling England

1. Tomalin, 1982, p. 23.

2. Ibid., p. 258.

3. When I give talks about Hudson's life and times, I like to relate this early anecdote and ask the audience to guess what this smell could be, then reveal the source at the end. All kinds of answers are proffered, but almost never the correct one.

4. Tomalin, 1982, p. 23.

5. Hudson, *Birds in London*, p. 56.

6. The Argentine capital had 178,000 inhabitants in 1869 – it was then the size of Southampton.

7. In Hudson's poem *The London Sparrow* published in the magazine *Merry England* in 1883.

8. Miller, p. 16. Cites Haymaker.

9. Roberts, 1924, p. 137. NB Sclater was a doctor, not a professor.

10. Ibid.

11. Shrubsall, 2007, p. 30.

12. Hudson, *The Book of a Naturalist*, p. 286.

13. I contacted the Zoological Society of London and was delighted to discover that they still have Hudson's letters. Archivist Sarah Broadhurst very helpfully sent me scans of all 16.

14. *On the Origin of Species*, 1859.

15. Dewar, p. 16.

16. Ibid., p. 23. Also Tomalin, 1982, p. 237.

17. Hudson, *The Book of a Naturalist*, p. 214.

18. Ibid.

19. Tomalin, 1982, p. 110. Also https://nonsenselit.com/2010/12/30/britains-audubon-and-edward-lear/

20. Hudson, *Idle Days in Patagonia*, p. 183.

21.   Tree, pp. 257–8.

22.   Roberts 1924, p. 29.

23.   Ibid.

24.   Ibid. I would love to see this article – perhaps it will surface again one day.

25.   Hudson later recalled a British Museum gentleman, working on bird anatomy, once saying to him: 'I wish that when you are out in the woods you would put a dozen or two of young blackbirds, fledged and unfledged, in a bottle of spirits and send them up to me!' Letter to Cunninghame Graham, July 1901.

26.   Shrubsall 2007, p. 150.

27.   Roberts, 1924, p. 30. Roberts also wrote that 'Hudson never said a word as to the why of his departure from the Argentine.' p. 27.

28.   Tomalin, 1982, p. 107.

29.   Roberts, 1924, p. 39.

30.   Hudson, *The Book of a Naturalist*, p. 189.

31.   Roberts, 1924, p. 29.

32.   Ibid.

33.   Ibid.

34.   Hudson's poem *A London Sparrow* was published in the magazine *Merry England* in 1883.

2. Salvations

1.   Shrubsall, 2007, p. 36.

2.   Hudson, *Afoot in England*, p. 33.

3.   Hudson, *Birds in London*, p. 121.

4.   Ibid., p. 169.

5.   Ibid., p. 170.

6.   Ibid.

7.   Roberts, 1924, p. 54.

8.   Shrubsall, 2007, p. 28.

9.   Hudson, *Afoot in England*, p. 33.

10.   Hudson, *The Book of a Naturalist*, p. 124.

11.   Hudson, *Afoot in England*, p. 34.

12.   Roberts, 1925, p. 255.

13.   Hudson, *A Crystal Age*, p. 1.

14.   Ibid., p. 8.

15.   Ibid., p. 9.

16.   Ibid., p. 17.

17.   Ibid., p. 21.

18.   Ibid., p. 97.

19.   Ibid., p. 86. *The Vicar of Bray* is an eighteenth-century satirical song, and the Bull of Bashan a Biblical reference.

20.   Ibid., p. 95.

21.   Ibid., p. 86.

22.   Shrubsall and Coustillas, p. 85.

23.   Clarke, p. 16.

24.   Shrubsall and Coustillas, p. 84.

25.   Ibid., p. 85.

26.   Ibid., p. 84.

27.   Hudson, *Nature in Downland*, p. 17.

28.   Ibid., p. 56.

29.   Jefferies, *Eye of the Beholder*, p. 33 The Hedgerow Sportsman.

30.   Ibid.

31.   Miller, p. 37.

32.   Hudson, *Afoot in England*, p. 76. The year of Hudson's church visit is uncertain.

33.   Ibid., p. 77.

34.   Ibid.

35.   Ibid.

36.   Ibid.

37.   Boase, p. 32

38.   Looker, p. 151.

3. The Bird Society

1.   Bird protection movements had already been appearing elsewhere. In January 1888, the Dicky Bird Society created by the *Newcastle Weekly Chronicle* had, it claimed, 140,000 members: it had been running for 12 years. Its child members signed the pledge 'I hereby promise to be kind to all living things, to protect them to the utmost of my power, to feed the birds in winter time, and never to take or destroy a nest. I also promise to get as many boys and girls as possible to join the Dicky Bird Society.' In January 1886, the Selborne League merged with the Plumage League as the Selborne Society for the Preservation of Birds, Plants and Pleasant Places.

2.   Boase, p. 55.

3.   *Bird Notes and News* no. 7, 1916. Obituary for Eliza Phillips.

4.   Ibid.

5.   *Bird Notes and News*, 1941 (centenary of Hudson's birth). Quoted in Samstag, and reproduced in Looker.

6.   *Bird Notes and News* no. 7, 1916. Obituary for Eliza Phillips.

7.   Looker, p. 152.

8.   *Bird Notes and News* no. 7, 1916. Obituary for Eliza Phillips.

9.   Shrubsall, 2007 p. 231.

10.   Clarke, p. 12.

11.   Ibid.

12.   Tomalin. 1982, p. 107.

13.   Shrubsall, 2007, p. 45.

14.   Boase and Blomfield, pers. comm.

15.   Tomalin, 1982, p. 147.

16.   Shrubsall, 2007, p. 783.

17.   Ibid., p. 39.

18.   Ibid., p. 69.

19.   Ibid., p. 43.

20.   Ibid.

21.   Ibid., pp. 44–5.

22.   Ibid., p. 45.

23.   Hudson, *Birds in London*, p. 105.

24.   Ibid.

25.   Shrubsall and Coustillas, p. 88.

26.   *The Times*, 23 March 1936, p. 18.

27.   Tschifelly, p. 243.

28.   Ibid., p. 199. Also https://blog.historicenvironment.scot/2018/01/adventurer-worth-remembering/

29.   Ibid., p. 232.

30.   Ibid., p. 233.

31.   Ibid., p. 240.

32.   Curle. Letter of March 1890.

33.   Ibid.

34.   Ibid.

35.   Ibid.

36.   Tomalin, 1982, p. 155.

37.   Ibid., p. 156.

38.   Curle, p. 91. Hudson also described this horse in *The Naturalist in La Plata*, 1892.

39.   Munro, 2022, is one of the biographies with more on this. Also Jauncey.

40.   Roberts, 1924, p. 71.

41.   Ibid.

42.   Ibid., p. 120.

43.   Tschifelly, p. 266.

44.   https://www.nicholasshakespeare.com/writing/w-h-hudson/

45.   Hudson, *Far Away and Long Ago*, Dent 1931 edition. Cunninghame Graham provided the preface.

46.   Ibid.

47.   Curle. Chester Square was then and remains one of the most expensive postcodes in London. In more recent times, it has been the residence of choice of rockstar Mick Jagger, one-time prime minister Margaret Thatcher and Russian oligarch Roman Abramovic.

48.   Roberts, 1925, p. 236.

49.   Tschifelly, p. 197.

50.   Watts and Davies, p. 223.

51.   Hudson, *Birds in London*, p. 225.

52.   Clarke, p. 12.

53.   Ibid. Hudson is referring to the Bird Society.

54.   Tschifelly, pp. 254 and 260.

55.   Curle.

56.   Ibid. Letter to his mother, 27 March 1893.

57.   Watts and Davies, p. 120.

58.   Letter to Cunninghame Graham, 29 March 1896.

59.   Tschifelly, p. 277. Watts and Davies, p. 121.

60.   Curle, p. 350.

61.   Tschifelly, p. 281.

62.   Shrubsall, 2007, p. 30.

63.   Ibid., p. 41.

64.   Morley Roberts certainly described him thus and set the tone, in his 1924 portrait, and the radical Edward Garnett enjoyed political jousts with Hudson.

65.   Tschifelly, p. 350.

66.   Ibid., p. 211.

67.   Donald, p. 243.

68.   Shrubsall, 2007, p. 44.

69.   Ibid., p. 50.

70.   Ibid. The Thunderer being *The Times*'s nickname.

71.   Ibid., p. 54.

72.   Ibid, pp. 50–1.

73.   Ibid.

74.   Ibid., p. 52.

75.   Ibid.

76.   Ibid., p. 74.

77.   Ibid., p. 77.

78.   Samstag, p. 57.

79.   Ibid., p. 70.

80.   Ibid., p. 73.

81.   Ibid., p. 74.

82.   Shrubsall, 2007, p. 40.

83.   Ibid.

84.   Portland, p. 47.

85.   Ibid., p. 139.

86.   www.nottshistory.org.uk/portland1907/portland8.htm

87.   Ibid.

88.   Ibid.

89.   Shrubsall, 2007, p. 39.

90.   Portland, p. 50.

91.   www.nottshistory.org.uk/portland1907/portland8.htm

92.   Ridley, p. 388.

93.   Boase, p. 99.

94.   Roberts, 1924, p. 67.

4. Branching Out

1.   Ibid., p. 68.

2.   Shrubsall, 2007, p. 31.

3.   https://community.rspb.org.uk//placestovisit/bemptoncliffs/b/bemptoncliffs-blog/posts/the-climmers-of-bempton

4.   Roberts, 1925, p. 313.

5.   https://onlinelibrary.wiley.com/doi/full/10.1111/j.1474-919X.2012.01274.x   Birkhead, T.R. and Gallivan, P.T. Alfred Newton's contribution to ornithology: a conservative quest for facts rather than grand theories.

6. Wollaston, p. 145.
7. Shrubsall and Coustillas, p. 119.
8. Roberts, 1925, p. 312.
9. Shrubsall and Coustillas, p. 119.
10. Shrubsall, 2007, p. 51.
11. Ibid., p. 54.
12. Ibid.
13. Ibid.
14. Thompson, p. 280.
15. Garnett, 1923, p. 127.
16. Hudson, *Birds in a Village*, p. 28.
17. Ibid.
18. Shrubsall, 2007, p. 112.
19. Ibid., p. 43.
20. Ibid.
21. Hudson, *Birds in a Village*, p. 12.
22. Shrubsall, 2007, p. 44.
23. Roberts, 1925, p. 16.
24. Ibid., p. 16.
25. Hudson, *Birds in a Village*, preface, p. v.
26. Roberts, 1925, p. 16.
27. Ibid.
28. Shrubsall, 2007, p. 360.
29. Hudson, *The Book of a Naturalist*, p. 107.
30. Shrubsall, 2007, p. 422.
31. Ibid., p. 422. In 1908, Hudson was still bemoaning the fact that his publisher refused to allow *British Birds* to be revised, as it was still selling.
32. Ibid., p. 78.
33. Garnett, E, 1923, p. 18.
34. Hudson, *The Book of a Naturalist*, p. 191.
35. Trevelyan, p. 39.
36. Ibid., p. 56.
37. Ibid., p. 57.
38. Ibid.
39. Grey, *The Charm of Birds*, p. 4.
40. Garnett, E., 1923, p. 16.
41. Shrubsall and Coustillas, p. 21.
42. Looker, p. 153.
43. Shrubsall and Coustillas, p. 21.
44. Ibid., p. 60.
45. Wilson, p. 186.
46. Shrubsall, 2007, p. 70.
47. Ibid.
48. Ibid.
49. Ibid., p. 71.

50.　Waterhouse, p. 22.

51.　Shrubsall, 2007, p. 44.

52.　Ibid., p. 50.

53.　Ibid., p. 62.

54.　Ibid.

55.　It is noteworthy that biographer Ruth Tomalin thought this was Hudson's paternal grandmother, and his friend Edward Garnett thought maternal.

56.　Garnett, E., 1923, p. 137. Writing in 1914.

57.　Hudson, *Nature in Downland*, p. 32.

58.　Shrubsall, 2007, p. 64.

59.　Shrubsall and Coustillas, p. 20.

60.　Shrubsall, 2007, p. 64.

61.　Ibid.

62.　Ibid., p. 145.

63.　Ibid.

64.　https://www.facebook.com/brightonirish/photos/this-month-marks-the-120th-anniversary-of-the-uncrowned-king-of-ireland-charles-/290274937669109/?paipv=0&eav=Afa-jNCDBBcsFcl2mntMuPx3vvA_31aAw3jrI1-b3VlDUofWg2L0C_jGoCQmyZqKKxE&_rdr

5. Saving London's birds

1.　Hudson, *Birds in London*, p. 13.

2.　Ibid., preface, p. v.

3.　Ibid., p. 113.

4.　Ibid., p. 14.

5.　Ibid., p. 26.

6.　Ibid., p. 27.

7.　Ibid., p. 169.

8.　Ibid., p. 2.

9.　Ibid., p. 4.

10.　Ibid., p. 62.

11.　Ibid.

12.　Ibid.

13.　Ibid., p. 64.

14.　Ibid.

15.　Ibid., p. 1.

16.　Ibid., p. 107.

17.　Ibid., p. 49.

18.　Doughty, p. 115.

19.　Shrubsall and Coustillas, p. 60.

20.　Tschifelly, p. 280.

21.　Curle, p. 35.

22.　Ibid., p. 324.

23.　Watts and Davies, p. 130.

24. Tschifelly, p. 307.
25. Curle, p. 40.
26. Ibid., p. 49.
27. Shrubsall, 2007, p. 105.
28. Watts and Davies, p. 124.
29. Watts, 1969.
30. Curle, p. 57.
31. Tschifelly, p. 320.
32. Ibid., p. 332.
33. Looker, p. 32.
34. Curle, p. 40.
35. Ibid., p. 57.
36. Tschifelly, p. 282.
37. Curle, p. 40.
38. Shrubsall, 2007, p. 154.
39. *The Country Diary of an Edwardian Lady* is said to have sold over a million copies in its first year.
40. Shrubsall, 2007, p. 202.
41. Ibid., p. 57.
42. Ibid., p. 59.
43. Ibid., p. 57.
44. Ibid.
45. Ibid., p. 59.
46. Ibid., p. 63.
47. Roper, p. 109.
48. *Birds of a Feather*, p. 64.
49. Ibid., p. 65.
50. Hudson, *A Traveller in Little Things*, p. 254.
51. http://hubbardplus.co.uk/hubbard/Emma_Evans/art_work/chestnut%20box%20top.html
52. Hudson, *A Traveller in Little Things*, p. 254.
53. Shrubsall, 2007, p. 169.
54. Letter in *The Times*, 14 April 1900, reproduced as the RSPB pamphlet *Kew Gardens and Old Deer Park*.
55. Ibid.
56. Shrubsall, 2007, p. 144.
57. Ibid., p. 145.
58. Hudson, *Birds in London*, p. 163.
59. *The Times*, 14 April 1900.
60. Shrubsall, 2007, p. 161.
61. Hudson, *Birds in London*, p. 56.
62. Ibid., p. 162.
63. Ibid., p. 166.
64. Ibid., p. 167. The RSPB counts Hudson's *Times* letter among its collection of pamphlets, this one called *Kew Gardens and Old Deer Park*, in which his letter was reprinted in full.

65.   Ibid.

66.   Ibid.

67.   Ibid., p. 172.

68.   Ibid., p. 174.

69.   Ibid., p. 175.

70.   Ibid., p. 176.

71.   Ibid., p. 181.

72.   Ibid., p. 180.

73.   Ibid., p. 199.

74.   The Kew archive has a copy of a letter Hudson sent to the director in December 1900.

6. Further afoot

1.   http://www.sussexhistory.co.uk/seaward-sussex/seaward-sussex%20-%200027.htm

2.   Hudson, *Nature in Downland*, p. 22. Hudson quotes letter LVI from White's *The Natural History of Selborne*.

3.   Shrubsall, 2007, p. 108.

4.   Ibid., p. 174. Also Hudson's Selborne article published as chapter 15 of *Birds and Man*, 1915 edition.

5.   Hudson, *Birds and Man*, p. 283.

6.   Ibid.

7.   Ibid., p. 291.

8.   Ibid., p. 300.

9.   Ibid.

10.   Hudson, *Nature in Downland*, p. 14.

11.   *Birds of a Feather*, p. 7.

12.   Hudson, *Nature in Downland*, p. 127.

13.   *Birds of a Feather*, p. 71.

14.   Hudson, *Nature in Downland*, p. 66.

15.   *Birds of a Feather*, p. 66.

16.   Shrubsall, 2007, p. 168.

17.   Hudson, *Nature in Downland*, p. 76.

18.   Ibid., p. 78. Hudson is quoting M.A. Lower, in *Contributions to Literature*.

19.   Ibid., p. 77.

20.   Hudson, *Nature in Downland*, p. 131. There may have been ortolan buntings on the menu on occasion, but the name probably covers a range of small songbirds served up as a delicacy.

21.   Ibid., p. 130.

22.   *Birds of a Feather*, p. 62.

23.   Ibid., p. 63.

24.   Shrubsall, 2007, p. 141.

25.   Ibid., p. 150.

26.   Ibid., p. 141.

27.   Ibid.

28.   Ibid., p. 107.

29.   Hudson, *Nature in Downland*, p. 6.

30.   Ibid., p. 5.

31.   Ibid., p. 163.

32.   Ibid.

33.   Shrubsall, 2007, p. 180.

34.   *Birds of a Feather*, p. 66.

35.   Hudson, *Nature in Downland*, p. 166.

36.   *Birds of a Feather*, p. 66.

37.   Hudson, *Nature in Downland*, p. 166.

38.   Ibid., p. 165.

39.   Ibid., p. 36.

40.   Ibid., p. 183.

41.   Ibid., p. 157.

42.   Ibid.

43.   Shrubsall, 2007, p. 147. MacCormick had his first exhibition at the Royal Academy of Art the previous year.

44.   This inn dates from 1500 and is still in operation today. I visited while in Chichester to speak to the local RSPB group. The inn is now a branch of a well-known Italian restaurant chain. Hudson returned in 1917 and found it and its 'two ancient ladies' unchanged. At first they didn't recognise him but when they did he 'noted a sudden strange chill towards me'. Shrubsall, 2007, p. 588.

45.   Shrubsall, p. 150.

46.   Ibid.

47.   Ibid., p. 152.

48.   Ibid.

49.   Ibid., p. 151. After the crushing disappointment with the barn owl in Sussex, it is worth mentioning that Hudson had more success in liberating a bullfinch in Hampshire, described in a letter to Eliza Phillips. *Birds of a Feather*, p. 69.

50.   Ibid.

51.   Ibid. Hudson must have returned to let Mrs Couzins know the sad news about the owl. His spirits cannot have been improved when news reached him of the death of John Ruskin. One critic thought *Nature in Downland* had 'a style not unworthy of Ruskin' (Shrubsall, 2007, p. 178), though typically Hudson considered the praise exaggerated.

52.   Gates, p. 170. Also published in leaflet no. 33 of the SPB.

53.   Ibid.

54.   Looker, p. 151.

55.   Curle, p. 59.

56.   Shrubsall, 2007, p. 338.

57.   Trevelyan, p. 39.

58.   Ibid., p. 380.

59.   Shrubsall, 2007, p. 187.

60.   Obituary in *The Speaker*, 3 March 1906.

61.   Trevelyan, p. 40.

62.   Roberts, 1925, p. 24.

63.   Waterhouse, p. 104.

64.   Shrubsall, 2007, p. 182.

65.   Ibid., p. 187.
66.   Ibid.
67.   Roberts, 1925, p. 24.
68.   Curle, p. 59.
69.   Ibid., p. 60.
70.   Shrubsall, 2007, p. 193.
71.   Ibid.
72.   Ibid.
73.   Ibid.
74.   Ibid.
75.   Hudson, *Hampshire Days*. In Chapter 12 this cottage summer is described in more detail.
76.   Shrubsall, 2007, p. 195.
77.   Ibid.
78.   Ibid.
79.   Ibid., p. 339.
80.   Trevelyan, p. 379.
81.   Ibid., p. 380.
82.   Roberts, 1925, p. 23.
83.   Ibid.
84.   https://fashioningfeathers.info/murderous-millinery/
85.   Roberts, 1925, p. 344.

7. Early Edwardians

1.   Doughty, p. 116.
2.   Curle, p. 60.
3.   Watts and Davies, p. 199.
4.   Ibid.
5.   Portland, p. 126.
6.   Hudson, *Hampshire Days*, p. 61.
7.   McGhie, p. 239.
8.   Shrubsall, 2007, p. 182.
9.   Hudson, *Birds and Man*, p. 241.
10.   Shrubsall, 2007, p. 211.
11.   *Birds of a Feather*, p. 81.
12.   nottshistory.org.uk /portland1907/portland8.htm
13.   Shrubsall, 2007, p. 385.
14.   Roberts, 1925, p. 43.
15.   Ibid., p. 30. Roberts has this as 1901 but I think it more likely 1903.
16.   There is a similarly curious lack of any mention in Hudson's books and letters of those other champions of conservation the Rothschilds, based not far from Woburn at Waddesdon Manor. At this time, in 1901, the Rothschild family were donating Wicken Fen to the National Trust, and Woodwalton Fen nature reserve was also being established. After I gave a talk about Hudson to students of Cambridge University, the question of any Rothschild links came up. See McGhie, below – Note 24.
17.   *Birds of a Feather*, p. 25.

18.   *Ibid.*
19.   Shrubsall, 2007, p. 222.
20.   *Birds of a Feather*, p. 33. Recalled on 23 January 1913.
21.   Roberts, 1924, p. 263.
22.   Shrubsall. 2007, p. 280.
23.   Ibid., p. 289.
24.   McGhie, pp. 238–39.
25.   *Saturday Review*, 24 November 1900.
26.   Ibid., Hudson had written in protest about albatross killing in his 1893 book *Birds in a Village*.
27.   Shrubsall, 2007, p. 199.
28.   *Saturday Review*, 24 November 1900.
29.   Ibid.
30.   Shrubsall 2007, p. 266.
31.   Ibid., p. 203.
32.   Ibid.
33.   Curle, p. 61.
34.   Ibid.
35.   Ibid.
36.   Ibid., p. 52.
37.   Ibid.
38.   Shrubsall, 2007, p. 206.
39.   Ibid.
40.   Watts and Davies, p. 124.
41.   Shrubsall and Coustillas, p. 88.
42.   Garnett, D., p. 108.
43.   Watts and Davies, p. 173. Smith, p. 107.
44.   www.guardian.com/books/2017/nov/17/the-uncommon-reader-a-life-of-edward-garnett-by-helen-smith-review
45.   Garnett, E., 1923, p. 2.
46.   Ibid., p. 3.
47.   Ibid., p. 1.
48.   Shrubsall, 2007, p. 368.
49.   Smith, p. 193.
50.   Ibid.
51.   Ibid., p. 151.
52.   Garnett, E., 1923, p. 13.
53.   Hudson frequently uses this term to describe remote rural places of which he was fond.
54.   Ibid. 2 July 1902.
55.   Garnett, E., 1923, p. 22.
56.   *Birds of a Feather*, p. 71.
57.   Shrubsall, 2007, p. 272.
58.   Ibid., p. 344.
59.   Ibid.
60.   *Birds of a Feather*, p. 63.

61.   Ibid.
62.   Ibid.
63.   Ibid.
64.   Roberts, 1925, p. 26.
65.   Shrubsall 2007, p. 238.
66.   *Birds of a Feather*, p. 15.
67.   Ibid.
68.   Hudson, *Dead Man's Plack and An Old Thorn*, Preamble p. 4.
69.   Hudson, *Hampshire Days*, p. 26.
70.   Hudson, *Dead Man's Plack and An Old Thorn*, Preamble p. 4.
71.   Hudson, *A Traveller in Little Things*, p. 256.
72.   Garnett, E., 1923, p. 22.
73.   Garnett, E., 1923, p. 44.
74.   Shrubsall, 2007, p. 227, also in *The Book of a Naturalist*.
75.   Ibid., p. 259.
76.   Ibid., p. 268.
77.   Ibid., p. 271.
78.   Ibid., p. 301.
79.   Ibid.
80.   *Birds of a Feather*, p. 83.
81.   Hudson, *Afoot in England*, p. 273.
82.   Ibid., p. 276.
83.   *Birds of a Feather*, p. 84.
84.   Ibid.
85.   Ibid.
86.   Hudson, *Afoot in England*, p. 271.
87.   Ibid., p. 290.
88.   Ibid., p. 270.
89.   Ibid., p. 291.
90.   Ibid., p. 293.
91.   *Bird Notes and News*, spring 1941, p. 91.
92.   Wilson, p. 173.
93.   Curle, p. 64.
94.   Shrubsall. 2007, p. 211.
95.   Shrubsall and Coustillas, p. 39. George Gissing gave his friend Hudson a presentation copy of this book, in 1891.
96.   Ibid., p. 217.
97.   Shrubsall, 2007, p. 410.
98.   Garnett, E., 1923, p. 38. Hudson had evidently invited Garnett to the meeting and later told him he 'didn't miss very much'.
99.   Shrubsall, 2007, p. 285.
100. Ibid., p. 333.
101. Ibid.
102. Ibid.
103. Ibid.

104.   Ibid., p. 336.
105.   Ibid., p. 342.
106.   *Bird Notes and News*, 1904, p. 61.
107.   Shrubsall and Coustillas, p. 69.
108.   Doughty, p. 116.
109.   Ibid.
110.   Ibid.
111.   Shrubsall, 2007, p. 231
112.   Ibid., p. 232.
113.   Ibid., p. 231.
114.   Ibid., p. 223.
115.   *Bird Notes and News*, 1916 obituary.
116.   Shrubsall, 2007, p. 271.
117.   Roberts, 1925, p. 67.
118.   Hudson, *Afoot in England*, p. 47.
119.   Roberts, 1925, p. 190.
120.   Shrubsall, 2007, p. 467.
121.   Ibid., p. 327.
122.   Ibid.
123.   Hudson, *The Book of a Naturalist*, p. 106.
124.   Shrubsall, 2007.
125.   Ibid., p. 392.
126.   *Birds of a Feather*, p. 87. Christmas 1911.
127.   Shrubsall and Coustillas, p. 64.
128.   https://www.odmp.org/officer/2150-game-warden-guy-m-bradley
129.   *Bird-Lore*, August 1905. Journal of the National Association of Audubon Societies.
130.   Curle, p. 88.
131.   Shrubsall, 2007, p. 402. No surviving trace exists of these relatives of Hudson, or any correspondence with them, that I am aware of.

8. Modernists

1.   Ibid., p. 321.
2.   Ibid., p. 330.
3.   Roberts, 1925, p. 48.
4.   Garnett, E., 1923, p. 52.
5.   Shrubsall, 2007, p. 316.
6.   Ibid., p. 329.
7.   Ibid., p. 345.
8.   Ibid., p. 427. Musicians Rowland Solomon and Carda Maclean.
9.   Roberts, 1925, p. 80.
10.   Ibid., and Looker, p. 87.
11.   Roberts, 1925, p. 82.
12.   Curle, p. 95.

13.   Shrubsall, 2007, p. 419.

14.   Ibid., p. 435.

15.   Ibid., p. 210.

16.   https://www.jstor.org/stable/4119491 Hudson's 1906 bird report for Kew.

17.   *Birds of a Feather*, p. 55.

18.   The RSPB archive has correspondence from 1900–1 resulting from an inquiry by Hudson into the nesting of the great crested grebe in the Pen Ponds, Richmond Park.

19.   Roberts, 1925, p. 29.

20.   Shrubsall, 2007, p. 348.

21.   Shrubsall and Coustillas, p. 67.

22.   *The Times*, 6 June 1905.

23.   Roberts, 1925, p. 57.

24.   Shrubsall, 2007, p. 388.

25.   *Worthing Herald*, clipping in the RSPB archive, undated.

26.   Shrubsall, 2007, p. 623.

27.   Ibid.

28.   *Birds of a Feather*, p. 56.

29.   Shrubsall, 2007, p. 266.

30.   Letter 28 June 1903, held in the RSPB archive.

31.   Curle, p. 67.

32.   *Birds of a Feather*, p. 66

33.   Shrubsall, 2007, p. 323.

34.   Ibid., p. 266.

35.   Curle, p. 57.

36.   Ibid.

37.   Ibid., p. 80.

38.   Tschifelly, p. 332.

39.   Curle, p. 81.

40.   Ibid.

41.   Ibid., p. 80.

42.   Watts and Davies, p. 224, and Curle, p. 89.

43.   Shrubsall, 2007, p. 403.

44.   Roberts, 1925, p. 81.

45.   Garnett, E., 1923, p. 78.

46.   Ibid.

47.   Curle, p. 91.

48.   Ibid.

49.   Ibid.

50.   Don Roberto's book about the Jesuit missionaries in Paraguay, *A Vanished Arcadia*, was published in 1902. It would influence the 1986 film *The Mission*, with Robert de Niro, Liam Neeson and Jeremy Irons heading a star-studded cast.

51.   Shrubsall, p. 38, Hudson had sent a brief letter to Fisher Unwin about his manuscript – working title *Mr Avel* – in 1892.

52.   Garnett, E., 1923, p. 58.

53.   Shrubsall, 2007, p. 328.

54.   Ibid.

55.   This satirical title had been a staunch supporter of the plumage campaign.

56.   Ibid., p. 336.

57.   https://brill.com/display/title/32617 Chambers, H. Centennial Essays on Joseph Conrad's *Chance*, p. 98, Brill, 2016.

58.   Shrubsall, 2007, p. 367.

59.   Garnett, E., 1923, p. 62.

60.   Hudson, *Nature in Downland*, p. 139.

61.   https://archives.blog.parliament.uk/2020/11/24/if-he-thinks-fit/ The 1736 Theatre Licensing Act had, nearly two centuries earlier, made the Lord Chamberlain censor-in-chief.

62.   Smith, p. 170.

63.   Roberts, 1925, p. 92.

64.   Ibid., p. 93.

65.   Garnett, E., 1923, p. 103.

66.   Ibid.

67.   Smith, p. 170.

68.   Garnett, E., p. 123.

69.   Smith, p. 173.

70.   Garnett, D., p. 107.

71.   Curle, p. 81.

72.   Shrubsall, 2007, p. 332.

73.   Ibid., p. 49.

74.   Ibid., p. 374.

75.   Roberts, 1925, p. 67.

76.   Ibid.

77.   Ibid., p. 78.

78.   Shrubsall, 2007, p. 27.

79.   Roberts, 1925, p. 80.

80.   Garnett, E., 1923, p. 74.

81.   https://edward-thomas-fellowship.org.uk

82.   Thomas, E. *In Pursuit of Spring*.

83.   Thomas, H. A Memory of W.H. Hudson, *The Times*, 27 August 1965.

84.   Miller, p. 117.

85.   Shrubsall and Coustillas, p. 100.

86.   Garnett, E., 1923, p. 135.

87.   Ibid., p. 134.

88.   Ibid., p. 135.

89.   Ibid., p. 125.

90.   Hollis, p. 29. Also Garnett, E., 1923, p. 118.

91.   Shrubsall, 2007, p. 479.

92.   Ibid., p. 527.

93.   Tomalin, 1982, p. 183.

94.   Smith, p. 213.

95.   Marrot, p. 288.

96. Garnett, E., 1923, p. 116.
97. https://www.theguardian.com/books/2017/nov/17/the-uncommon-reader-a-life-of-edward-garnett-by-helen-smith-review
98. Garnett, E., 1923, p. 133.
99. Ibid.
100. Garnett, D., p. 105.
101. Ibid.
102. Shrubsall, 2007, p. 334.
103. Ibid., p. 425.
104. Roberts, 1925, p. 86.
105. Ibid.
106. Shrubsall, 2007, p. 403.
107. Ibid., p. 404.
108. Ibid., p. 407.
109. Ibid., p. 403.
110. Ibid., p. 407.
111. Ibid., p. 425.
112. Garnett, D., p. 33.
113. Ibid., p. 31.
114. Ibid., p. 26.
115. Ibid.
116. Ibid., p. 28.
117. Garnett, E., 1923, p. 66.
118. Shrubsall, 2007, p. 368.
119. Ibid., p. 384.
120. Garnett, E., 1923, p. 66.
121. Shrubsall, 2007, p. 368.
122. Ibid.
123. Garnett, D., p. 30.
124. Ibid., p. 65.
125. Garnett, E. 1923, p. 134.
126. Garnett, D., p. 31.
127. Ibid.
128. https://www.countrylife.co.uk/property/country-houses-for-sale-and-property-news/bloomsbury-house-in-cambridgeshire-30324#:~:text=Described%20as%20'The%20most%20beautiful,Lady%20into%

9. Later Edwardians

1. Brooke, M., 1934, p. 290.
2. Ibid., p. 236.
3. Shrubsall, p. 357. Although the dates of Hudson's letters sometimes had to be estimated, in a letter about first meeting Brooke he makes reference to the wedding of her (second) son taking place the next day, which was in 1904.
4. Brooke, M., 1934, pp. 275–6.

5. Ibid, pp. 276–7.
6. Shrubsall, 2007, p. 463. Letter to Austin Harrison.
7. Ibid., p. 357.
8. Brooke, M., 1934, p. 8.
9. Ibid., p. 278.
10. Ibid.
11. Shrubsall, 2007, pp. 355–7.
12. Brooke, M., 1934, p. 34.
13. *Birds of a Feather*, p. 40.
14. Shrubsall, 2007, p. 450.
15. Ibid., p. 364.
16. Brooke, M., 1934, p. 269.
17. Brett, S., p. 28.
18. Hignett., p. 30.
19. Brooke, S., p. 27
20. Ibid., p. 28.
21. Ibid., p. 29.
22. Hignett, p. 34.
23. Ibid.
24. Seymour, 1992, p. 287.
25. Books by Brooke, S., Brooke, M., Eade and Hignett.
26. Hignett, p. 34.
27. Ibid., p. 35.
28. Eade, p. 47.
29. Brooke, M., 1934, p. 279.
30. Hignett, p. 36.
31. Brooke, M., 1934, p. 279.
32. Ibid.
33. Ibid.
34. Brooke, M., 1934, p. 238.
35. Hignett, p. 35.
36. Ibid.
37. Ibid., p. 36. Hignett also notes that in 1918 Dorothy Brett mentioned to Lady Ottoline Morrell that 'the old Ranee said she heard I was having the same affair! with you as I had with her', p. 85.
38. Shrubsall, 2007, p. 439.
39. Brooke, S., p. 48.
40. https://www.youtube.com/watch?v=-pzdg8Z4xk8
41. Hignett, p. 46.
42. Russell, p. 77.
43. Garnett, E., 1923, p. 76.
44. Roberts, 1925, p. 66.
45. *The Speaker*, 3 March 1906.
46. Garnett, E., 1923, p. 78.
47. Trevelyan, p. 152.

48. Ibid., p. 159.
49. Ibid.
50. Ibid., p. 160.
51. Ibid., p. 159.
52. Shrubsall and Coustillas, p. 66.
53. Shrubsall, 2007, p. 395.
54. Ibid., p. 563.
55. Roberts, 1925, p. 315.
56. Garnett, E., 1923, p. 73.
57. Ibid., p. 75.
58. Roberts, 1925, p. 248.
59. Shrubsall, 2007, p. 392.
60. Curle, p. 90.
61. Garnett, E., 1923, p. 74.
62. *Birds of a Feather*, p. 30.
63. Garnett, E., 1923, p. 75.
64. Roberts, 1925, p. 114.
65. Ibid., p. 75.
66. Hudson, *The Land's End*, pp. 54 and 198.
67. Roberts, 1925, p. 155.
68. Hudson, *The Land's End*, p. 202.
69. Ibid., p. 204.
70. Ibid., p. 207.
71. Ibid.
72. Ibid., p. 203
73. Roberts, 1925, p. 96.
74. Ibid., p. 98.
75. Ibid., p. 256.

10. Wider Horizons

1. Shrubsall, 2007, p. 62.
2. Roberts, 1925, p. 302.
3. Ibid., p. 79.
4. Shrubsall, 2007, p. 652.
5. Ibid., pp. 689–90.
6. Ibid.
7. Ibid., p. 766.
8. Looker, p. 153. Etta Lemon later noted: 'as far as I can remember he never went to Fallodon'.
9. Shrubsall, 2007, p. 362.
10. Roberts, 1925, p. 52.
11. Shrubsall and Coustillas, p. 65.
12. Roberts, 1925, p. 53.

13. Ibid., p. 52.
14. Hudson, *Adventures Among Birds*, p. 113.
15. Ibid.
16. Ibid.
17. Shrubsall, 2007, p. 625.
18. As for Wales, I don't think Hudson ever got there. In 1913, Edward Garnett suggested he might consider wintering in south Wales, on the Pembrokeshire coast. 'Of course I would like that coast if only for its bird life,' Hudson replied, 'but I see no prospect of getting to places I want to visit any more.'
19. Roberts, 1925, p. 47.
20. Miller, p. 119.
21. Hudson, *Adventures Among Birds*, p. 95.
22. Ibid., p. 99.
23. Shrubsall, 2007, p. 437.
24. Hudson, *Adventures Among Birds*, p. 99.
25. Ibid.
26. Shrubsall, 2007, p. 436.
27. Hudson, *Adventures Among Birds*, p. 100.
28. Ibid., p. 102.
29. Shrubsall and Coustillas, p. 112.
30. Hudson, *Adventures Among Birds*, p. 104.
31. Roberts, 1925, p. 47.
32. *Birds of a Feather*, p. 25.
33. https://www.researchgate.net/publication/229068728_The_decline_of_the_Ring_Ouzel_in_Britain – the species' range size has decreased by 43% in last 40 years alone. BTO.
34. Roberts, 1925, p. 46. I had to suppress a chuckle at the reference to Buxton in the index of *Men, Books and Birds* – the compilation of Hudson's letters to Morley Roberts. It says, simply 'Buxton, unpleasant'.
35. Hudson, *Afoot in England*, p. 67.
36. *Birds of a Feather*, p. 63.
37. Ibid.
38. Shrubsall and Coustillas, p. 60.
39. Shrubsall, 2007, p. 95.
40. Hudson, *Afoot in England*, p. 53.
41. Garnett, E., 1923, p. 106.
42. Ibid., p. 109.
43. Hudson, *Birds in London*, p. 187.
44. Hudson, *Adventures Among Birds*, p. 100.
45. Garnett, E., 1923, p. 109.
46. Ibid. 105.
47. Ibid.
48. Roberts, 1925, p. 101.
49. Shrubsall and Coustillas, p. 109.
50. Garnett, E., 1923, p. 112.
51. Ibid., p. 113.

52.   *Birds of a Feather*, p. 28.
53.   Shrubsall, 2007, p. 472.
54.   Hudson may have forgotten at this point that the first book he ever bought as a boy and an early inspiration was Scots poet James Thomson's *Seasons* (1726). Tomalin, 1982, p. 40.
55.   Ibid., p. 469.
56.   Hudson, *Birds of La Plata*, p. 12. Curle wrote the introduction.
57.   Shrubsall, 2007, p. 82.
58.   Ibid., p. 280.
59.   Hudson, *A Shepherd's Life*, p. 41.
60.   Shrubsall, 2007, p. 216.
61.   Ibid., p. 344.
62.   Ibid., p. 267.
63.   Ibid., p. 239.
64.   Hudson, *A Shepherd's Life*, p. 266.
65.   Shrubsall, 2007, p. 295.
66.   Hudson, *A Shepherd's Life*, p. 123.
67.   Shrubsall, 2007, pp. 288 and 384.
68.   Curle, p. 73.
69.   Shrubsall, 2007, p. 345.
70.   Ibid., p. 341.
71.   Ibid., p. 267.
72.   Ibid., p. 306.
73.   Garnett, E., 1923, p. 73.
74.   Tomalin, 1982, p. 131.
75.   Ibid., p. 111.
76.   Hudson, *Birds and Man*, p. 258.
77.   Ibid.
78.   Garnett, E., 1923, p. 65.
79.   Roberts, 1925, p. 98.
80.   Shrubsall, 2007, p. 280.
81.   Roberts, 1925, p. 98.
82.   Ibid., p. 95. Transcribed as Telbury but Hudson had probably written Tetbury, which is correct.
83.   *Birds of a Feather*, p. 22.
84.   Garnett, E., 1923, p. 68.
85.   Ibid., p. 61.
86.   Ibid.

11. Shadows of War

1.   Roberts, 1925, p. 108.
2.   Ibid.
3.   Ibid.
4.   Shrubsall and Coustillas, p. 76.

5.   Boase, p. 198. Hudson had resigned his membership in 1908.

6.   https://api.parliament.uk/historic-hansard/commons/1910/jul/19/ plumage-prohibition-of-sale-or-exchange

7.   Roberts, 1925, p. 111.

8.   Ibid., p. 109. In the Bible, Ananias was struck down dead for lying. The Heavenly Twins is a reference from Greek mythology.

9.   https://www.gov.uk/government/history/past-foreign-secretaries/edward-grey

10. Shrubsall, 2007 p. 449.

11. Garnett, E., 1923, p. 139.

12. Samstag, pp. 9–10. RSPB pamphlet no. 69 published 1912.

13. Watts and Davies, p. 262.

14. Shrubsall, 2007, p. 485.

15. *Feathers and Facts – A Reply to the Feather Trade*. RSPB pamphlet.

16. Ibid.

17. Garnett, E., 1923, p. 114.

18. Watts and Davies, pp. 230–2.

19. Garnett, E., 1923, p. 118.

20. Roberts, 1925, p. 114.

21. *Birds of a Feather*, p. 27.

22. Garnett, E., 1923, p. 127.

23. Ibid., p. 126.

24. *Birds of a Feather*, p. 36.

25. Doughty, p. 57.

26. Shrubsall, 2007, p. 476.

27. Garnett E., 1923, p. 129.

28. Ibid.

29. Shrubsall and Coustillas, p. 98.

30. Shrubsall, 2007, p. 237.

31. Ibid., p. 481.

32. Ibid.

33. https://fortnightlyreview.co.uk/2013/05/tagore-in-london/ – William Rothenstein on Rabindranath Tagore in London.

34. Ibid.

35. Ibid.

36. Watts and Davies, p. 260.

37. https://fortnightlyreview.co.uk/2013/05/tagore-in-london/ William Rothenstein on Rabindranath Tagore in London.

38. https://www.academia.edu/49991515/One_Brother_in_India_Very_ Remarkable_T_E_s_Favourite_Brother_Will

39. Garnett, E., 1923, p. 130.

40. Ibid. Hudson may have meant style rather than sty, and been lost in transcription.

41. Ibid., p. 142.

42. Garnett, E., 1923, p. 160.

43. Ibid., p. 221.

44. Ibid., p. 7.

45. Smith, p. 245.

46.   Boase, p. 51.

47.   Ibid.

48.   https://britishlibrary.typepad.co.uk//untoldlives/2014/06/franz-ferdinand-shooting-and-shopping-in-england.html

49.   http://www.nottshistory.org.uk/portland1907/portland8.htm

50.   https://www.bbc.co.uk/news/uk-england-nottinghamshire-25008184 – Could Franz Ferdinand Welbeck gun accident have halted WWI?'

51.   https://www.gov.uk/government/history/past-foreign-secretaries/edward-grey

52.   Greville, 1929, p. 104. https://archive.org/stream/in.ernet.dli.2015.82520/2015.82520.Lifes-Ebb-And-Flow_djvu.txt

53.   https://zoologyweblog.blogspot.com/2019/06/feathers-fly-in-london-why-on-earth-was.html Why on Earth was Chalmers Mitchell of London Zoo on the wrong side of a conservation battle of the early 1900s?

54.   Ibid.

55.   Ibid.

56.   Shrubsall, 2007, p. 555.

57.   Ibid., p. 485.

58.   Roberts, 1925, p. 121.

59.   https://www.yorkpress.co.uk/news/11099770.march-25/ Archive 1914.

60.   Garnett, E., 1923, p. 140.

61.   Shrubsall, 2007, p. 487.

62.   Ibid., p. 489.

63.   Brooke, M., 1934, p. 278.

64.   Ibid.

65.   Ibid.

66.   Hudson, *Lost British Birds*, p. 36.

67.   *Birds of a Feather*, p. 44.

68.   Hudson, *The Book of a Naturalist*, p. 1, ch. 1, and Roberts, 1925, p. 160.

69.   Hudson, *Birds and Man*, p. 249.

70.   Garnett, E., 1923, p. 145.

71.   https://www.nationalarchives.gov.uk/pathways/firstworldwar/first_world_war/p_archduke_assassination.htm

72.   Shrubsall, 2007, p. 494.

73.   Marrot, p. 394, Diary entry, 14 July 1914.

74.   Ibid., p. 395.

75.   Ibid.

76.   Ibid.

77.   https://blog.nationalarchives.gov.uk/lamps-going-europe/

12. 'The Lamps Are Going Out'

1.   Roberts, 1925, p. 124.

2.   Ibid.

3.   Shrubsall, 2007, p. 499.

4.   Roberts, 1925, p. 124.

5.   Shrubsall, 2007, p. 496.

6.   Brooke, M., 1934, p. 281.
7.   Roberts, 1925, p. 124.
8.   Garnett, E., 1923, p. 145.
9.   Roberts, 1925, p. 129.
10. Garnett, E., 1923, p. 145.
11. Ibid., p. 145.
12. Garnett, E., 1923, p. 162.
13. *Birds of a Feather*, p. 43.
14. Ibid.
15. Garnett, E., 1923, p. 146.
16. Roberts, 1925, p. 126.
17. Garnett, E., 1923, p. 146.
18. Roberts, 1925, p. 128.
19. Ibid., p. 129.
20. Roberts, 1925, p. 129. Curiously, General Sir Smith-Dorrien rented Grey Friars from Ranee Margaret late in 1917. Shrubsall, 2007, p. 585.
21. Shrubsall, 2007, p. 514.
22. Roberts, 1925, p. 134.
23. Ibid., p. 130.
24. Shrubsall, 2007, p. 522.
25. Ibid., p. 523.
26. Hudson, *The Book of a Naturalist*, ch. 1. See also Roberts, 1925, p. 160.
27. *Birds of a Feather*, p. 43.
28. Roberts, 1925, p. 315.
29. Ibid., p. 188.
30. *Birds of a Feather*, p. 43.
31. Ibid.
32. Roberts, 1925, p. 131
33. https://www.sothebys.com/en/auctions/ecatalogue/2008/english-literature-history-childrens-books-illustrations-l08405/lot.214.html
34. Shrubsall, 2007, p. 524.
35. Marrot, p. 460.
36. Shrubsall, 2007, p. 513.
37. Ibid.
38. I made several enquiries to Forbes that went unanswered. Perhaps this lead has not yet gone cold.
39. Hudson, *Green Mansions*. Galsworthy provided the Introduction.
40. Garnett, E., 1923, p. 149.
41. Shrubsall, 2007, p. 528.
42. Ibid., p. 153.
43. Roberts, 1925, p. 136.
44. Garnett, E., 1923, p. 149.
45. Shrubsall, 2007, p. 550.
46. Roberts, 1925, p. 137.
47. Shrubsall, 2007, p. 528.

48.   Marrot, p. 418.
49.   Ibid.
50.   Roberts, 1925, p. 139.
51.   Shrubsall, 2007, p. 558.
52.   Marrot, pp. 544–6.
53.   Roberts, 1925, p. 139.
54.   Ibid., p. 140.
55.   Ibid., p. 139.
56.   Ibid.
57.   Ibid., p. 154.
58.   Shrubsall, 2007, p. 579.
59.   Ibid., p. 533.
60.   Ibid., p. 562.
61.   Ibid., p. 543.
62.   Roberts, 1925, p. 142.
63.   Ibid.
64.   Ibid.
65.   Roberts, 1924, p. 110.
66.   Garnett, E., 1923, p. 150.
67.   Roberts, 1925, p. 147.
68.   https://fortnightlyreview.co.uk/2013/05/tagore-in-london/
69.   Shrubsall, 2007, p. 539.
70.   Ibid., p. 543.
71.   *Bird Notes and News* 1916.
72.   Ibid., p. 561.
73.   Ibid., p. 565.
74.   Ibid., p. 552.
75.   Tschiffely, p. 364.
76.   Ibid.
77.   Ibid., p. 371.
78.   Curle, page not recorded.
79.   Ibid., pp. 360, 369, and Watts and Davies, p. 237.
80.   Curle, November 1915.
81.   Garnett, E., 1923, p. 137.
82.   Ibid.
83.   He was also known to the neighbours as Domingo, having been born on St Dominic's Day.
84.   Hudson, *Far Away and Long Ago*, p. 67.
85.   Watts and Davies, pp. 276.
86.   Ibid., pp. 374–7.
87.   Garnett, E., 1923, p. 143.
88.   Tschifelly, p. 375.
89.   Curle, p. 99.
90.   Ibid.
91.   Tschifelly, p. 369.

92.   Shrubsall, 2007, p. 525.
93.   Shrubsall and Coustillas, p. 102.
94.   Ibid., p. 572.
95.   Shrubsall, 2007, p. 573.
96.   Garnett, E., 1923, p. 169.
97.   Roberts, 1925, p. 166.
98.   Shrubsall and Coustillas, p. 103.
99.   Roberts, 1925, p. 166.
100.   Ibid.
101.   Thomas, H. A Memory of W.H. Hudson, *The Times*, 27 August 1965.
102.   Roberts, 1925, p. 237.
103.   Shrubsall, 2007, pp. 623, 632, 678.
104.   Shrubsall and Coustillas, p. 104.
105.   Ibid.
106.   Shrubsall, 2007, p. 563.
107.   Marrot, p. 436.
108.   Ibid., p. 563.
109.   Shrubsall, 2007, p. 574.
110.   Garnett, E., 1923, p. 161.
111.   Roberts, 1925, p. 151.
112.   Ibid., p. 152.
113.   Ibid.
114.   Brooke, M., 1934, p. 25.
115.   Eade, p. 44.
116.   Ibid., p. 41.
117.   Shrubsall, 2007, p. 585.
118.   Ibid., p. 591.
119.   Ibid., p. 593.
120.   Ibid., p. 591.
121.   Ibid., p. 593.
122.   Roberts, 1925, p. 167.
123.   Ibid., p. 168.
124.   Ibid., p. 170.
125.   Ibid., p. 182.
126.   Ibid.
127.   Ibid., p. 248.
128.   Ibid., 1925, p. 182.
129.   Garnett, E., 1923, p. 184.
130.   Roberts, 1925, p. 246.
131.   Shrubsall, 2007, p. 555.
132.   Doughty, p. 134.
133.   https://fieldsports-journal.com/fieldsports/fish/politics-by-other-means
134.   Trevelyan, p. 342.
135.   Roberts, 1925, p. 175.
136.   Ibid.

137. Ibid., p. 176.

138. Ibid.

139. Garnett, E., 1923, p. 184.

140. Shrubsall, 2007, p. 542.

141. Roberts, 1925, p. 186.

142. *Birds of a Feather*, p. 46.

143. Shrubsall, 2007, p. 817.

144. Roberts, 1925, p. 191.

145. Ibid., p. 193.

146. Ibid., p. 194.

147. Ibid., p. 193.

148. Ibid.

149. *Birds of a Feather*, p. 46.

150. Ibid.

151. Shrubsall, 2007, p. 625.

152. Ibid., p. 627.

153. Ibid., p. 629.

154. Roberts, 1925, p. 204.

155. The global flu pandemic killed 150,000 Britons, including two of my dad's uncles, after whom he was named, who survived the trenches only to die the same month the war ended. Worldwide, in just two years it infected 500 million people – a third of the world population – and killed 50 million.

156. Ibid., p. 181.

157. Garnett, E., 1923, p. 185.

158. Marrot, p. 436.

159. Ibid.

160. Shrubsall, 2007, p. 604.

161. Marrot, p. 437.

162. Roberts, 1925, p. 200.

163. Ibid., p. 203.

164. Garnett, E., 1923, pp. 182–4, 186–9.

165. Virginia Woolf's review of *Far Away and Long Ago* was published anonymously in the *Times Literary Supplement* in 1918.

166. https://www.gutenberg.org/files/19691/19691-h/19691-h.htm

167. Looker, p. 57. Also Hudson, *A Hind in Richmond Park*. The Dent 1929 edition includes adverts at the back with testimonies including this one.

13. Picking up the Pieces

1. Doughty, pp. 134–5.

2. Ibid.

3. Shrubsall, 2007, p. 682.

4. Roberts, 1924.

5. Shrubsall, 2007, p. 626.

6. Ibid., pp. 693 and 697.

7.   Ibid., p. 704.
8.   Tschifelly, p. 380.
9.   Roberts, 1925., p. 218.
10.   Trevelyan, p. 349.
11.   Roberts, 1925, pp. 215.
12.   Curle, 22 October (about 1919).
13.   Roberts, 1925., p. 230.
14.   Ibid., p. 235.
15.   Curle. 22 October (about 1919).
16.   Ibid.
17.   Ibid., p. 98.
18.   Trevelyan, p. 350.
19.   Ibid., pp. 351–2.
20.   Ibid.
21.   Roberts, 1925, p. 246.
22.   https://www.heritageandhistory.com/contents1a/2008/10/the-president-visits-carlisle/?doing_wp_cron=1670428119.0188579559326171875000
23.   Roberts, 1925, p. 100.
24.   Hudson, *Birds in Town and Village*, preface, p. v.
25.   Ibid.
26.   Hudson, *A Traveller in Little Things*, Dent 1924. Reviews quoted at end of book.
27.   Ibid.
28.   Shrubsall, 2007, p. 686.
29.   Roberts, 1925, p. 272.
30.   Hudson, *Birds of La Plata*, preface, p. xiv.
31.   Garnett, D., p. 32.
32.   Garnett, E., 1923, p. 208.
33.   *Birds of a Feather*, p. 52.
34.   Ibid.
35.   Roberts, 1925, p. 79.
36.   *Birds of a Feather*, p. 52.
37.   Roberts, 1925, p. 230.
38.   Shrubsall, 2007, p. 674.
39.   *Birds of a Feather*, p. 52.
40.   Ibid., p. 51.
41.   Shrubsall, 2007, p. 755.
42.   Roberts, 1925, p. 281.
43.   Shrubsall, 2007, p. 651.
44.   Ibid., p. 822.
45.   Hudson, *Lost British Birds*, foreword, p. ix.
46.   Ibid.
47.   Roberts, 1925, p. 238.
48.   Ibid., p. 239
49.   Ibid., p. 238.
50.   Ibid., p. 267.

51.   *Birds of a Feather*, p. 53.
52.   Roberts, 1925, p. 238.
53.   Ibid.
54.   Shrubsall, 2007, p. 702.
55.   Ibid., p. 456.
56.   Ibid., p. 650.
57.   Ibid.
58.   Ibid.
59.   Ibid.
60.   Roberts. 1924.
61.   Garnett, E., 1923, p. 209.
62.   Shrubsall, 2007, p. 694.
63.   Ibid.
64.   Ibid.
65.   Ibid., p. 777.

14. Swansongs

1.   CinemaTreasures.org /theaters/25235
2.   Shrubsall, 2007, p. 705.
3.   *Green Mansions* the movie was finally released in 1959, starring Audrey Hepburn as Rima, and Anthony Perkins as Smith, the explorer.
4.   Roberts, 1925, p. 317.
5.   Ibid.
6.   Ibid., p. 318.
7.   Shrubsall, 2007, p. 816.
8.   Roberts, 1925, p. 277.
9.   Shrubsall, 2007, pp. 720–1.
10.   Roberts, 1925, p. 293.
11.   Ibid., p. 300.
12.   Hudson, *Afoot in England*, p. 291.
13.   The Plumage Act would be law from April 1922, more than three decades on from the first meetings in Manchester and Notting Hill. But was it – and the devil in its detail – the law that the wild bird protectors wanted? It was certainly a compromise, to overcome those who would continue to block it. There was the fact that not all imported plumage was banned, and that a list of exempt species was appended and would be subject to continued review and wrangling. The existence of permitted plumage did as expected present problems for customs officials with limited ornithological training. And with no ban on ownership or sale of plumage, because such a ban was thought to be unworkable and unenforceable, unscrupulous dealers could continue to claim that their bird wares pre-dated the new law.

There were many further seizures of plumage – including nearly 140,000 grebe skins in 1924, as these 'furs' continued to be promoted by the fashion industry. But there were few prosecutions – they averaged just one per year up to 1936.

The campaign effort on plumage law did not stop in 1922. Seven more bills to ban ownership and sale were introduced in Parliament between 1922 and 1940. All of them failed. On a happier note, amid much less publicity there was another victory for the environment and wild birds in 1922 when the Oil in Navigable Waters Act was passed, effective from 1923. There would now be fines for ships discharging oil within 3 miles of the coast.

Slowly but surely, the global trade in bits of birds for costume adornment died out. Don Roberto would report in 1926 that egret feathers had no value in a city where they once were traded. One thing's for sure – fashionable people don't as a rule wear dead kittiwakes on their heads anymore.

After 1919 there were four attempts in the next eight years to introduce a Wild Birds Protection Bill. The first three failed owing to a lack of parliamentary time, and the fourth as too many amendments were tabled. The Wild Birds Protection Act was finally converted into law in 1954, another 32 years after the Plumage Act, and the death of Hudson. It was also the year that RSPB president Winifred Cavendish-Bentinck, Duchess of Portland, passed away. Tireless leader Etta Lemon had died the previous year, aged 92, having worked for the society until the age of 79 – she would have stayed on for longer if she'd had her way. Both had given more than 60 years' dedicated service to the cause.

14.   Samstag, p. 153.
15.   Shrubsall, 2007, p. 728.
16.   Ibid., p. 737.
17.   Ibid., p. 744.
18.   Ibid., p. 742.
19.   *Birds of a Feather*, p. 55.
20.   Roberts, 1925, p. 313.
21.   Ibid., p. 328.
22.   Ibid., p. 185.
23.   Shrubsall, 2007, pp. 739, 760, 756 and 759.
24.   *Birds of a Feather*, pp. 55–6.
25.   Ibid., p. 56.
26.   Roberts, 1925, p. 315.
27.   Ibid.
28.   Shrubsall, 2007, p. 762.
29.   Roberts, 1925, p. 136.
30.   Ibid.
31.   Ibid., p. 341.
32.   Ibid.
33.   Ibid., p. 332.
34.   Shrubsall, 2007, p. 763.
35.   Ibid., p. 764.
36.   Ibid.
37.   Ibid.
38.   Roberts, 1925, p. 342.
39.   Ibid., p. 345.
40.   Ibid.
41.   Ibid., p. 347.
42.   Hudson, *Lost British Birds*, foreword, p. ix.
43.   Ibid.
44.   Shrubsall, 2007, p. 341.

45.　Roberts, 1925, p. 254.

46.　Hudson, *Lost British Birds*, foreword, p. vii.

47.　Ibid., p. viii.

48.　Ibid., p. x.

49.　Shrubsall, 2007, p. 777.

50.　Garnett, E., 1923, p. 216.

51.　https://www.birdlife.org/news/2022/10/18/an-omnibus-scrapbook-of-birdlife100-global-congress-and-centenary/

52.　Garnett, E., 1923, p. 216.

53.　Roberts, 1925, p. 347.

54.　Ibid., p. 351.

55.　Ibid., p. 349.

56.　Curle, p. 112.

57.　Shrubsall, 2007, p. 773.

58.　Ibid., p. 779.

59.　Ibid., p. 782.

60.　Roberts, 1925, p. 350.

61.　Ibid.

62.　Ibid.

63.　Thomas, H. *A Memory of W.H. Hudson*. Fleece Press, Wakefield, 1984.

64.　Roberts, 1925, p. 349.

65.　Ibid., p. 785.

A last word

1.　Shrubsall, 2007, pp. 821–2.

W.H. Hudson's books and pamphlets

Hudson's collected works were published soon after his death, in 24 volumes, an astonishing output, in the circumstances of his life.
Image courtesy of Ian Ward and King's Lynn magazine.

1895	*British Birds*
1898	*Birds in London*
1900	*Nature in Downland*
1901	*Birds and Man*
1902	*El Ombú* (later *South American Sketches*)
1903	*Hampshire Days*
1904	*Green Mansions: A Romance of the Tropical Forest*
1905	*A Little Boy Lost*
1908	*The Land's End. A Naturalist's Impressions in West Cornwall*
1909	*Afoot in England*
1910	*A Shepherd's Life: Impressions of the South Wiltshire Downs*
1913	*Adventures Among Birds*
1916	*Tales of the Pampas*
1918	*Far Away and Long Ago: A History of My Early Life*
1919	*The Book of a Naturalist*
1919	*Birds in Town and Village*
1920	*Birds of La Plata* (revised from 1887 minus Sclater's part)
1920	*Dead Man's Plack* and *An Old Thorn*
1921	*A Traveller in Little Things*
1922	*A Hind in Richmond Park*
1922–23	*The Collected Works*, in 24 volumes
1923	*Rare, Vanishing and Lost British Birds* (expanded from Hudson's 1894 *Lost British Birds* pamphlet)
1923	*Ralph Herne* (originally serialised as a short story in 1888)
1930	*South American Romances*: bringing together *The Purple Land, Green Mansions, El Ombú*
1946	*Tales of the Gauchos* (an anthology of Hudson stories about gaucho life and the wonders of nature)

Bibliography

Arocena, F. (2003) *William Henry Hudson: Life, Literature and Science*. USA: McFarland and Co.

Baldwin, S. (1926) *On England – And Other Addresses*. London: Penguin.

Bates, H.E. (1950) *Edward Garnett*. London: Max Parrish.

Boase, T. (2018) *Mrs Pankhurst's Purple Feather: Fashion, Fury and Feminism – Women's Fight for Change*. London: Aurum Press.

Anand, S. (2009) *Daisy: The Life and Loves of the Countess of Warwick*. London: Piatkus.

Blunt, W. (1920) *My Diaries: Being a Personal Narrative of Events, 1888–1914*. London: Martin Secker.

Brockway, P. (2010) *Sir Edward Grey: More than a Politician*. Winchester: Winchester College.

Brooke, M. (1913) *My Life in Sarawak*. Oxford: Oxford Paperbacks.

Brooke, M. (1934) *Good Morning and Good Night*. London: Constable & Co. Ltd.

Brooke, S. (1970) *Queen of the Head-Hunters: Sylvia, Lady Brooke The Ranee of Sarawak*. Oxford: Oxford University Press.

Clarke, R. (2004) *Pioneers of Conservation; The Selborne Society and the (Royal) SPB*. Working Paper. London: The Selborne Society and Birkbeck College CEPAR.

Creighton, L. (1907) *Dorothy Grey*. London: Spottiswoode & Co Ltd. Privately printed.

Curle, R. (ed.) (1941) *W.H. Hudson's Letters to R.B. Cunninghame Graham with a Few to Cunninghame Graham's Mother, Mrs Bontine*. London: Golden Cockerel Press.

Cunninghame Graham, G. (2018) *Rhymes from a World Unknown*. Palala Press.

Cunninghame Graham, J. (2004) *Gaucho Laird: The Life of R.B. 'Don Roberto' Cunninghame Graham*. The Long Riders Guild Press.

Cunninghame Graham, R. (1930) *The Horses of the Conquest*. London: William Heinemann Ltd.

Dewar, D.W. (ed.) (1951) *Letters on the Ornithology of Buenos Ayres by W.H. Hudson*. New York: Cornell University Press.

Donald, D. (2020) *Women Against Cruelty – Protection of Animals in Nineteenth-Century Britain*. Manchester: Manchester University Press.

Eade, P. (2007) *Sylvia, Queen of the Headhunters: An Outrageous Englishwoman and Her Lost Kingdom*. London: Weidenfeld & Nicolson.

Epstein, J. (2006) *Let There Be Sculpture*. London: Hesperides Press.

Faulkner West, H. (1932) *Robert Bontine Cunninghame Graham: His Life and Works*. London: Cranley and Day.

Faulkner West, H. (1958) *William Henry Hudson's Diary Concerning his Voyage from Buenos Aires to Southampton on the Ebro*. Hanover, New Hampshire: Westholme Publications.

Fisher, J. (1966). *The Shell Bird Book*. Norwich: Ebury Press and Michael Joseph.

Fitter, R. (1949) *London's Birds*. London: Collins.

Ford, F.M. (1937) *Portraits from Life*. New York: Houghton Mifflin.

Ford, F.M. (1938) *Mightier than the Sword*. New York: Houghton Mifflin.

Frederick, J.T. (1972) *William Henry Hudson*. New York: Twayne.

Friedman, T. (1988) *Epstein's Rima: The Hyde Park Atrocity; Creation and Controversy*. Leeds: Henry Moore Centre.

Garnett, D. (1979) *Great Friends: Portraits of Seventeen Writers*. Macmillan.

Garnett, D. (ed.) (1941) *Letters of T.E. Lawrence*. London: World Books.

Garnett, E. (ed.) (1923) *153 Letters from W.H. Hudson*. London: Nonesuch Press.

Gates, B.T. (2002) *In Nature's Name: An Anthology of Women's Writing and Illustration 1780–1930*. Chicago: University of Chicago Press.

Gosse, P. (1949) *Traveller's Rest*. London: Cassell.

Greville, F.E.M. (1929) *Life's Ebb and Flow*. University of Michigan: W. Morrow.

Greville, F.E.M. (1936) *Nature's Quest*. London: John Murray.

Guthrie, J. (1920) *Edward Thomas's Letters to W.H. Hudson*. London: Mercury.

Hamilton, R. (1946) *W.H. Hudson: The Vision of Earth*. London: J.M. Dent.

Hardwick, J. (1990) *An Immodest Violet. The Life of Violet Hunt*. London: Andre Deutsch.

Haymaker, R.E. (1954). *From Pampas to Hedgerows and Downs: A Study of W.H. Hudson*. New York: Record Press.

Hignett, S. (1985) *Brett: From Bloomsbury to New Mexico*. New York: Watts.

Hudson, W.H. (see the list of Hudson's books.)

Hunt, V. (1900) *The Flurried Years*. London: Hurst & Blackett.

Jameson, C.M. (2013) *Looking for the Goshawk*. London: Bloomsbury.

Jefferies, R. (1987) *Eye of the Beholder*. Hampshire: Ashford Press.

Karney, R. (1993) *The Life of Audrey Hepburn*. London: Bloomsbury.

Looker, S. (ed.) (1947) *W.H. Hudson: A Tribute by Various Writers*. Worthing: Aldridge Brothers.

Marrot, H.V. (1935) *The Life and Letters of John Galsworthy*. London and Toronto: Heinemann Ltd.

Massingham, H.J. (1941) *Remembrance*. London: Batsford.

McGhie, H. (2018) *Henry Dresser and Victorian Ornithology: Birds, Books and Business*. Manchester: Manchester University Press.

Miller, D.L.S. (1985) *W.H. Hudson and the Elusive Paradise*. PhD thesis. London: Royal Holloway College.

Munro, L. (2017) *An Eagle in a Henhouse – Selected Political Speeches and Writings of R.B. Cunninghame Graham*. Turriff: Ayrton Publishing Ltd.

Munro, L. (2019) *R.B. Cunninghame Graham's contribution to the political and literary life of Scotland: party, prose, and the political aesthetic*. Glasgow: https://theses.gla.ac.uk/73017/1/2019munrophd.pdf

Munro, L. (2022) *R.B. Cunninghame Graham and Scotland: Party, Prose and Political Aesthetic*. Edinburgh: Edinburgh University Press.

Nicholson, M. (1926) *Birds in England: An Account of the State of Our Bird-Life and a Criticism of Bird Protection*. London: Chapman and Hall.

Payne, J.R. (1977) *W.H. Hudson: A Bibliography*. Folkestone. Wm Dawson and Sons Ltd.

Portland, Duke of. (1937) *Men, Women and Things: Memories of the Duke of Portland*. London: Faber and Faber Ltd.

Reynolds, F. (2016) *The Fight for Beauty: Our Path to a Better Future*. London: Oneworld Publications.

Ridley, J. (2012) *Bertie: A Life of Edward VII*. London: Chatto and Windus.

Roberts, M. (1924) *W. H. Hudson, A Portrait*. London: Eveleigh Nash & Grayson Limited.

Roberts, M. (1925) *Men, Books and Birds*. London: Grayson & Grayson.

Roper, F. (2018) *Victorian Hangover: Childhood Memories 1899–1920*. Privately published.

Rothenstein, W. (1931) *Men and Memories: Recollections of William Rothenstein, 1900–1922*. London: Faber.

Rothenstein, W. (1923) *Twenty-Four Portraits*. London: Chatto and Windus.

RSPB. (1964) *Birds and Green Places: A Selection from the Writings of W.H. Hudson*. Sandy, Bedfordshire.

Russell, B. (1958) *Portraits from Memory and other Essays*. London: Allen & Unwin Ltd.

Samstag, T. (1988) *For The Love of Birds: The Story of the RSPB*. Sandy: RSPB. (The official history of the RSPB covering the years 1889–1988.)

Seymour, M. (2004) *A Ring of Conspirators: Henry James and his Inner Circle, 1895–1915*. New York: Simon & Schuster.

Seymour, M. (1992). *Ottoline Morrell: Life on the Grand Scale*. London: Hodder and Stoughton.

Sheaill, J. (1976) *Nature in Trust: The History of Nature Conservation in Britain*. Glasgow: Blackie & Sons Ltd.

Shrubsall, D. (1978) *W.H. Hudson, Writer and Naturalist*. Tisbury, Wiltshire: Compton Press Ltd.

Shrubsall, D. (ed.) (1981) *Birds of a Feather – Unpublished Letters of W.H. Hudson*. Salisbury, Wiltshire: Moonraker Press.

Shrubsall, D. (ed.) (2007) *The Unpublished Letters of W.H. Hudson – The First Literary Environmentalist 1841–1922*. New York: Edwin Mellen Press.

Shrubsall, D. and Coustillas, P. (eds) (1985) *Landscapes and Literati – Unpublished Letters of W.H. Hudson and George Gissing*. Salisbury, Wiltshire: Michael Russell (Publishing) Ltd.

Smith, H. (2018) *The Uncommon Reader: A Life of Edward Garnett, Mentor and Editor of Literary Genius*. New York: Farrar, Straus and Giroux.

Thomas, E. (1922) *Cloud Castle and Other Papers*. London: Duckworth.

Thomas, E. (1978) *Richard Jefferies*. London and Boston: Faber and Faber.

Thomas, H. (1984) *A Memory of W.H. Hudson*. Wakefield: Fleece Press.

Tippet, B. (2004) *W.H. Hudson in Hampshire*. Hampshire County Council.

Thompson, F. (1945) *Lark Rise to Candleford*. London: Oxford University Press.

Tomalin, R. (1954) *W.H. Hudson: A Biography*. London: H.F. and G. Witherby Ltd.

Tomalin, R. (1982) *W.H. Hudson: A Biography*. London: Faber.

Tree, I. (2004) *The Bird Man: The Extraordinary Story of John Gould*. London: Ebury Publishing.

Trevelyan, G.M. (1937) *Grey of Fallodon*. London: Longmans.

Tschifelly, A.F. (1937) *Don Roberto: The Life of R.B. Cunninghame Graham*. London: Heinemann.

Waterhouse, M. (Ed.) (1999) *The Cottage Book – The Undiscovered Country Diary of an Edwardian Statesman*. By Sir Edward and Lady Grey. London: Victor Gollancz.

Wallace, A.R. (1905) *My Life: A Record of Events and Opinions*. London: Chapman and Hall.

Watts, C. and Davies, L. (1979) *Cunninghame Graham: A Critical Biography*. Cambridge: Cambridge University Press.

West, H.F. (1958) *William Henry Hudson's Diary Concerning His Voyage from Buenos Aires to Southampton on the Ebro from 1 April 1874 to 3 May 1874 Written to His*

Brother Albert Merriam Hudson with Notes By Doctor Jorge Cesares of Buenos Aires. Hanover, New Hampshire: Westholm Publications.

West, H.F. (1958) *For a Hudson Biographer.* Hanover, New Hampshire: Westholm Publications.

Wilson, G.F. (1922) *A Bibliography of the Writings of W.H. Hudson.* London: The Bookman's Journal.

Wilson, J. (2016) *Living in the Sound of the Wind: A Personal Quest For W.H. Hudson, Naturalist And Writer From The River Plate.* London: Constable.

Wollaston, A.F.R. (1921) *Professor Alfred Newton.* London: John Murray.

Index

References to figures and photographs appear in *italic* type; references to 'Hudson' in subheadings are to William Henry Hudson.

laws to protect birds
inadequate 143
law-breaking 96
Lear, Edward 12
Lelant, Cornwall 244–6
see also Badger's Holt cottage, Lelant,
Cornwall
Lemel, Isidor 34
Lemon, Etta (née Smith) 27, 284
annual Hudson Lecture proposal 292
Bird Society's aims 48
conference in Provence 78
death 330n13
as honorary secretary 34–5, 141
Hudson as chairman 36, 50
Hudson attending Eliza Phillips's
meetings 30–1
Hudson not visiting Fallodon 71
Hudson wanting to enlist 249
Hudson's legacy to Bird Society 282,
286–7
Hudson's letter to The Times April
1900 87
joining the BOU 220
letters to ladies wearing plumage 28
Linda Gardiner as assistant 140
'Our Duties to Wild Animals' 101–2
Lemon, Frank 48, 144
letters x–xi, 8–9, 16–17, 158, 293
Liberal Party 72, 192, 221
'Life in a Pinewood' (Hudson essay)
233–4
Lincoln's Inn Fields 77
Lindsay, David, 27th Earl of Crawford and
10th Earl of Balcarres 87, 155
Linnaean Society 109
A Linnet for Sixpence (Hudson) 146
A Literary Pilgrimage in England
(Guthrie) 255
A Little Boy Lost (Hudson) 160, 176,
274, 279
Lofft, Capel 138–9
London 4–7
August 1914 238
August 1915 241
markets 77
population 7
treatment of birds 75–7
winters 194
see also parks
London Zoo 119–20
Longman (publisher) 136
Looe, Cornwall 258

Looking for the Goshawk (Jameson) 176
Lost British Birds (BS pamphlet) 46, 61, 84,
143, 233, 275, 283, 284–5, 287–9
Lucas, E.V. 160
Lusitania, RMS 250

MacCormick, Arthur 100
MacDougall, Jessie 261–4
Maidenhead, Berkshire 224
Malmesbury, Wiltshire 214, 218
Malony (Hudson's grandmother) 73
Man of Devon (Galsworthy) 170
Manchester Guardian 272
Marland, Eric 63
Marlborough Place, Brighton 96
marriage of the Hudsons 16–17
Martin, Robert Montgomerie 30
Martin, Wiltshire (Winterbourne Bishop)
215, 216–17
Marvell, Andrew 81
Massingham, H.J. 291
Maud Merryweather (Hudson nom de
plume) 19
Maxwell, Sir Herbert 60, 101
McKay, Tommy 158
McKenna, Reginald 291
memoirs 260
Men, Women and Things (Duke of
Portland) 228–9
Midway Cottage, Cookham, Berkshire
64–5, 273
military use of plumage 77–8, 108, 113
The Mission (film) 40
Mitchell, Peter Chalmers 120–1, 229–30
Mont Blanc restaurant, Gerrard Street
125, 127, 129, 278
circle of writers 130, 159–60, 173
literary luncheon club 191
writers and censorship 161
Montcuq, France 173
Morocco 79, 80
Morris, William 32
motor cars 212–13
My Moor (Hubbard) 85

National Association of Audubon
Societies 290
National Physical Laboratory 86
National Portrait Gallery 151, 277
natterjack toad 174
Natural History Museum 229
Natural History of Selborne (White) 91–2
The Naturalist in La Plata 47, 158

Milton Keynes UK
Ingram Content Group UK Ltd.
UKHW041819181023
430868UK00002B/26

9 781784 273286